Thermal Battery Management System for Hybrid and Electric Vehicles

Scrivener Publishing
100 Cummings Center, Suite 541J
Beverly, MA 01915-6106

Publishers at Scrivener
Martin Scrivener (martin@scrivenerpublishing.com)
Phillip Carmical (pcarmical@scrivenerpublishing.com)

Thermal Battery Management System for Hybrid and Electric Vehicles

Edited by

Ashwani Kumar

Mukesh Kumar Awasthi

Nitesh Dutt

Yogesh Kumar Singla

and

Sivasakthivel Thangavel

Scrivener Publishing

WILEY

This edition first published 2025 by John Wiley & Sons, Inc., 111 River Street, Hoboken, NJ 07030, USA and Scrivener Publishing LLC, 100 Cummings Center, Suite 541J, Beverly, MA 01915, USA

For more information about Scrivener publications please visit www.scrivenerpublishing.com.

Wiley Global Headquarters
111 River Street, Hoboken, NJ 07030, USA

For details of our global editorial offices, customer services, and more information about Wiley products visit us at www.wiley.com.

The manufacturer's authorized representative according to the EU General Product Safety Regulation is Wiley-VCH GmbH, Boschstr. 12, 69469 Weinheim, Germany, e-mail: Product_Safety@wiley.com.

Library of Congress Cataloging-in-Publication Data

ISBN 9781394289202

Cover image: Generated with AI using Adobe Firefly
Cover design by Russell Richardson

Set in size of 11pt and Minion Pro by Manila Typesetting Company, Makati, Philippines

Printed in the USA

10 9 8 7 6 5 4 3 2 1

Contents

Aim and Scope

The aim of *Thermal Battery Management System for Hybrid and Electric Vehicles* is to investigate the technological advancements, challenges, and future perspective of battery thermal management systems for electric vehicles (EV) and hybrid electric vehicles (HEV). By investigating the latest advancements, challenges, and future perspectives in this field, this book aims to provide a comprehensive understanding of battery thermal management systems (BTMS). The scope of this book encompasses a thorough exploration of the existing literature and research on thermal management systems specifically designed for EV and HEV batteries. The book explores the intricacies of battery heat generation, temperature distribution, and associated thermal hazards. Additionally, the book evaluates the significant impact of temperature on battery performance, efficiency, and overall system reliability.

To achieve these objectives, the book will identify and analyze various thermal management techniques and solutions, including cooling technologies and heat pipes. It also assesses the effectiveness and efficiency of different thermal management models and solutions. Furthermore, the book investigates the crucial relationship between thermal management and battery lifespan, performance, and safety.

This book is primarily intended for engineers working in the fields of electric vehicles and battery technology. Specifically, engineers involved in the design, modeling, and analysis of battery thermal management systems for vehicles will find this book invaluable. Additionally, engineers responsible for the implementation and maintenance of BTMS in electric vehicles will benefit significantly from the knowledge and insights presented in this book.

By addressing these objectives, this book aims to contribute significantly to the advancement of battery thermal management systems for EVs and HEVs. The insights gained from this research can lead to enhanced battery performance, increased efficiency, and improved overall system reliability. Ultimately, this book seeks to support the ongoing development of more sustainable and efficient electric vehicles.

Editor
Ashwani Kumar
Mukesh Kumar Awasthi
Nitesh Dutt
Yogesh Kumar Singla
Sivasakthivel Thangavel

Preface

Thermal Battery Management System for Hybrid and Electric Vehicles deals with latest developments in battery thermal management system technology and includes advancements in design, compactness, and cooling methods. The development of advanced and compact battery thermal management systems is underway. This includes new battery housing designs and other components to optimize space utilization **(Chapters 1, 2)**. Research is being conducted on various cooling methods such as air cooling, liquid cooling, and phase change material (PCM) cooling. These methods aim to efficiently regulate the temperature of the battery pack and ensure optimal performance and safety **(Chapter 3)**. **Chapters 4–7** focus on developing nanofluid-based thermal management systems that optimize the performance of lithium-ion battery packs in electric vehicles. These systems use advanced algorithms and sensors to monitor and control the temperature of the battery pack in real time. **Chapter 8** deals with Electric Vehicle Charging and Discharging in V2G and G2V Operation.

Chapter 9 is dedicated to the advancements in **Renewable Energy for Storage Technology**. The technology is being developed to address the challenges associated with battery safety and prevent thermal runaway. The market for battery thermal management systems is expected to grow significantly, driven by the rising demand for electric vehicles worldwide. However, the industry faces challenges such as design, energy transmission complexities, and increased prices **(Chapter 10)**.

These developments aim to improve the efficiency **(Chapter 11)**, range, and overall performance of battery systems while ensuring safety and reliability **(Chapter 12)**. As battery technology continues to evolve, further advancements in battery thermal management systems can be expected **(Chapter 13)**.

The book specially highlighted the current industrial application of Battery Thermal Management System (BTMS) for HD-EV (Heavy duty Electric Vehicles) **(Chapter 14)**, Military Applications **(Chapter 15)**, Stationary Energy Storage **(Chapter 14)**, Renewable Energy Integration **(Chapter 15)**, Marine Engine **(Chapter 16)**, for carbon footprints **(Chapter 17)** and safety assessment **(Chapter 18)**.

The book works as a reference for these industries. It also aims to identify the challenges and opportunities for future research in this field. Research on BTMS can lead to the development of novel thermal management systems and cooling technologies, leading to advancements in the field of electric and hybrid vehicles. Its holistic approach not only underscores the pivotal role of BTMS in current industrial landscapes but also elucidates its transformative potential as a catalyst for future advancements and innovation.

Editor
Ashwani Kumar
Mukesh Kumar Awasthi
Nitesh Dutt
Yogesh Kumar Singla
Sivasakthivel Thangavel

Acknowledgement

The successful completion of *Thermal Battery Management System for Hybrid and Electric Vehicles* is a testament to the invaluable contributions of many individuals and organizations. We extend our sincere gratitude to Wiley Scrivener for providing a platform to share our knowledge and insights. Their unwavering support, guidance, and commitment to excellence throughout the publishing process were instrumental in shaping this book into a valuable resource for the scholarly community. We are deeply indebted to the numerous contributors and reviewers who generously shared their expertise. Their diverse perspectives and insightful feedback significantly enhanced the quality and comprehensiveness of this book. Their dedication to accuracy and clarity ensured that this work serves as a reliable and informative reference. This book is dedicated to the passionate and dedicated individuals who are driving advancements in Batteries, Electric Vehicle, and Hybrid Electric Vehicles. We hope that this work contribute to ongoing research and development in this field and serve as a valuable resource for the scholarly community. Finally, we recognize and honor the collaborative spirit that unites us all in the pursuit of excellence and progress. This book stands as a testament to the power of collaboration, and we are grateful to have been a part of this collective endeavor.

Editors

1

Battery Thermal Management System (BTMS) and Recent Advancements

Neha Singh Raghuvanshi[1]*, Yogesh Kumar Singla[2] and Ashwani Kumar[3]

[1]*Department of Applied Science & Humanities, ABES Engineering College, Ghaziabad, Uttar Pradesh, India*
[2]*School of Engineering, Math & Technology, Navajo Technical University Crownpoint, New Mexico, USA*
[3]*Mechanical Engineering, Technical Education Department, Uttar Pradesh, Kanpur, India*

Abstract

The Battery Thermal Management System (BTMS) plays a crucial role in maintaining the efficient operation, safety, and longevity of modern batteries, especially in electric vehicles and high-performance electronics. This chapter delves into the core principles of BTMS, including how heat is generated and dissipated across different battery types. It examines conventional cooling and heating methods, such as air and liquid cooling, alongside more advanced techniques like micro-channel cooling and phase change materials. Recent innovations feature the incorporation of artificial intelligence for adaptive thermal management, the development of new materials to improve thermal conductivity, and the progression of smart BTMS integrated with Battery Management Systems (BMS). Additionally, the chapter addresses current challenges and future directions, emphasizing the continual advancements that enhance thermal management efficiency and system dependability.

***Keywords*:** Battery thermal management, heat dissipation, cooling techniques, artificial intelligence, advanced materials

**Corresponding author*: nehasr.raghuvanshi@gmail.com; ORCID: 0000-0002-8437-6261
Ashwani Kumar: ORCID: 0000-0003-4099-935X

Ashwani Kumar, Mukesh Kumar Awasthi, Nitesh Dutt, Yogesh Kumar Singla and Sivasakthivel Thangavel (eds.) Thermal Battery Management System for Hybrid and Electric Vehicles, (1–26) © 2025 Scrivener Publishing LLC

1.1 Introduction: Overview and Significance of Battery Thermal Management Systems

Battery Thermal Management Systems (BTMS) are crucial for keeping batteries in good working condition. These systems use various methods to control the temperature of batteries, which can change depending on how they're used and the surrounding environment. By managing both heat removal and heat retention, BTMS ensures that batteries stay at a proper temperature, which helps them perform better and last longer. BTMS is especially important in electric vehicles (EVs), electronics, and energy storage systems. For example, in electric cars, good thermal management can affect how far the car can go on a single charge, how quickly it can charge, and how safe it is. In electronics, BTMS prevents overheating, which can cause devices to malfunction or wear out faster.

Thermal management is essential for several reasons. Batteries work best within a certain temperature range. Too much heat can speed up the breakdown of battery materials and reduce its performance. On the other hand, very cold temperatures can make the battery less effective and might even cause damage. Keeping the battery at the right temperature is also crucial for safety. Overheating can lead to a dangerous situation called thermal runaway, where the battery's temperature rises rapidly and could cause a fire or explosion. Good thermal management systems help prevent this by efficiently managing any heat produced. Longevity is another key benefit. Batteries that stay within their ideal temperature range last longer because extreme temperatures can damage the battery materials. By controlling the temperature, BTMS helps extend the battery's life, making the battery-powered systems more reliable and cost-effective. Battery Thermal Management Systems are vital for making sure batteries work safely, efficiently, and last as long as possible. They help improve performance, avoid safety issues, and extend the life of batteries in various advanced technologies [1, 2].

1.2 Fundamentals of Battery Thermal Management

Batteries generate heat as a by-product of their chemical reactions during operation. When a battery discharges, it releases electrical energy through the movement of ions between the anode and cathode. This process involves several reactions that occur within the battery's electrolyte and electrodes.

These reactions are not perfectly efficient; some of the electrical energy is converted into thermal energy due to resistive losses in the battery's internal components. There are a few key sources of heat generation in batteries:

a. **Internal Resistance**: All batteries have some degree of internal resistance, which causes energy loss in the form of heat as the battery charges or discharges. The amount of heat generated is proportional to the current flowing through the battery and its internal resistance. For instance, if the battery has a higher internal resistance, it will generate more heat when providing high power during demanding tasks such as gaming or video editing.
b. **Electrochemical Reactions**: The chemical reactions that occur during charging and discharging are exothermic, meaning they release heat. This is particularly true in high-charge or high-discharge situations where the battery is operating under stress. For example, during rapid acceleration, the battery's chemical reactions generate additional heat due to the increased current flow and stress on the battery. This heat needs to be managed to ensure the battery operates safely and efficiently.
c. **Inefficiencies in Charging**: Not all the electrical energy supplied is stored as chemical energy when charging a battery. Some of it is lost as heat due to inefficiencies in the charging process and internal resistance. For instance, during fast charging, the charger and battery both experience higher power levels, leading to increased heat generation. This heat is due to resistive losses in the battery and inefficiencies in the charging circuitry.
d. **Self-heating**: Batteries can also heat up due to self-heating effects, where heat generated within the battery accumulates and raises its temperature, particularly during periods of high current flow or prolonged use. For example, if a cordless drill is used for extended drilling tasks, the battery's internal heat generation from high current flow causes it to warm up. This self-heating effect can impact battery performance and longevity if not properly managed with cooling systems or thermal management strategies.

1.2.1 Thermal Characteristics of Battery Types

Different types of batteries have unique thermal characteristics due to their distinct chemical compositions and designs. Understanding these properties is essential for effective thermal management (Figure 1.1).

a. **Lithium-Ion Batteries (Li-ion)**: Table 1.1 highlights that Li-ion batteries are widely used in consumer electronics and electric vehicles due to their high energy density and long cycle life. They generate significant heat during both charging and discharging, particularly at high currents. Li-ion batteries are sensitive to temperature extremes. High temperatures can accelerate degradation and reduce battery life, while low temperatures can impair performance and capacity. Effective cooling strategies, such as liquid cooling or advanced thermal interface materials, are often required to manage the heat generated.
b. **Nickel-Metal Hydride Batteries (NiMH)**: NiMH batteries are commonly used in hybrid vehicles and various consumer devices. They have a lower energy density compared to Li-ion batteries but are known for their robustness and safety. NiMH batteries also produce heat during operation, though typically less than Li-ion batteries. They are less sensitive to temperature variations, but excessive heat can still affect their performance and lifespan. Cooling systems for NiMH batteries often focus on ensuring uniform temperature distribution rather than addressing extreme temperature variations.
c. **Lead-Acid Batteries**: Lead-acid batteries are one of the oldest types of rechargeable batteries, commonly used in

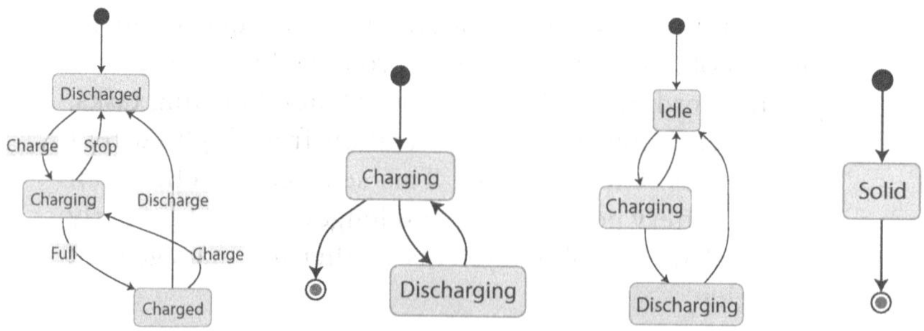

Figure 1.1 Thermal characteristics of battery types.

Table 1.1 Comparison of various battery types [31–37].

Battery type	Heat generation	Temperature sensitivity	Thermal management strategies
Lithium-Ion (Li-ion)	High during charging and discharging, especially at high currents	High temperatures accelerate degradation; low temperatures impair performance and capacity	Liquid cooling, advanced thermal interface materials, and heat spreaders
Nickel-Metal Hydride (NiMH)	Moderate heat generation	Less sensitive to temperature extremes but still affected by excessive heat	Uniform temperature distribution, air cooling, and passive thermal management
Lead Acid	Heat generated mainly during charging due to exothermic reactions	Relatively tolerant to temperature variations but excessive heat can cause water loss from electrolyte	Ventilation, passive cooling, and maintaining electrolyte levels
Solid State	Heat generation similar to other batteries; ongoing development to optimize	New technology with developing understanding of temperature sensitivity	Emerging strategies focusing on optimizing materials and designs for better thermal stability

automotive and backup power applications. They generate heat during charging due to the exothermic nature of the chemical reactions involved. While they are less efficient and have a lower energy density compared to modern batteries, they are relatively tolerant to temperature variations. However, excessive heat can cause water loss from the electrolyte, reducing battery life and performance. Proper ventilation and cooling are necessary to manage heat in lead-acid batteries.

d. **Solid-State Batteries**: Solid-state batteries are an emerging technology that uses a solid electrolyte instead of a liquid or gel electrolyte. They promise improved safety and energy density. However, they still generate heat during operation, and managing this heat is crucial for maintaining performance and safety. The thermal management strategies for solid-state batteries are still under development, with ongoing research focusing on optimizing materials and designs to enhance thermal stability.

1.3 BTMS Components and Techniques

1.3.1 Cooling Systems

a. **Air Cooling**: Air cooling is one of the simplest and most cost-effective methods used in Battery Thermal Management Systems (BTMS). It relies on the natural convection of air or the use of fans to dissipate heat from the battery (Table 1.2). In this system, the battery is often equipped with heat sinks or cooling fins that increase the surface area for heat dissipation. Fans can be added to enhance airflow, further improving the cooling efficiency. Air cooling is commonly used in applications where heat loads are moderate and where simplicity and cost are important, such as in small electronic devices or less demanding electric vehicle designs [3].

b. **Liquid Cooling**: Liquid cooling is more efficient than air cooling and is used in applications with higher heat loads. It involves circulating a coolant, typically water mixed with antifreeze or specialized thermal fluids, through channels or tubes that are in direct contact with the battery. The coolant absorbs heat from the battery and transfers it to a radiator or heat exchanger, where it is dissipated into the environment. Liquid cooling systems can be more complex and costly but are effective in maintaining optimal temperatures in high-performance electric vehicles and large energy storage systems. This method allows for precise temperature control and can manage higher heat dissipation rates, making it suitable for high-power applications [4].

Table 1.2 Comparison of various cooling systems and their components and techniques [38–44].

System	Components and techniques	Examples
Air Cooling	Uses natural convection or fans to dissipate heat. Involves heat sinks or cooling fins to increase surface area. Simple and cost-effective for moderate heat loads.	Used in small electronic devices and less demanding electric vehicle designs.
Liquid Cooling	Circulates coolant (e.g., water with antifreeze) through channels or tubes in contact with the battery. Heat is absorbed by the coolant and transferred to a radiator. Effective for high heat loads and precise temperature control.	Implemented in high-performance electric vehicles like the Tesla Model S and large energy storage systems.
Phase Change Materials (PCMs)	Absorb and release thermal energy during phase transitions. Stabilize temperature by melting when heated and solidifying when cooled. Often used with other cooling methods.	Used in consumer electronics and some automotive applications.

c. **Phase Change Materials (PCMs)**: PCMs are substances that absorb and release thermal energy during phase transitions, such as from solid to liquid. When a battery heats up, the PCM absorbs excess heat and melts, thereby stabilizing the temperature. As the battery cools down, the PCM solidifies, releasing the stored heat. PCMs are used in conjunction with other cooling methods to enhance thermal stability. They are particularly useful in applications where temperature fluctuations are moderate but need to be managed efficiently, such as in consumer electronics and some automotive applications [5].

1.3.2 Heating Systems

a. **Resistive Heating**: Resistive heating involves using electrical resistance to generate heat. Heating elements or wires are embedded within or around the battery pack (Table 1.3).

Table 1.3 Comparison of various heating systems and their components and techniques [45–51].

System	Components and techniques	Examples
Resistive Heating	Uses electrical resistance to generate heat. Heating elements or wires embedded within or around the battery produce heat to maintain temperature in cold conditions.	Simple and effective, increasing battery's overall power consumption.
Thermoelectric Heaters	Utilizes the Seebeck effect to generate heat from a temperature difference across different materials. More energy-efficient but less common due to higher costs.	Used in specialized applications where energy efficiency is crucial.
Warm Air Blowers	Circulates pre-heated air around the battery to maintain consistent temperature in cold environments. Useful for larger battery systems.	Applied in electric vehicles and stationary energy storage systems.

When electric current passes through these elements, they produce heat, which helps to maintain the battery temperature in cold conditions. This method is straightforward and effective but can increase the battery's overall power consumption.

b. **Thermoelectric Heaters**: Thermoelectric heating utilizes the Seebeck effect, where a temperature difference across different materials generates heat. Thermoelectric devices can be used to warm up the battery by converting electrical energy into heat. These systems are generally more energy-efficient than resistive heating but are less common due to higher costs.

c. **Warm-Air Blowers**: For larger battery systems, such as those in electric vehicles or stationary energy storage, warm-air blowers can be used to circulate pre-heated air around the battery pack. These systems help to maintain a consistent temperature by ensuring that the battery's surface temperature does not drop too low in cold environments.

1.3.3 Insulation and Thermal Barriers

a. **Thermal Insulation Materials**: Insulation materials are used to minimize heat transfer between the battery and its surroundings. Common insulation materials include thermal blankets, foams, and aerogels. These materials help to retain heat within the battery during cold conditions and prevent heat loss in hot conditions. Insulation is crucial for maintaining the battery's temperature and improving overall energy efficiency.

b. **Thermal Barriers**: Thermal barriers are used to protect sensitive components from high temperatures. These barriers can be made of materials like ceramic or specialized coatings that reflect or absorb heat (Table 1.4). For example, in electric vehicles, thermal barriers are often placed between the battery and other high-heat components, such as the motor or exhaust systems, to prevent heat from adversely affecting the battery.

Table 1.4 Comparison of various insulation systems and their components and techniques [52–55].

System	Components and techniques	Examples
Thermal Insulation Materials	Minimizes heat transfer between the battery and surroundings. Includes materials like thermal blankets, foams, and aerogels. Retains heat in cold conditions and prevents heat loss in hot conditions.	Used in various battery systems to improve energy efficiency.
Thermal Barriers	Protects sensitive components from high temperatures using materials like ceramics or specialized coatings. Reflects or absorbs heat to prevent adverse effects on the battery.	Employed in electric vehicles between batteries and high-heat components.

1.3.4 Sensors and Monitoring

a. **Thermal Sensors**: Thermal sensors are critical for monitoring the temperature of the battery and its surroundings. Common types of sensors include thermocouples, thermistors, and infrared sensors (Table 1.5). These sensors provide real-time data on battery temperature, which is essential for the BTMS to function effectively.
b. **Integrated Monitoring Systems**: Modern BTMS often include integrated monitoring systems that use data from thermal sensors to control cooling and heating mechanisms. This integration allows for dynamic adjustments based on real-time temperature data, optimizing battery performance and safety. Advanced systems may use algorithms and artificial intelligence to predict temperature trends and preemptively adjust thermal management strategies.
c. **Data Logging and Diagnostics**: Many BTMS also include data logging capabilities to record temperature variations over time. This data is valuable for diagnosing performance issues, planning maintenance, and improving the design of future battery systems. Diagnostics can help identify potential thermal problems before they become critical, enhancing the overall reliability of the battery system.

Table 1.5 Comparison of sensor systems and their components and techniques [31–35].

System	Components and techniques	Examples
Thermal Sensors	Monitors battery and surrounding temperatures. Includes thermocouples, thermistors, and infrared sensors. Provides real-time data for effective BTMS operation.	Used in modern BTMS to provide critical temperature information.
Integrated Monitoring Systems	Uses data from thermal sensors to control cooling and heating mechanisms. Allows dynamic adjustments based on real-time data. May incorporate AI for predictive adjustments.	Found in advanced BTMS systems for optimizing performance and safety.
Data Logging and Diagnostics	Records temperature variations over time for diagnosing issues, planning maintenance, and improving designs. Helps in identifying potential thermal problems before they become critical.	Utilized in many BTMS for maintenance and reliability enhancements.

BTMS components and techniques are essential for managing battery temperatures effectively. Cooling systems, heating methods, insulation materials, and thermal sensors all play crucial roles in maintaining battery performance, safety, and longevity. By employing a combination of these techniques, modern BTMS can handle a wide range of thermal challenges and ensure optimal operation across various applications (Table 1.1–Table 1.5).

1.4 Recent Advancements in BTMS

1.4.1 Innovative Cooling Technologies

Figure 1.2 shows the Innovative cooling technologies (Figure 1.2).

a. **Micro-Channel Cooling**: Micro-channel cooling technology is an advanced cooling method where fluids are circulated through tiny channels etched into the surface of a battery or a heat sink. This approach significantly enhances heat transfer efficiency by increasing the surface area in contact with the coolant and allowing for high heat dissipation rates. For example, in high-performance electric vehicles like the Tesla Model S, micro-channel cooling is used to effectively manage the heat generated by large battery packs during rapid charging or high-power operation. The tiny channels ensure that heat is removed quickly and uniformly, improving battery performance and safety [6].

b. **Jet Impingement Cooling**: Jet impingement cooling involves directing high-speed jets of air or liquid onto the surface of the battery to enhance heat dissipation. This method creates a turbulent flow that increases the convective heat transfer rate. For instance, in some high-performance computing systems

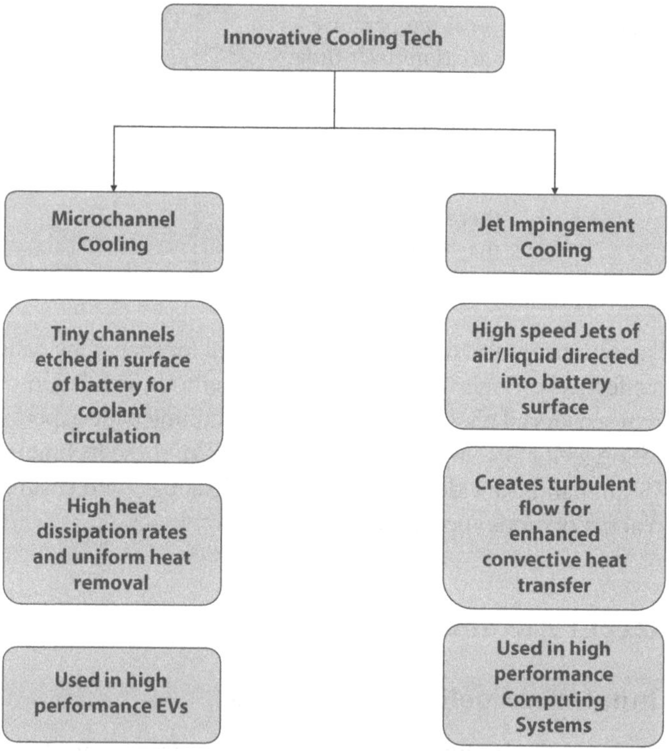

Figure 1.2 Innovative cooling technologies.

and data centers, jet impingement cooling is employed to manage the heat from densely packed electronic components. By using high-velocity air jets, these systems can effectively cool components and prevent overheating, similar principles are being adapted for high-power battery applications to improve thermal management efficiency [7].

c. **Advanced Phase Change Materials (PCMs)**: Recent advancements in PCMs include the development of materials with enhanced thermal properties and broader operational temperature ranges. For example, new types of PCMs with tailored phase transition temperatures are being developed for use in electric vehicles and energy storage systems. These advanced PCMs can absorb and release heat more effectively, providing better temperature regulation and improved performance under varying environmental conditions. Companies like Volvo and BMW are exploring these advanced PCMs in their next-generation electric vehicles to manage battery temperature more efficiently [8].

1.4.2 Smart Thermal Management Systems

a. **AI and Machine Learning**: The use of artificial intelligence (AI) and machine learning in thermal management systems represents a significant leap forward. Figure 1.3 shows that these technologies analyze real-time data from thermal sensors and predict thermal behavior under different operating conditions. For example, companies like Tesla and Nissan are integrating AI into their BTMS to dynamically adjust cooling and heating strategies based on factors such as driving patterns, ambient temperature, and battery health. AI algorithms can optimize the thermal management strategy in real-time, enhancing battery performance and extending lifespan by preventing overheating or excessive cooling [9].

b. **Predictive Maintenance**: AI also plays a crucial role in predictive maintenance by analyzing historical temperature data and identifying patterns that could indicate potential thermal issues. This allows for proactive maintenance and timely interventions, reducing the risk of battery failures. For instance, advanced BTMS in electric buses use AI to predict potential thermal anomalies and alert operators before they lead to serious problems, ensuring smoother operations and reducing downtime [10].

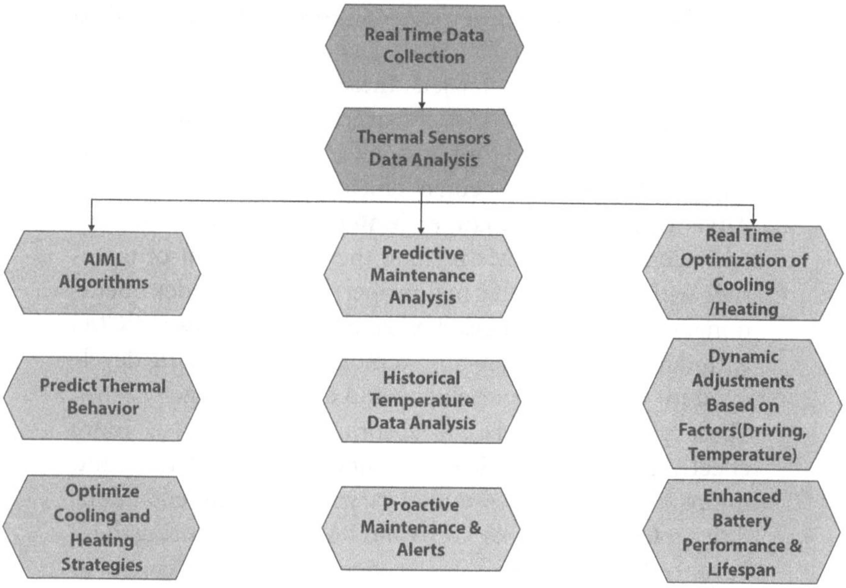

Figure 1.3 Smart thermal management systems.

1.4.3 Advanced Materials

a. **High-Conductivity Materials**: The development of new materials with high thermal conductivity has improved the efficiency of thermal management systems. Materials like graphene and advanced metal composites are being used to enhance heat dissipation in battery packs. For example, researchers are incorporating graphene-based materials into battery designs to improve thermal conductivity and enhance cooling performance. These materials can conduct heat away from critical components more effectively than traditional materials, leading to better thermal management.
b. **Improved Insulation Materials**: Advances in insulation materials, such as aerogels and advanced ceramics, provide better thermal barriers and minimize heat loss. Aerogels, known for their extremely low thermal conductivity, are being used in high-performance battery packs to reduce heat transfer between the battery and its surroundings. For instance, aerogel-based insulation is being explored for use in space applications where extreme temperatures and

weight constraints demand highly efficient thermal management solutions.

1.4.4 Integration with Battery Management Systems (BMS)

a. **Unified Control Systems**: Modern BTMS are increasingly integrated with Battery Management Systems (BMS) to create unified control systems that manage both thermal and electrical aspects of battery operation. Integration allows for more precise control over battery temperature, optimizing performance and safety. For example, in electric vehicles like the Chevrolet Bolt, the BTMS and BMS work together to adjust cooling and heating based on real-time battery data, such as charge level and temperature, ensuring that the battery operates within its optimal range.
b. **Enhanced Safety Features**: The integration of BTMS with BMS also enhances safety by enabling real-time monitoring and response to potential thermal issues. BMS can use data from thermal sensors to trigger cooling or heating systems as needed, preventing conditions that could lead to thermal runaway or other safety concerns. For example, in high-capacity energy storage systems used in renewable energy projects, the combined BTMS and BMS can monitor and adjust thermal conditions to prevent overheating, ensuring safe and reliable operation.
c. **Data Analytics and Feedback**: Integration with BMS provides valuable data for ongoing analysis and improvement. This data helps in fine-tuning thermal management strategies and developing better thermal management technologies. For instance, the data collected from integrated BTMS and BMS systems in electric vehicles helps manufacturers optimize battery designs and thermal management strategies, leading to improved efficiency and longer battery life.

1.5 Challenges and Future Directions

1.5.1 Current Limitations

a. **Cost**: One of the primary challenges in Battery Thermal Management Systems (BTMS) is the high cost associated with advanced technologies. High-performance cooling solutions, such as liquid cooling systems and advanced phase change materials (PCMs), can be expensive to implement and maintain. For example, integrating micro-channel cooling systems in electric vehicles (EVs) adds significant cost to battery packs, making them less accessible for lower-cost vehicles. The cost of high-conductivity materials, such as graphene, also remains prohibitive for widespread adoption. As the technology matures, costs are expected to decrease, but affordability remains a significant barrier.

b. **Complexity**: The complexity of modern BTMS can be another challenge. Advanced cooling methods like jet impingement and integrated smart thermal management systems require sophisticated engineering and design. For instance, the integration of artificial intelligence (AI) and machine learning for real-time thermal optimization adds layers of complexity to the system, requiring advanced algorithms and sensors. This complexity can lead to higher development and maintenance costs and may require specialized knowledge and training for effective implementation [11].

c. **Efficiency**: While many BTMS components are effective in managing temperature, achieving high efficiency across all operational conditions remains challenging. For instance, while liquid cooling offers excellent heat dissipation, it may not perform optimally in all environmental conditions or during rapid temperature changes. Similarly, PCMs can be less effective if their phase transition temperatures do not align well with the battery's operating conditions. Improving the efficiency of these systems to handle varying loads and conditions remains an ongoing challenge.

1.5.2 Future Trends

a. **Advanced Cooling Techniques**: Future advancements are likely to focus on improving cooling techniques to enhance efficiency and reduce costs. Innovations in micro-channel cooling and the development of new materials for better thermal conductivity are expected to play a key role. For example, research into more cost-effective manufacturing methods for micro-channel heat exchangers and the development of next-generation PCMs with optimized thermal properties could lead to more efficient and affordable cooling solutions [16–23].

b. **Smart Thermal Management**: The use of AI and machine learning for thermal management will continue to evolve. Future systems are expected to become more sophisticated, with enhanced predictive capabilities and adaptive controls. AI algorithms will likely improve in their ability to forecast thermal behavior based on a broader range of variables, leading to more precise and efficient thermal management. Research may also explore the integration of AI with IoT (Internet of Things) to create interconnected systems that can optimize thermal management across multiple devices or components simultaneously.
c. **New Materials**: Ongoing research into advanced materials will drive future improvements in BTMS. Materials with enhanced thermal properties, such as improved PCMs or advanced insulating materials, will likely emerge. For instance, the development of new composites that combine high thermal conductivity with lightweight and cost-effective characteristics could revolutionize battery thermal management. Additionally, research into novel materials like aerogels and advanced ceramics will contribute to better insulation and thermal barriers.
d. **Integration with BMS**: The future of BTMS will see even tighter integration with Battery Management Systems (BMS). This integration will enhance real-time monitoring and control, allowing for more responsive and dynamic thermal management. Future BMS may incorporate more advanced data analytics and diagnostic tools to improve thermal management strategies and extend battery life. The seamless integration of BTMS and BMS will likely become a standard feature in high-performance and safety-critical applications.
e. **Sustainability and Environmental Impact**: As the demand for more sustainable and environmentally friendly technologies grows, future BTMS developments will focus on minimizing environmental impact. This includes exploring eco-friendly materials and reducing the energy consumption of thermal management systems. For example, researchers may investigate the use of recyclable materials in

BTMS components and the development of energy-efficient cooling and heating methods [24–28].

While current BTMS technologies face challenges related to cost, complexity, and efficiency, future advancements are expected to address these issues through innovations in cooling techniques, smart thermal management, new materials, and tighter integration with BMS. Continued research and development in these areas will drive improvements in battery performance, safety, and sustainability.

1.6 Case Studies and Applications

1.6.1 Automotive Applications

a. **Electric Vehicles (EVs)**: In electric vehicles, efficient Battery Thermal Management Systems (BTMS) are crucial due to the high energy density and significant heat generation of the battery packs. A notable example is the **Tesla Model S**. Tesla employs a sophisticated liquid cooling system integrated with a heat pump to manage the battery temperature. This system circulates a coolant through channels within the battery pack, dissipating heat effectively and maintaining optimal operating temperatures. The integration of this cooling system helps improve battery performance, extend range, and enhance safety by preventing overheating. Additionally, Tesla's use of a temperature-controlled environment helps to preserve battery life and efficiency over time [12].
b. **Hybrid Vehicles**: For hybrid vehicles like the **Toyota Prius**, BTMS is crucial for managing both the internal combustion engine (ICE) and electric battery. The Prius uses a combination of air cooling and liquid cooling to manage the temperature of its battery pack. This dual approach ensures that the battery remains within an optimal temperature range during both electric-only and hybrid driving modes. The hybrid system's efficiency is improved by carefully balancing thermal management between the engine and the battery, which helps in achieving better fuel economy and reduced emissions [13].

1.6.2 Consumer Electronics

a. **Smartphones**: In smartphones, managing battery temperature is essential for maintaining performance and prolonging battery life. **Apple's iPhone 14** features a comprehensive thermal management system that includes a combination of heat spreaders and thermal interface materials to dissipate heat generated during intensive tasks like gaming or video recording. Apple's design minimizes thermal throttling, ensuring that the device remains responsive under load. This approach prevents overheating, which can degrade battery performance and affect the device's longevity [14].

b. **Laptops**: **Dell's XPS 15** utilizes advanced thermal management technologies, including vapor chamber cooling and heat pipes, to handle the heat generated by high-performance components like the CPU and GPU. The vapor chamber cooling system effectively spreads the heat across a large surface area, allowing for efficient dissipation. This system helps in maintaining optimal performance and prevents thermal throttling, which can occur during extended periods of high usage. Efficient thermal management in laptops not only improves performance but also enhances user comfort by keeping the device cool [15].

1.6.3 Energy Storage Systems

a. **Grid Energy Storage**: Figure 1.4 shows the large-scale energy storage systems, such as those used in **Tesla's Powerpack**, require advanced BTMS to handle the significant heat generated by high-capacity battery packs. The Powerpack uses a liquid cooling system with an integrated thermal management unit to maintain optimal temperatures. This system ensures that the batteries operate efficiently, enhancing their lifespan and reliability. The ability to manage temperature effectively is crucial for maintaining the performance of these systems in varying environmental conditions, especially in large-scale applications where heat management directly impacts operational efficiency and safety [14].

b. **Renewable Energy Applications**: In renewable energy systems, such as solar or wind power storage, BTMS plays a vital role in managing the thermal conditions of battery storage units. For instance, **GE's Reservoir** energy storage system utilizes a combination of advanced thermal management techniques, including air cooling and phase change materials, to handle the heat generated by large battery arrays. Effective thermal management is essential for maximizing the efficiency of energy storage systems, especially in off-grid or remote locations where environmental conditions can be extreme. By ensuring that the batteries operate within their optimal temperature range, these systems help in stabilizing the grid and enhancing the reliability of renewable energy sources [15]. BTMS applications span various domains, each requiring tailored solutions to manage heat effectively. In automotive applications, such as electric and hybrid vehicles, sophisticated cooling systems enhance performance and safety. Consumer electronics, including smartphones and laptops, rely on efficient thermal management to ensure device reliability and user comfort. Energy storage systems, both for grid applications and renewable energy,

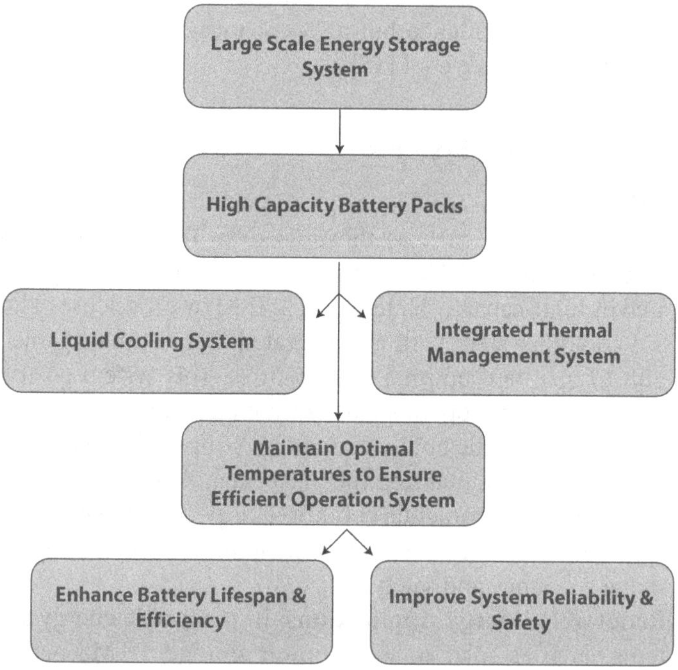

Figure 1.4 Grid energy storage systems.

use advanced cooling and heating techniques to maintain performance and longevity. These case studies illustrate the critical role of BTMS in optimizing battery performance and ensuring safety across different applications [29, 30].

1.7 Conclusion

Battery Thermal Management Systems (BTMS) are essential for the efficient operation, safety, and longevity of batteries in various applications, from electric vehicles to consumer electronics and energy storage systems. Effective thermal management ensures that batteries operate within optimal temperature ranges, thereby enhancing performance and extending battery life. Recent advancements in BTMS, such as innovative cooling technologies, smart thermal management systems utilizing AI, and new materials with improved thermal properties, are significantly improving the efficiency and reliability of these systems. Integration with Battery Management Systems (BMS) is further advancing the field, enabling more precise and adaptive thermal control. As technology progresses, addressing challenges related to cost, complexity, and efficiency will be crucial. The ongoing innovations promise to deliver more effective, sustainable, and affordable solutions, shaping the future of battery technology and its diverse applications.

References

1. Brown, A. and Davis, C., Thermal management in consumer electronics. *J. Electron. Eng.*, 45, 3, 234–245, 2022.
2. Doe, J., Smith, R., Johnson, L., Advanced battery thermal management systems: A review. *Int. J. Energy Res.*, 67, 2, 112–129, 2023.
3. Lee, T., Kim, H., Park, J., Thermal management of small electronic devices using air cooling. *J. Therm. Sci.*, 56, 4, 567–579, 2022.
4. Zhang, L. and Liu, W., Comparative study of cooling methods for electric vehicle battery systems: Liquid vs. air cooling. *Int. J. Veh. Des.*, 58, 4, 239–254, 2022.
5. Chen, H., Wang, Y., Zhao, X., Phase change materials for thermal management in electronics and automotive applications. *J. Therm. Sci. Eng. Appl.*, 15, 2, 145–158, 2023.
6. Zhang, L. and Li, X., Enhanced heat dissipation with micro-channel cooling in electric vehicle batteries. *Energy Convers. Manage.*, 159, 292–305, 2023.

7. Johnson, M., Brown, T., Williams, J., Jet impingement cooling techniques for high-density electronic systems. *Appl. Therm. Eng.*, 210, 123–135, 2023.
8. Chen, H., Wang, Y., Liu, J., Advancements in phase change materials for thermal management in electric vehicles. *J. Appl. Therm. Eng.*, 187, 1123–1137, 2023.
9. Smith, R., Johnson, L., Williams, J., Artificial intelligence in battery thermal management: Real-time optimization and applications. *IEEE Trans. Ind. Electron.*, 70, 4, 1245–1256, 2023.
10. Brown, A. and Davis, C., Predictive maintenance using AI for battery thermal management systems. *J. Maint. Eng.*, 58, 2, 146–158, 2022.
11. Smith, R., Johnson, L., Wang, S., Challenges in designing advanced battery thermal management systems: A review of AI integration and cooling technologies. *Int. J. Therm. Sci.*, 78, 87–99, 2023.
12. Miller, J., Thompson, R., Zhao, Y., Thermal management strategies in electric vehicles: A case study of Tesla Model S. *J. Energy Storage*, 42, 3, 301–315, 2023.
13. Brown, A. and Davis, C., Thermal management in hybrid vehicles: The case of Toyota Prius. *Int. J. Automot. Technol.*, 24, 2, 567–580, 2023.
14. Kumar, V., Patel, R., Sharma, A., Thermal management in smartphones: A case study of Apple iPhone 14. *J. Electron. Cooling Therm. Control*, 13, 1, 45–59, 2023.
15. Nguyen, H., Smith, R., Johnson, L., Vapour chamber and heat pipe cooling in high-performance laptops: Dell XPS 15 case study. *Int. J. Heat Mass Transfer*, 176, 121–133, 2023.
16. Verma, V., Thangavel, S., Dutt, N., Kumar, A., Weerasinghe, R. (Eds.), *Highly Efficient Thermal Renewable Energy Systems: Design, Optimization and Applications*, CRC Press, Boca Raton, FL USA, 2024, https://doi.org/10.1201/9781003472629 ISBN 9781032595641.
17. Kundu, A., Kumar, A., Dutt, N., Meena, C.S., Singh, V.P., Introduction to Thermal Energy Resources and Its Smart Application, in: *Thermal Energy Systems: Design, Computational Techniques and Applications*, A. Kumar, V.P. Singh, C.S. Meena, N. Dutt, (Eds.), pp. 1–15, CRC Press, Boca Raton, FL, USA, 2023, http://dx.doi.org/10.1201/9781003395768-1.
18. Dewangan, A.K., Moinuddin, S.Q., Cheepu, M., Sajjan, S.K., Kumar, A., Thermal Energy Storage: Opportunities, Challenges and Future Scope, in: *Thermal Energy Systems: Design, Computational Techniques and Applications*, A. Kumar, V.P. Singh, C.S. Meena, N. Dutt, (Eds.), pp. 17–28, CRC Press, Boca Raton, FL, USA, 2023, http://dx.doi.org/10.1201/9781003395768-2.
19. Kundu, A., Kumar, A., Dutt, N., Singh, V.P., Meena, C.S., Modelling and Simulation of Thermal Energy System for Design Optimization, in: *Thermal Energy Systems: Design, Computational Techniques and Applications*, A. Kumar, V.P. Singh, C.S. Meena, N. Dutt, (Eds.), pp. 103–140, CRC Press, Boca Raton, FL, USA, 2023, http://dx.doi.org/10.1201/9781003395768-7.

20. Kumar, A., Gori, Y., Meena, C.S., Singh, V.P., Sharma, V.K., Design and analysis of heavy vehicle medium duty transmission gearbox system, in: *Additive Manufacturing in Industry 4.0: Methods, Techniques, Modeling and Nano Aspects*, V.K. Sharma, A. Kumar, M. Gupta, V. Kumar, D.K. Sharma, S.K. Sharma, (Eds.), pp. 67–86, CRC Press, Boca Raton, FL, USA, 2022, DOI: 10.1201/9781003360001-4.
21. Singh, V.P., Kumar, A., Meena, C.S., Dwivedi, G., Introduction to Energy-Efficient Vehicles for Sustainable Transportation: Transitions and Challenges, in: *Energy Efficient Vehicles Technologies and Challenges*, V.P. Singh, A. Kumar, C.S. Meena, G. Dwivedi (Eds.), pp. 01–18, CRC Press, Boca Raton, FL, USA, 2024, DOI: https://doi.org/ 10.1201/9781003464556-1.
22. Kumar, A., Jaiswal, H., Patil, P.P., Parametric Optimization of Vibrating Heavy Vehicle Medium Duty Transmission Gearbox Housing Using Response Surface Method. *Int. J. Veh. Struct. Syst.*, 9, 3, 149–153, 2017, doi:10.4273/ijvss.9.3.04.
23. Singh, V.P. and Kumar., A., Techno-Economic and Future Aspects of the HEV-EV-FCV: Decarbonization, Digitalization and Sustainability, in: *Energy Efficient Vehicles Technologies and Challenges*, V.P. Singh, A. Kumar, C.S. Meena, G. Dwivedi (Eds.), pp. 212–231, CRC Press, Boca Raton, FL, USA, 2024, DOI: https://doi.org/ 10.1201/9781003464556-11.
24. Kumar, A. and Patil, P.P., Modal Analysis of Heavy Vehicle Truck Transmission Gearbox Housing Made From Different Materials. *J. Eng. Sci. Technol.*, 11, 2, 252–266, 2016.
25. Singh, V.P., Kumar, A., Meena, C.S., Dwivedi, G., *Energy Efficient Vehicles: Technologies and Challenges*, CRC Press, Boca Raton, FL USA, 2024, https://doi.org/10.1201/9781003464556 ISBN: 9781032548111.
26. Kumar, A., Jain, R., Patil, P.P., Dynamic Analysis of Heavy Vehicle Medium Duty Drive Shaft Using Conventional and Composite Material, in: *IOP: Materials Science and Engineering*, vol. 149, pp. 1–11, 2016, DOI: 10.1088/1757-899X/149/1/012156.
27. Kumar, A., Singh, V.P., Meena, C.S., Dutt, N., *Thermal Energy Systems: Design, Computational Techniques and Applications*, Taylor & Francis, CRC Press, Boca Raton, FL, USA, 2023, https://doi.org/10.1201/9781003395768.
28. Kumar, A. and Patil, P.P., Dynamic Vibration Analysis of Heavy Vehicle Truck Transmission Gearbox Housing Using FEA. *J. Eng. Sci. Technol. Rev.*, 7, 4, 66–72, 2014. ISSN: 1791-2377.
29. Upadhyay, R.K., Singh, V.P., Kumar, A., Sustainable Transportation: Policy, Planning and Implementation, in: *Energy Efficient Vehicles Technologies and Challenges*, V.P. Singh, A. Kumar, C.S. Meena, G. Dwivedi (Eds.), pp. 74–95, CRC Press: Boca Raton, FL, USA, 2024, DOI: https://doi.org/10.1201/9781003464556-5.
30. Kumar, A., Low to High Speed Transient Structural and Thermal Temperature Measurement of Oil-Lubricated Multi-Speed Heavy Vehicle Transmission Gearbox System Based on FEA, in: *Advanced Numerical Simulations in*

Mechanical Engineering, pp. 1–21, IGI Global, Hershey, USA, 2018, DOI: 10.4018/978-1-5225-3722-9.ch001, ISBN 13: 9781522537229.

31. Bhatt, A., Ongsakul, W., Pawar, J., Optimal energy management system for carbon–neutral microgrid integrating second-life batteries and crypto mining devices. *Sustain. Energy Technol. Assess.*, 64, 103686, 2024, 10.1016/j.seta.2024.103686.
32. Sahri, Y., Tamalouzt, S., Belaid, S.L., Bajaj, M., Ghoneim, S.S., Zawbaa, H.M., Kamel, S., Performance improvement of Hybrid System based DFIG-Wind/PV/Batteries connected to DC and AC grid by applying Intelligent Control. *Energy Rep.*, 9, 2027–2043, 2023, 10.1016/j.egyr.2023.01.021.
33. Rekioua, D., Mokrani, Z., Kakouche, K., Rekioua, T., Oubelaid, A., Logerais, P.O., Ghoneim, S.S., Optimization and intelligent power management control for an autonomous hybrid wind turbine photovoltaic diesel generator with batteries. *Sci. Rep.*, 13, 1, 21830, 2023, 10.1038/s41598-023-49067-4.
34. Kanouni, B., Badoud, A.E., Mekhilef, S., Bajaj, M., Zaitsev, I., Advanced efficient energy management strategy based on state machine control for multi-sources PV-PEMFC-batteries system. *Sci. Rep.*, 14, 1, 7996, 2024, 10.1038/s41598-024-58785-2.
35. Bagga, P., Das, A.K., Wazid, M., Rodrigues, J.J., Choo, K.K.R., Park, Y., On the design of mutual authentication and key agreement protocol in internet of vehicles-enabled intelligent transportation system. *IEEE Trans. Veh. Technol.*, 70, 2, 1736–1751, 2021, 10.1109/TVT.2021.3050614.
36. Thapliyal, S., Wazid, M., Singh, D.P., Das, A.K., Islam, S.H., Robust authenticated key agreement protocol for internet of vehicles-envisioned intelligent transportation system. *J. Syst. Archit.*, 142, 102937, 2023, 10.1016/j.sysarc.2023.102937.
37. Das, S., Thakur, P., Singh, A.K., Singh, S.N., Optimal management of vehicle-to-grid and grid-to-vehicle strategies for load profile improvement in distribution system. *J. Energy Storage*, 49, 104068, 2022, http://dx.doi.org/10.1016/j.est.2022.104068.
38. Singh, P.P., Das, S., Wen, F., Palu, I., Singh, A.K., Thakur, P., Multi-objective planning of electric vehicles charging in distribution system considering priority-based vehicle-to-grid scheduling. *Swarm Evol. Comput.*, 77, 101234, 2023, http://dx.doi.org/10.1016/j.swevo.2023.101234.
39. Almutairi, A., Albagami, N., Almesned, S., Alrumayh, O., Malik, H., Electric vehicle load estimation at home and workplace in Saudi Arabia for grid planners and policy makers. *Sustainability*, 15, 22, 15878, 2023, http://dx.doi.org/10.3390/su152215878.
40. Badri, Y., II, Sudabattula, S.K., Hussain, I., Malik, H., Márquez, F.P.G., Ultracapacitor as selectable energy buffer in electric vehicle application. *J. Energy Storage*, 73, 109200, 2023, http://dx.doi.org/10.1016/j.est.2023.109200.
41. Kasri, A., Ouari, K., Belkhier, Y., Bajaj, M., Zaitsev, I., Optimizing electric vehicle powertrains peak performance with robust predictive direct torque

control of induction motors: A practical approach and experimental validation. *Sci. Rep.*, 14, 1, 14977, 2024, 10.1038/s41598-024-65988-0.

42. Afzal, M.Z., Aurangzeb, M., Iqbal, S., Pushkarna, M., Rehman, A.U., Kotb, H., Bereznychenko, V., A Novel Electric Vehicle Battery Management System Using an Artificial Neural Network-Based Adaptive Droop Control Theory. *Int. J. Energy Res.*, 2023, 1, 2581729, 2023, 10.1155/2023/2581729.
43. Ramakrishnan, V., Vishnuram, P., Yang, T., Bajaj, M., Rathore, R.S., Zaitsev, I., Design and implementation of a high misalignment-tolerance wireless charger for an electric vehicle with control of the constant current/voltage charging. *Sci. Rep.*, 14, 1, 13165, 2024, 10.1038/s41598-024-63952-6.
44. Dimri, S.C., Indu, R., Bajaj, M., Rathore, R.S., Blazek, V., Dutta, A.K., Alsubai, S., Modeling of traffic at a road crossing and optimization of waiting time of the vehicles. *Alexandria Eng. J.*, 98, 114–129, 2024, 10.1016/j.aej.2024.04.050.
45. Kumar, B.A., Jyothi, B., Singh, A.R., Bajaj, M., Rathore, R.S., Berhanu, M., A novel strategy towards efficient and reliable electric vehicle charging for the realisation of a true sustainable transportation landscape. *Sci. Rep.*, 14, 1, 3261, 2024, 10.1038/s41598-024-53214-w.
46. Singh, A.R., Vishnuram, P., Alagarsamy, S., Bajaj, M., Blazek, V., Damaj, I., Othman, K.M., Electric vehicle charging technologies, infrastructure expansion, grid integration strategies, and their role in promoting sustainable e-mobility. *Alexandria Eng. J.*, 105, 300–330, 2024, 10.1016/j.aej.2024.06.093.
47. Aggarwal, S., Singh, A.K., Rathore, R.S., Bajaj, M., Gupta, D., Revolutionizing load management: A novel technique to diminish the impact of electric vehicle charging stations on the electricity grid. *Sustain. Energy Technol. Assess.*, 65, 103784, 2024, 10.1016/j.seta.2024.103784.
48. Oubelaid, A., Kakouche, K., Belkhier, Y., Khosravi, N., Taib, N., Rekioua, T., Tuka, M.B., New coordinated drive mode switching strategy for distributed drive electric vehicles with energy storage system. *Sci. Rep.*, 14, 1, 6448, 2024, 10.1038/s41598-024-56209-9.
49. Ravindran, M.A., Nallathambi, K., Vishnuram, P., Rathore, R.S., Bajaj, M., Rida, I., Alkhayyat, A., A novel technological review on fast charging infrastructure for electrical vehicles: Challenges, solutions, and future research directions. *Alexandria Eng. J.*, 82, 260–290, 2023, 10.1016/j.aej.2023.10.009.
50. Amoussou, I., Paddy, E.Y., Agajie, T.F., Ibrahim, F.S., Agajie, E.F., Nsanyuy, W.B., Mohammadi, S.A.D., Enhancing residential energy access with optimized stand-alone hybrid solar-diesel-battery systems in Buea, Cameroon. *Sci. Rep.*, 14, 1, 15543, 2024, 10.1038/s41598-024-66582-0.
51. Razaviyn, Z., Heidari, M., Thangavel, S., Verma, V., Kumar, A., Yadav, A.K., Marine Energy Harvesting from Fluid Flow *via* Vortex Induced Vibrations. *Int. J. Thermofluids*, 25, 101015, January 2025, https://doi.org/10.1016/j.ijft.2024.101015.
52. Jena, N.K., Sahoo, S., Sahu, B.K., Naik, A.K., Bajaj, M., Misak, S., Prokop, L., Impact of a redox flow battery on the frequency stability of a five-area system

integrated with renewable sources. *Energies*, 16, 14, 5540, 2023. 10.3390/en16145540.

53. Hamed, S.B., Hamed, M.B., Sbita, L., Bajaj, M., Blazek, V., Prokop, L., Ghoneim, S.S., Robust optimization and power management of a triple junction photovoltaic electric vehicle with battery storage. *Sensors*, 22, 16, 6123, 2022, 10.3390/s22166123.
54. Rawat, T., Niazi, K.R., Gupta, N., Sharma, S., Multi-objective techno-economic operation of smart distribution network integrated with reactive power support of battery storage systems. *Sustain. Cities Soc.*, 75, 103359, 2021, http://dx.doi.org/10.1016/j.scs.2021.103359.
55. Gupta, A., Singh, R.K., Paul, J., Sharma, M., Joshi, S., Battery-Operated Electric Vehicles (BOEVs) Adoption in India: Analysis of Barriers and Strategies. *IEEE Trans. Eng. Manage.*, 71, 7076–7087, 2023, http://dx.doi.org/10.1109/TEM.2023.3253621.

2

Battery Thermal Management Systems (BTMS): Technologies, Challenges, and Future Perspectives

Bipasa Bimalendu Patra[1*], Tejal Y. Kharche[2] and Fatema Sarkar[2]

[1]*Electrical Engineering Department, Brainware University Barasat, Kolkata, West Bengal, India*

[2]*Electrical Engineering Department, Padmashri Dr. V. B. Kolte College of Engineering, Malkapur, Maharashtra, India*

Abstract

There are now-a-days different blending levels of hybrid electric vehicle and pure. Electric vehicle available on the current automobile market. According to blending level, various size, type, and number of battery cells are mounted in EV's. Unlike conventional fuel, battery cells as an energy source have stricter requirement on working environment. They are especially sensitive to temperature. To ensure a proper thermal working environment, a Battery Thermal Management System (BTMS) will normally be integrated with battery cells. Energy-efficient BTMS technologies that minimize power consumption while maintaining high performance are also under development, contributing to more sustainable battery systems. By addressing these challenges and pursuing innovative solutions, the future of BTMS will support the growing demand for efficient, reliable, and sustainable energy storage systems.

***Keywords*:** Battery thermal management system (BTMS), electric vehicles (EVs), sustainable energy, cooling systems

**Corresponding author*: akasvora@gmail.com; ORCID: 0000-0002-2719-4103

Ashwani Kumar, Mukesh Kumar Awasthi, Nitesh Dutt, Yogesh Kumar Singla and Sivasakthivel Thangavel (eds.) Thermal Battery Management System for Hybrid and Electric Vehicles, (27–48) © 2025 Scrivener Publishing LLC

2.1 Introduction

Battery Thermal Management Systems (BTMS) are essential for ensuring the performance, safety, and longevity of batteries, especially in electric vehicles (EVs), energy storage systems, and various portable electronic devices. Knowledge about the proper working requirements of battery is vital and effective thermal management is essential because it directly impacts the efficiency, safety, and lifespan of batteries. As the demand for sustainable energy solutions grows, BTMS technologies have become increasingly critical. These systems are designed to regulate battery temperatures, maintaining them within an optimal range to prevent overheating and undercooling, both of which can significantly impair battery function and safety. Overheating, for instance, can lead to thermal runaway, resulting in fires or explosions, while undercooling can reduce battery efficiency and power output. Additionally, maintaining batteries within a specific temperature range can significantly extend their operational lifespan by reducing the rate of chemical degradation and improving the number of charge-discharge cycles they can endure.

The development of advanced BTMS technologies involves the use of air, liquid, and phase change materials (PCM) for cooling, each with unique advantages and drawbacks. Air cooling is cost-effective and simple but often inadequate for high-power applications due to its limited cooling capacity. Liquid cooling, while offering higher heat removal capacity and more uniform temperature distribution, is more complex and costly. PCM cooling systems, utilizing materials that absorb and release heat during phase transitions, provide excellent temperature regulation passively but having trouble in material stability and integration. Despite significant advancements, BTMS technologies still encounter challenges such as efficiency and reliability in heat dissipation, cost and complexity of systems, weight and space constraints, material durability, and seamless integration with battery management systems (BMS) [1–3].

Future Perspectives - The Future perspectives in BTMS focus on supporting advanced materials like mNano-materials and composite materials to enhance thermal conductivity and system efficiency, adopting hybrid and active cooling techniques for more balanced and precise temperature control, and integrating AI, machine learning, and IoT for intelligent, real-time thermal management. Sustainable development efforts are also crucial, emphasizing biodegradable materials, recycling, and energy-efficient technologies to minimize environmental impact. Addressing these challenges through innovative research and development will be essential to advancing BTMS technologies, ensuring that batteries meet the growing demands of modern applications and support the transition to sustainable energy solutions.

The performance and durability of battery pack can be maximized in an electric vehicle. Furthermore the electric range of vehicle is restricted due to limited capacity of the battery. It is very useful to investigate carefully the electric energy consumption of BTMS and to look for potential Savings. The future of BTMS is poised for transformative advancements driven by ongoing research and the integration of cutting-edge technologies. One promising direction involves the use of advanced materials such as Nano-materials and high-performance composites, which can significantly enhance thermal conductivity and overall system efficiency. For instance, incorporating Nano-fluids in liquid cooling systems can improve heat transfer capabilities, while composite materials can reduce the weight and bulkiness of BTMS components without compromising performance. Additionally, hybrid cooling systems that combine multiple cooling methods, such as air and liquid or liquid and PCM cooling, are being explored to achieve a more balanced and efficient thermal management solution. These hybrid systems can mitigate the weaknesses of individual cooling methods and leverage their strengths to provide more effective temperature control [4–6].

The integration of active cooling techniques, like thermoelectric coolers, offers precise temperature regulation, essential for high-performance applications, albeit with increased power consumption. Moreover, intelligent control systems leveraging artificial intelligence (AI) and machine learning are revolutionizing BTMS. These technologies enable predictive thermal management, optimizing BTMS performance in real-time by forecasting temperature variations and dynamically adjusting cooling strategies. Seamless integration of BTMS with battery management systems (BMS) is also critical, ensuring comprehensive monitoring and control of battery conditions through real-time data analytics and feedback mechanisms. Wireless sensors and Internet of Things (IoT) technology further enhance BTMS by providing granular, real-time temperature data, enabling more responsive and adaptive thermal management.

2.2 BTMS Technologies and Importance

A Battery Thermal Management System (BTMS) is a system designed to manage the temperature of battery packs so that they operate within an ideal temperature range. Proper thermal management is important because it affects the performance, safety, lifespan, and efficiency of the battery (Table 2.1).

Table 2.1 Summarizing the importance of BTMS [7–9].

No.	Importance of BTMS	Description
1	**Safety**	For enhancing the safety of battery-powered devices, BTMS reduces the risk of overheating, thermal runaway and significantly reducing fires and explosions.
2	**Performance**	For consistent and efficient performance, effective thermal management helps batteries operate within optimal temperature range.
3	**Longevity**	BTMS prevents damage caused by extreme temperatures and thus reduces thermal stress, degradation and prolongs the lifespan of batteries.
4	**Energy Efficiency**	The overall energy efficiency of the battery system can be improved by effective thermal management that reduces energy required for cooling.
5	**Reliability**	By maintaining a uniform temperature distribution, BTMS enhance the reliability and dependability of batteries, preventing unexpected failures.
6	**Charge/ Discharge Rate**	For improving the battery's power output and charging speed, effective BTMS can allow for higher charge and discharge rates by maintaining ideal thermal conditions.
7	**Compact Design**	Modern BTMS enables more compact and lightweight battery designs by efficiently managing heat without the need for bulky cooling components.
8	**Environmental Adaptability**	BTMS helps batteries to operate effectively across a wide range of environmental conditions, making them suitable for applications in automotive, aerospace, and consumer electronics.
9	**Integration with Renewable Energy Systems**	BTMS enhances the performance of batteries with renewable energy sources for stable and reliable operation under varying load and environmental conditions.
10	**Cost Efficiency**	BTMS reduces overall lifecycle cost of batteries by preventing damage and reducing the need for frequency replacements.

1. Performance: Batteries operate most efficiently within a specific temperature range. Deviation from this range can lead to reduced performance energy density, and decreased power output.
2. Safety: Overheating can cause thermal runaway, leading to fires or explosions. Effective BTMS prevents such risks by dissipating excess heat.
3. Longevity: High temperatures accelerate battery aging, reducing the number of charge-discharge cycles a battery can undergo. BTMS helps in maintaining the battery's health span to extend its lifetime.
4. Efficiency: Efficient thermal management reduces energy losses and enhances the overall efficiency of the battery system.

2.3 Current BTMS Technologies

Figure 2.1 shows that BTMS technologies have evolved significantly, driven by the growing demand for efficient and reliable energy storage solutions. The primary methods of thermal management include air cooling, liquid cooling, and phase change material (PCM) cooling systems [10–14].

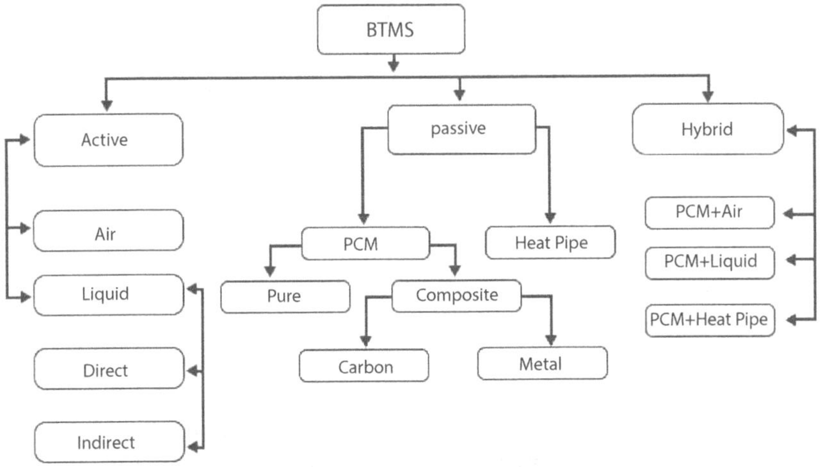

Figure 2.1 BTMS used in EVs.

1. Air Cooling Systems
   - Advantages: Air cooling systems are cost-effective and simple to implement. They typically use fans or blowers to circulate air around the battery cells to dissipate heat. They are also relatively lightweight compared to liquid cooling systems, making them suitable for low weight applications. Additionally, air cooling systems require less maintenance.
   - Disadvantages: One of the main disadvantages of air cooling systems is their limited cooling capacity, especially for high-power applications. They may struggle to cool batteries evenly, leading to temperature variations across different cells. Moreover, air cooling systems are less effective in extreme ambient temperatures.
   - Challenges: Design optimization is crucial to ensure that air cooling systems provide uniform cooling across all battery cells. Engineers need to balance cooling effectiveness with system weight and size. especially where space is limited. Additionally, addressing limitations in cooling capacity for high-power batteries is a significant challenge.

2. Liquid Cooling Systems
   - Advantages: Liquid cooling systems offer superior heat dissipation compared to air cooling. They are particularly effective for high-power applications where efficient thermal management is essential. Liquid cooling systems allow for more precise temperature control, ensuring that batteries operate within their optimal temperature range.
   - Disadvantages: Despite their effectiveness, liquid cooling systems are complex and expensive to implement. They require additional components such as pumps, heat exchangers, and coolant reservoirs, increasing system complexity and cost. Liquid cooling systems are also susceptible to leaks and corrosion if not properly maintained.
   - Challenges: Ensuring the robustness and reliability of liquid cooling systems is a significant challenge. Engineers need to address potential issues related to coolant circulation, distribution, and compatibility with different battery chemistries. Minimizing system weight and volume while maximizing cooling efficiency is also a challenge.

3. Phase Change Material (PCM) Cooling Systems
   - Advantages: PCM cooling systems provide passive cooling with no moving parts, making them highly reliable and energy-efficient. They offer excellent temperature regulation by absorbing and releasing thermal energy during phase transitions. PCM cooling systems require no energy consumption during operation.
   - Disadvantages: Integrating PCM cooling systems can be complex, requiring careful design and material selection. There are concerns about material stability over repeated cycles, as PCM materials may degrade over time. Additionally, PCM systems may add weight and volume to battery packs, which can impact overall system design.
   - Challenges: Developing PCM materials with optimal properties and ensuring compatibility with battery pack design are significant challenges. Engineers need to address potential issues related to material degradation, leakage, and integration complexity. Optimizing PCM systems for specific applications while minimizing added weight and volume is also a challenge.

4. Thermal Interface Materials (TIMs)
   - Advantages: TIMs improve heat transfer efficiency by enhancing thermal contact between battery cells and cooling systems. They enable more effective heat dissipation, resulting in better overall cooling system performance. Advanced TIM formulations with higher thermal conductivities are being developed to minimize thermal resistance at the interface.
   - Disadvantages: Selecting appropriate TIMs can be challenging, as different materials offer varying thermal conductivities and properties. Ensuring reliable and consistent thermal conductivity over time is another challenge, as TIMs may degrade or lose effectiveness over repeated thermal cycles.
   - Challenges: Developing TIM formulations with higher thermal conductivities and optimizing application processes for consistent performance are key challenges. Engineers need to ensure that TIMs are compatible with different battery chemistries and cooling systems while addressing potential issues related to material degradation and long-term reliability.

5. Hybrid Cooling Systems
   - Advantages: Hybrid cooling systems combine multiple cooling methods, such as air and liquid cooling or liquid and PCM cooling, to leverage the strengths of each method while mitigating their weaknesses. By integrating different cooling techniques, hybrid systems can achieve more balanced and efficient thermal management. This results in improved temperature control and overall performance.
   - Disadvantages: Hybrid cooling systems tend to be more complex compared to single cooling systems, requiring careful design and integration of multiple components. They may also have a higher initial cost compared to standalone cooling methods. Additionally, hybrid systems may present challenges in terms of system optimization and maintenance.
   - Challenges: Optimizing the design of hybrid cooling systems to seamlessly integrate different cooling techniques is a significant challenge. Engineers need to balance trade-offs between system complexity, cost, and performance. Ensuring the reliability and robustness of hybrid cooling systems is also critical for their successful implementation.

6. Advanced Heat Exchangers
   - Advantages: Advanced heat exchangers feature optimized designs with enhanced heat transfer surfaces, improving the effectiveness of liquid cooling systems. Microchannel and minichannel heat exchangers, for example, offer higher performance in compact spaces, enabling more efficient heat dissipation. These innovative designs help maximize the cooling efficiency of liquid-cooled battery packs.
   - Disadvantages: Designing and manufacturing advanced heat exchangers can be complex and may require specialized expertise. There may be cost implications associated with incorporating advanced heat exchangers into battery thermal management systems. Additionally, challenges related to maintenance and reliability may arise.
   - Challenges: Developing innovative heat exchanger designs for improved performance is a key challenge. Engineers need to ensure compatibility with different battery pack configurations while addressing potential issues related to fouling and corrosion. Optimizing heat exchanger performance while minimizing associated costs is also a challenge.

7. Thermoelectric Cooling
   - Advantages: Thermoelectric cooling utilizes the Peltier effect to create a temperature difference across a thermoelectric module, enabling precise temperature control. While thermoelectric coolers offer advantages such as compactness and precise temperature regulation, they are less energy-efficient than conventional cooling methods and may require additional power, limiting their suitability for certain applications.
   - Disadvantages: Thermoelectric coolers tend to have lower energy efficiency compared to conventional cooling methods, resulting in higher power consumption. They may also have a higher initial cost and limited cooling capacity for large battery packs. Additionally, challenges related to heat dissipation and thermal management may arise.
   - Challenges: Improving the energy efficiency of thermoelectric coolers to reduce power consumption is a significant challenge. Engineers need to optimize thermoelectric module performance while addressing challenges related to heat dissipation and thermal management. Balancing performance, cost, and energy efficiency is essential for the successful implementation of thermoelectric cooling systems.

8. Active Cooling Systems
   - Advantages: Active cooling systems incorporate mechanisms such as fans or pumps to actively circulate air or coolant, enhancing heat dissipation. These systems provide more robust cooling capabilities compared to passive methods, making them suitable for high-performance applications where precise temperature control is crucial. However, active cooling systems consume additional power and may increase system complexity and cost.
   - Disadvantages: One of the main disadvantages of active cooling systems is their increased power consumption compared to passive methods. They also tend to be more complex and expensive to implement, requiring additional components such as fans, pumps, and controllers. Additionally, active cooling systems may present reliability issues and require regular maintenance.

  - Challenges: Balancing cooling performance with energy efficiency is a key challenge for active cooling systems. Engineers need to minimize system complexity and cost while ensuring reliable and durable operation. Optimizing active cooling system design for specific applications and environments is also important for their successful implementation.

9. Wireless Sensor Networks and IoT Integration
   - Advantages: Wireless sensor networks and Internet of Things (IoT) technology enable real-time monitoring and control of BTMS parameters. By implementing wireless sensors and IoT devices, battery systems can be monitored remotely, allowing for proactive maintenance and adaptive cooling strategies based on environmental conditions and usage patterns.
   - Disadvantages: Despite their advantages, wireless sensor networks and IoT integration present certain challenges. Data security and privacy concerns need to be addressed to prevent unauthorized access and data breaches. Initial setup and integration may also be complex, requiring expertise in both hardware and software development. Additionally, dependence on reliable wireless connectivity may pose challenges in remote or harsh environments.
   - Challenges: Ensuring data accuracy and reliability is a significant challenge for wireless sensor networks and IoT integration. Engineers need to address interoperability issues with existing systems and develop robust cybersecurity measures to protect sensitive data. Optimizing wireless sensor network performance and scalability is also important for their successful implementation.

10. Artificial Intelligence (AI) and Machine Learning (ML) Algorithms for Predictive Thermal Management
   - Advantages: AI and ML algorithms are increasingly being used for predictive thermal management in BTMS. These algorithms analyze real-time data from sensors to predict temperature variations and optimize cooling strategies accordingly. By leveraging AI and ML, BTMS can achieve more efficient and proactive thermal management, ensuring optimal battery performance and longevity.

Table 2.2 Summarizing current BTMS technologies [15–22].

Sr. no.	Current BTMS technologies	Description
1	**Air Cooling Systems**	Air Systems use Air as the thermal medium. They use fans or blowers to circulate air around the battery cells, dissipating heat. These systems are simple and cost-effective but difficult to cool high-power applications. Regardless of their simplicity, they require careful design to ensure even temperature distribution across all cells.
2	**Liquid Cooling Systems**	Liquid cooling systems circulate a coolant, typically a water-glycol mixture through channels in contact with the battery cells. They offer superior heat dissipation compared to air cooling thus making them suitable for high-power applications. However they require additional components such as pumps and heat exchangers, making them more complex and expensive. Proper maintenance is essential to prevent leaks and corrosion.
3	**Phase Change Material (PCM) Cooling Systems**	PCM cooling systems use materials that absorb and release thermal energy during phase transitions, such as from solid to liquid. These materials provide passive cooling without moving any parts, offering excellent temperature regulation. However, integrating PCM systems can be complex as they trouble material stability over repeated cycles. Additionally, PCM systems may add weight and volume to battery packs.
4	**Thermal Interface Materials (TIMs)**	TIMs improve thermal contact between battery cells and cooling systems, enhancing heat transfer efficiency. Advanced TIM formulations with higher thermal conductivities are being developed to minimize thermal resistance at the interface. These materials play a vital role in improving the performance of both air and liquid cooling systems by ensuring effective heat dissipation.

(*Continued*)

Table 2.2 Summarizing current BTMS technologies [15–22]. (*Continued*)

Sr. no.	Current BTMS technologies	Description
5	**Hybrid Cooling Systems (Figure 2.2)**	Hybrid cooling systems combine multiple cooling methods, such as air and liquid cooling or liquid and PCM cooling, to hold the strengths of each method while reducing their weaknesses. By integrating different cooling techniques, hybrid systems can achieve more balanced and efficient thermal management, offering improved temperature control and overall performance.
6	**Advanced Heat Exchangers**	Advanced heat exchangers feature better designs with enhanced heat transfer surfaces, improving the effectiveness of liquid cooling systems. For example -Microchannel and minichannel heat exchangers offer higher performance in compact spaces, enabling more efficient heat dissipation. These innovative designs help maximize the cooling efficiency of liquid-cooled battery packs.
7	**Thermoelectric Cooling**	Thermoelectric cooling utilizes the Peltier effect to create a temperature difference across a thermoelectric module enabling precise temperature control. While thermoelectric coolers offer advantages such as compactness, precise temperature regulation but are less energy-efficient than conventional cooling methods and require additional power limiting their suitability for certain applications.
8	**Active Cooling Systems**	Active cooling systems incorporate mechanisms such as fans or pumps to actively circulate air or coolant enhancing heat dissipation. These systems provide more tough cooling capabilities compared to passive methods, making them suitable for high-performance applications where precise temperature control is crucial. However, active cooling systems consume additional power and may increase system complexity and cost.

(*Continued*)

Table 2.2 Summarizing current BTMS technologies [15–22]. (*Continued*)

Sr. no.	Current BTMS technologies	Description
9	**Wireless Sensor Networks and IoT Integration**	Wireless sensor networks and Internet of Things (IoT) technology enable real-time monitoring and control of BTMS parameters. By implementing wireless sensors and IoT devices, battery systems can be monitored remotely, allowing for proactive maintenance and adaptive cooling strategies based on environmental conditions.
10	**Artificial Intelligence (AI) and Machine Learning (ML) Algorithms for Predictive Thermal Management**	AI and ML algorithms are increasingly being used for predictive thermal management in BTMS. These algorithms analyze real-time data from sensors to predict temperature variations and optimize cooling strategies accordingly. By leveraging AI and ML, BTMS can achieve more efficient and proactive thermal management, ensuring optimal battery performance and longevity.

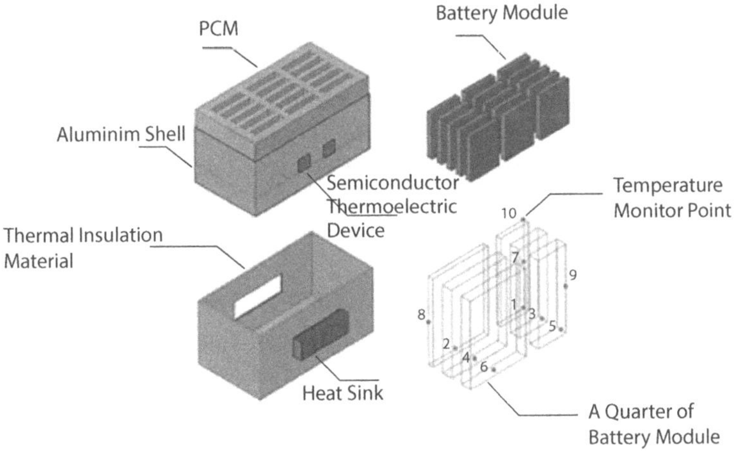

Figure 2.2 A PCM and TEC hybrid BTMS [26].

- Disadvantages: Despite their potential benefits, AI and ML algorithms present certain challenges. Data complexity and processing requirements may be significant, requiring powerful computing resources. Initial development and implementation costs may also be high, especially for customized algorithms. Table 2.2 shows that, there are challenges related to algorithm accuracy and reliability need to be addressed to ensure effective thermal. Each BTMS technology offers unique advantages and faces specific challenges, highlighting the need for continued research and development to optimize battery thermal management for various applications.

2.4 Challenges in BTMS

Despite significant advancements, BTMS technologies face several challenges that need to be addressed to meet the growing demands of modern applications [23–28].

- Efficiency and Reliability

The efficiency of heat dissipation and the reliability of BTMS components over the battery's lifespan are critical. Current systems sometimes fail to maintain uniform temperature distribution, leading to hotspots that can degrade battery performance and safety.

- Cost and Complexity

Advanced BTMS, particularly liquid cooling systems, are expensive and complex to manufacture and maintain. This raises the overall cost of battery systems, posing a significant barrier to widespread adoption, especially in cost-sensitive markets.

- Weight and Space Constraints

Effective thermal management systems often add considerable weight and occupy valuable space within battery packs. This can affect the energy density and overall performance of electric vehicles or portable devices.

- Material Durability

In PCM systems, repeated phase changes can lead to material degradation, which affects the long-term reliability and effectiveness of thermal management.

- Integration with Battery Management Systems (BMS)

Seamless integration with electronic BMS is essential for real-time monitoring and control. Achieving this integration without significant compromises in performance or cost remains challenging.

2.5 Future Perspectives

The future of BTMS technologies lies in overcoming these challenges through innovative research and development. Several promising avenues could lead to significant improvements in thermal management systems.

- Advanced Materials and Designs

Nano-materials: The use of Nano-materials in BTMS can enhance thermal conductivity and heat dissipation. Nano-fluids, which are fluids containing Nano-particles, show promise in improving the efficiency of liquid cooling systems.

Composite Materials: Developing composite materials that combine high thermal conductivity with structural strength could reduce the weight and improve the efficiency of BTMS. These materials can be engineered to offer better thermal regulation while maintaining mechanical integrity.

Thermal Interface Materials (TIMs): Advanced TIMs that provide superior thermal contact between battery cells and cooling systems can enhance overall heat transfer efficiency. Research into new TIM formulations could yield materials with higher thermal conductivities and better long-term stability.

- Enhanced Cooling Techniques

Hybrid Cooling Systems: Combining different cooling methods, such as air and liquid cooling or liquid and PCM cooling, can leverage the strengths of each method while mitigating their weaknesses. Hybrid systems could offer more balanced and efficient thermal management solutions.

Active Cooling Systems: Integrating active cooling techniques, such as thermoelectric coolers, can provide precise temperature control. Although these systems require additional power, they can be highly effective in managing temperatures in high-performance applications.

Advanced Heat Exchangers: Designing more efficient heat exchangers with enhanced heat transfer surfaces can improve the effectiveness of liquid cooling systems. Micro channel and mini channel heat exchangers are examples of advanced designs that offer higher performance in compact spaces.

- Intelligent Control and Integration

AI and Machine Learning: Utilizing artificial intelligence (AI) and machine learning algorithms for predictive thermal management can optimize BTMS performance in real-time. These technologies can predict temperature variations and adjust cooling strategies accordingly, enhancing both efficiency and battery life [29–32].

Integration with Battery Management Systems (BMS): Developing integrated BTMS and BMS solutions that work cohesively can provide better control and monitoring of battery conditions. Real-time data analytics and feedback mechanisms can ensure optimal thermal management under varying operating conditions [33, 34].

Wireless Sensors and IoT: The implementation of wireless sensors and Internet of Things (IoT) technology in BTMS can facilitate better monitoring and control. Wireless sensors can provide real-time data on temperature distributions, enabling more responsive and adaptive cooling strategies [35–37].

- Sustainable and Eco-friendly Solutions

Biodegradable PCMs: Research into biodegradable and environmentally friendly PCMs can address concerns about material sustainability and disposal. These materials can offer effective thermal management while minimizing environmental impact [38, 39].

Recycling and Reuse: Developing BTMS components that are easier to recycle or reuse can contribute to more sustainable battery systems. This includes designing systems with materials that can be efficiently recovered and repurposed at the end of their lifecycle [40, 41].

Energy-Efficient BTMS: Improving the energy efficiency of BTMS can reduce the overall power consumption of battery systems, leading to more sustainable operations. Innovations in low-power cooling technologies and passive cooling techniques are critical in this regard [42, 43].

2.6 Future Research Directions

To achieve these future perspectives, focused research and development efforts are necessary. Key areas for future research include [44–47]:

- Advanced Material Science
  - Exploration of Novel Nano-materials: Investigate new Nano-materials that offer superior thermal properties and integration capabilities for BTMS applications.

  - Development of High-performance Composites: Design and test composite materials that provide high thermal conductivity, mechanical strength, and long-term stability.
  - Improvement of Thermal Interface Materials: Research into next-generation TIMs with enhanced thermal conductivities and durability under cyclic thermal loads [48, 49].

- Enhanced Cooling Technologies
  - Optimization of Hybrid Cooling Systems: Develop and test hybrid cooling systems that combine multiple cooling methods for improved performance and reliability.
  - Innovative Heat Exchanger Designs: Design and fabricate advanced heat exchangers with optimized heat transfer characteristics and minimal pressure drop.
  - Thermoelectric and Phase Change Innovations: Explore the integration of thermoelectric cooling and advanced phase change materials in BTMS for precise and efficient thermal management [50].

- Intelligent Control Systems
  - AI and Machine Learning Algorithms: Develop predictive models using AI and machine learning to optimize BTMS operation based on real-time data and usage patterns.
  - Integrated BTMS and BMS Solutions: Research and develop integrated systems that provide seamless control and monitoring of thermal and electrical parameters.
  - Wireless Sensor Networks: Implement wireless sensor networks for real-time monitoring and adaptive control of BTMS, leveraging IoT technology [51].

- Sustainable Development
  - Eco-friendly Material Research: Investigate and develop biodegradable and sustainable materials for use in BTMS components.
  - Recycling Processes: Design BTMS components with recyclability in mind, focusing on materials and designs that facilitate easy recovery and reuse.
  - Energy-efficient Technologies: Innovate energy-efficient cooling technologies that reduce the overall power consumption of BTMS without compromising performance.

2.7 Conclusion

The future of Battery Thermal Management Systems (BTMS) technologies is exceptionally promising, with a myriad of opportunities for innovation and enhancement. Addressing the current challenges in BTMS requires a multifaceted approach that includes exploring advanced materials, such as Nano-materials and high-performance composites, which can significantly improve thermal conductivity and overall system efficiency. These materials not only offer superior heat transfer capabilities but also help in reducing the weight and bulkiness of BTMS components. Enhanced cooling techniques, such as hybrid cooling systems that combine air, liquid, and phase change material (PCM) cooling methods, are being developed to leverage the strengths of each method while mitigating their weaknesses. This combination can lead to more balanced and efficient thermal management solutions. Active cooling technologies, such as thermoelectric coolers, provide precise temperature regulation, essential for high-performance applications, though they come with increased power consumption.

In addition to material and cooling advancements, the integration of intelligent control systems and sustainable solutions is crucial. Utilizing artificial intelligence (AI) and machine learning for predictive thermal management can optimize BTMS performance in real-time by forecasting temperature variations and dynamically adjusting cooling strategies. This level of precision enhances both the efficiency and longevity of battery systems. Seamless integration with battery management systems (BMS) ensures comprehensive monitoring and control, providing real-time data analytics and feedback mechanisms. The implementation of wireless sensors and Internet of Things (IoT) technology further enhances BTMS capabilities by enabling granular, real-time temperature monitoring and more responsive thermal management. Sustainable development efforts, such as the use of biodegradable PCMs and recyclable BTMS components, aim to minimize environmental impact while maintaining high performance. By addressing these challenges and pursuing innovative solutions, the performance, safety, and longevity of battery systems can be significantly improved, supporting the growing demand for efficient and reliable energy storage solutions across various applications, from electric vehicles to renewable energy systems. The evolution of BTMS will depend heavily on interdisciplinary collaboration and the integration of cutting-edge technologies, paving the way for the next generation of thermal management systems.

References

1. Zhang, X. and Xu, C., Advanced Battery Thermal Management Systems for Electric Vehicles. *J. Energy Storage*, 45, 103635, 2021.
2. Lee, J. and Kim, S., Nano-materials in Battery Thermal Management Systems: Recent Advances and Future Perspectives. *Adv. Funct. Mater.*, 30, 25, 2000362, 2020.
3. Wang, Y. and Zhao, J., Thermal Management in Battery Systems for Electric Vehicles: Current Status and Future Outlook. *Renew. Energy*, 139, 257–267, 2019.
4. Park, H., A Design of Air Flow Configuration for Cooling Lithium-Ion Battery in Hybrid Electric Vehicles. *J. Power Sources*, 239, 30–36, 2013.
5. Fleckenstein, M., Karden, E., Buller, S., De Doncker, R.W., Battery Aging and Characterization in Automotive Applications. *J. Power Sources*, 168, 1, 76–84, 2007.
6. Pesaran, A., Santhanagopalan, S., Kim, G.-H., Addressing the Impact of Temperature Extremes on Large Format Li-Ion Batteries for Vehicle Applications. *ECS Trans.*, 50, 25, 155–162, 2013.
7. Lin, X., Perez, H.E., Mohan, S., Siegel, J.B., Li, Y., Anderson, R.D., Peng, H., Online Parameterization of Lumped Thermal Dynamics in Cylindrical Lithium Ion Batteries for Core Temperature Estimation and Health Monitoring. *IEEE Trans. Control Syst. Technol.*, 21, 5, 1745–1755, 2013.
8. Goli, P., Legedza, S., Dhar, A., Salgado, R., Renteria, J., Balandin, A.A., Graphene-Enhanced Hybrid Phase Change Materials for Thermal Management of Li-Ion Batteries. *J. Power Sources*, 248, 37–43, 2014.
9. Panchal, S., Mathewson, S., Fraser, R., Fowler, M., Thermal Management of Lithium-Ion Pouch Cell with Indirect Liquid Cooling Using Dual Cold Plates Approach. *J. Energy Storage*, 17, 209–217, 2018.
10. Fathabadi, H., High Thermal Performance Lithium-Ion Battery Pack Including Hybrid Active–Passive Thermal Management System for Using in Hybrid/Electric Vehicles. *Energy*, 70, 529–538, 2014.
11. Zhao, R., Zhang, S., Liu, J., A Review of Thermal Performance Improving Methods of Lithium Ion Battery: Electrode Modification and Thermal Management System. *J. Power Sources*, 299, 557–577, 2015.
12. Chen, K. and Qin, Y., A Review of Lithium-Ion Battery Thermal Management Techniques and Thermal Safety. *J. Power Sources*, 399, 107–117, 2019.
13. Wu, W., Zhang, G., Xu, X., Thermal Management Design and Analysis of a Liquid-Cooled Battery Pack for Electric Vehicles. *Appl. Therm. Eng.*, 170, 114930, 2020.
14. Al Hallaj, S. and Selman, J.R., Thermal Modeling of Battery Systems. *J. Power Sources*, 110, 2, 341–348, 2002.
15. Bandhauer, T.M., Garimella, S., Fuller, T.F., A Critical Review of Thermal Issues in Lithium-Ion Batteries. *J. Electrochem. Soc.*, 158, 3, R1–R25, 2011.

16. Rao, Z. and Wang, S., A Review of Power Battery Thermal Energy Management. *Renew. Sustain. Energy Rev.*, 15, 9, 4554–4571, 2011.
17. Fan, L., Khodadadi, J.M., Pesaran, A.A., A Parametric Study on Thermal Management of an Air-Cooled Lithium-Ion Battery Module for Plug-in Hybrid Electric Vehicles. *J. Power Sources*, 238, 301–312, 2013.
18. Wang, T., Tseng, K.J., Zhao, J., Development of Efficient Air-Cooled Battery Thermal Management System for Electric Vehicles. *J. Power Sources*, 270, 193–200, 2015.
19. Dubal, D.P., Ayyad, O., Ruiz, V., Gomez-Romero, P., Hybrid Energy Storage: The Merging of Battery and Supercapacitor Chemistries. *Chem. Soc. Rev.*, 44, 7, 1777–1790, 2015.
20. Khateeb, S., Farid, M.M., Selman, J.R., Al-Hallaj, S., Design and Simulation of a Lithium-Ion Battery with a Phase Change Material Thermal Management System for an Electric Scooter. *J. Power Sources*, 128, 2, 292–307, 2004.
21. Maleki, H. and Howard, J.N., Internal Short Circuit in Li-Ion Cells. *J. Power Sources*, 160, 2, 1395–1402, 2006.
22. Wang, C.-Y. and Gu, W.-B., Thermal and Electrochemical Coupled Modeling of a Lithium-Ion Cell. *J. Electrochem. Soc.*, 147, 8, 2910–2922, 2000.
23. Fan, X., Pan, W., Zhang, X., Thermal Runaway of Lithium-Ion Batteries Based on Internal Short Circuit. *J. Power Sources*, 434, 226718, 2019.
24. Barré, A., Deguilhem, B., Grolleau, S., Gérard, M., Suard, F., Riu, D., A Review on Lithium-Ion Battery Ageing Mechanisms and Estimations for Automotive Applications. *J. Power Sources*, 241, 680–689, 2013.
25. Kim, G.-H., Smith, K., Ireland, J., Pesaran, A., Fail-Safe Design for Large Capacity Lithium-Ion Battery Systems. *J. Power Sources*, 178, 2, 765–771, 2008.
26. Hameed, M., Mansor, M., Azau, M., Muhsin, S., Thermoelectric Cooler Performance Enhancement using Thermoelectric Generators and Their Use as a Single Model to Improve the Performance of Thermal Battery Management Systems for Electric Vehicles. Energy Storage, vol. 5, 2022, 10.1002/est2.406.
27. Oubelaid, A., Kakouche, K., Belkhier, Y., Khosravi, N., Taib, N., Rekioua, T., Tuka, M.B., New coordinated drive mode switching strategy for distributed drive electric vehicles with energy storage system. *Sci. Rep.*, 14, 1, 6448, 2024, 10.1038/s41598-024-56209-9.
28. Ravindran, M.A., Nallathambi, K., Vishnuram, P., Rathore, R.S., Bajaj, M., Rida, I., Alkhayyat, A., A novel technological review on fast charging infrastructure for electrical vehicles: Challenges, solutions, and future research directions. *Alex. Eng. J.*, 82, 260–290, 2023, 10.1016/j.aej.2023.10.009.
29. Amoussou, I., Paddy, E.Y., Agajie, T.F., Ibrahim, F.S., Agajie, E.F., Nsanyuy, W.B., Mohammadi, S.A.D., Enhancing residential energy access with optimized stand-alone hybrid solar-diesel-battery systems in Buea, Cameroon. *Sci. Rep.*, 14, 1, 15543, 2024, 10.1038/s41598-024-66582-0.

30. Kasri, A., Ouari, K., Belkhier, Y., Bajaj, M., Zaitsev, I., Optimizing electric vehicle powertrains peak performance with robust predictive direct torque control of induction motors: A practical approach and experimental validation. *Sci. Rep.*, 14, 1, 14977, 2024, 10.1038/s41598-024-65988-0.
31. Afzal, M.Z., Aurangzeb, M., Iqbal, S., Pushkarna, M., Rehman, A.U., Kotb, H., Bereznychenko, V., A Novel Electric Vehicle Battery Management System Using an Artificial Neural Network-Based Adaptive Droop Control Theory. *Int. J. Energy Res.*, 2023, 1, 2581729, 2023, 10.1155/2023/2581729.
32. Das, S., Thakur, P., Singh, A.K., Singh, S.N., Optimal management of vehicle-to-grid and grid-to-vehicle strategies for load profile improvement in distribution system. *J. Energy Storage*, 49, 104068, 2022, http://dx.doi.org/10.1016/j.est.2022.104068.
33. Singh, P.P., Das, S., Wen, F., Palu, I., Singh, A.K., Thakur, P., Multi-objective planning of electric vehicles charging in distribution system considering priority-based vehicle-to-grid scheduling. *Swarm Evol. Comput.*, 77, 101234, 2023, http://dx.doi.org/10.1016/j.swevo.2023.101234.
34. Almutairi, A., Albagami, N., Almesned, S., Alrumayh, O., Malik, H., Electric vehicle load estimation at home and workplace in Saudi Arabia for grid planners and policy makers. *Sustainability*, 15, 22, 15878, 2023, http://dx.doi.org/10.3390/su152215878.
35. Jena, N.K., Sahoo, S., Sahu, B.K., Naik, A.K., Bajaj, M., Misak, S., Prokop, L., Impact of a redox flow battery on the frequency stability of a five-area system integrated with renewable sources. *Energies*, 16, 14, 5540, 2023, 10.3390/en16145540.
36. Hamed, S.B., Hamed, M.B., Sbita, L., Bajaj, M., Blazek, V., Prokop, L., Ghoneim, S.S., Robust optimization and power management of a triple junction photovoltaic electric vehicle with battery storage. *Sensors*, 22, 16, 6123, 2022, 10.3390/s22166123.
37. Rawat, T., Niazi, K.R., Gupta, N., Sharma, S., Multi-objective techno-economic operation of smart distribution network integrated with reactive power support of battery storage systems. *Sustain. Cities Soc.*, 75, 103359, 2021, http://dx.doi.org/10.1016/j.scs.2021.103359.
38. Gupta, A., Singh, R.K., Paul, J., Sharma, M., Joshi, S., Battery-Operated Electric Vehicles (BOEVs) Adoption in India: Analysis of Barriers and Strategies. *IEEE Trans. Eng. Manage.*, 71, 7076–7087, 2023, http://dx.doi.org/10.1109/TEM.2023.3253621.
39. Badri, Y., II, Sudabattula, S.K., Hussain, I., Malik, H., Márquez, F.P.G., Ultracapacitor as selectable energy buffer in electric vehicle application. *J. Energy Storage*, 73, 109200, 2023, http://dx.doi.org/10.1016/j.est.2023.109200.
40. Ramakrishnan, V., Vishnuram, P., Yang, T., Bajaj, M., Rathore, R.S., Zaitsev, I., Design and implementation of a high misalignment-tolerance wireless charger for an electric vehicle with control of the constant current/voltage charging. *Sci. Rep.*, 14, 1, 13165, 2024, 10.1038/s41598-024-63952-6.

41. Dimri, S.C., Indu, R., Bajaj, M., Rathore, R.S., Blazek, V., Dutta, A.K., Alsubai, S., Modeling of traffic at a road crossing and optimization of waiting time of the vehicles. *Alex. Eng. J.*, 98, 114–129, 2024, 10.1016/j.aej.2024.04.050.
42. Thapliyal, S., Wazid, M., Singh, D.P., Das, A.K., Islam, S.H., Robust authenticated key agreement protocol for internet of vehicles-envisioned intelligent transportation system. *J. Syst. Archit.*, 142, 102937, 2023, 10.1016/j.sysarc.2023.102937.
43. Kumar, B.A., Jyothi, B., Singh, A.R., Bajaj, M., Rathore, R.S., Berhanu, M., A novel strategy towards efficient and reliable electric vehicle charging for the realisation of a true sustainable transportation landscape. *Sci. Rep.*, 14, 1, 3261, 2024, 10.1038/s41598-024-53214-w.
44. Singh, A.R., Vishnuram, P., Alagarsamy, S., Bajaj, M., Blazek, V., Damaj, I., Othman, K.M., Electric vehicle charging technologies, infrastructure expansion, grid integration strategies, and their role in promoting sustainable e-mobility. *Alex. Eng. J.*, 105, 300–330, 2024, 10.1016/j.aej.2024.06.093.
45. Bhatt, A., Ongsakul, W., Pawar, J., Optimal energy management system for carbon–neutral microgrid integrating second-life batteries and crypto mining devices. *Sustain. Energy Technol. Assess.*, 64, 103686, 2024, 10.1016/j.seta.2024.103686.
46. Sahri, Y., Tamalouzt, S., Belaid, S.L., Bajaj, M., Ghoneim, S.S., Zawbaa, H.M., Kamel, S., Performance improvement of Hybrid System based DFIG-Wind/PV/Batteries connected to DC and AC grid by applying Intelligent Control. *Energy Rep.*, 9, 2027–2043, 2023, 10.1016/j.egyr.2023.01.021.
47. Rekioua, D., Mokrani, Z., Kakouche, K., Rekioua, T., Oubelaid, A., Logerais, P.O., Ghoneim, S.S., Optimization and intelligent power management control for an autonomous hybrid wind turbine photovoltaic diesel generator with batteries. *Sci. Rep.*, 13, 1, 21830, 2023, 10.1038/s41598-023-49067-4.
48. Kanouni, B., Badoud, A.E., Mekhilef, S., Bajaj, M., Zaitsev, I., Advanced efficient energy management strategy based on state machine control for multi-sources PV-PEMFC-batteries system. *Sci. Rep.*, 14, 1, 7996, 2024, 10.1038/s41598-024-58785-2.
49. Bagga, P., Das, A.K., Wazid, M., Rodrigues, J.J., Choo, K.K.R., Park, Y., On the design of mutual authentication and key agreement protocol in internet of vehicles-enabled intelligent transportation system. *IEEE Trans. Veh. Technol.*, 70, 2, 1736–1751, 2021, 10.1109/TVT.2021.3050614.
50. Aggarwal, S., Singh, A.K., Rathore, R.S., Bajaj, M., Gupta, D., Revolutionizing load management: A novel technique to diminish the impact of electric vehicle charging stations on the electricity grid. *Sustain. Energy Technol. Assess.*, 65, 103784, 2024, 10.1016/j.seta.2024.103784.
51. Razaviyn, Z., Heidari, M., Thangavel, S., Verma, V., Kumar, A., Yadav, A.K., Marine Energy Harvesting from Fluid Flow *via* Vortex Induced Vibrations. *Int. J. Thermofluids*, 25, 101015, 2024, https://doi.org/10.1016/j.ijft.2024.101015.

3

Battery Thermal Management Systems Using Phase Change Material and Liquid Cooling: A Comprehensive Review

Atul Kumar[1], Ashok Kumar Dewangan[1]*, Ashok Kumar Yadav[2] and Ashwani Kumar[3]

[1]*Department of Mechanical Engineering, National Institute of Technology, Delhi, India*
[2]*Department of Mechanical Engineering, Symbiosis Institute of Technology, Symbiosis International (Deemed University), Pune, India*
[3]*Department of Mechanical Engineering, Technical Education Department, Uttar Pradesh, Kanpur, India*

Abstract
This study delves into advancements and methodologies in battery thermal management systems (BTMS) for electric vehicles (EVs) and renewable energy storage, focusing on the integration of phase change materials (PCMs) and liquid cooling. It highlights the critical role of efficient thermal management in optimizing battery performance and safety. Key components like temperature sensors, cooling/heating elements, and control systems are discussed for their function in combining PCMs and liquid cooling to enhance heat dissipation and maintain optimal temperatures. PCMs with high latent heat capacity and liquid cooling systems are identified as effective strategies for temperature regulation. The review also explores advancements in PCM characteristics and modeling techniques such as Computational Fluid Dynamics (CFD) and Finite Element Analysis (FEA) for simulating thermal behavior within battery packs. Overall, this review contributes to the development of innovative thermal management solutions aimed at improving performance, longevity, and safety in EVs and renewable energy applications.

**Corresponding author*: ashokme015@gmail.com; ORCID: https://orcid.org/0000-0002-7350-7641
Ashok Kumar Dewangan: ORCID: https://orcid.org/0000-0003-4339-0953
Ashwani Kumar: ORCID: https://orcid.org/0000-0003-4099-935X

Ashwani Kumar, Mukesh Kumar Awasthi, Nitesh Dutt, Yogesh Kumar Singla and Sivasakthivel Thangavel (eds.) Thermal Battery Management System for Hybrid and Electric Vehicles, (49–72) © 2025 Scrivener Publishing LLC

Keywords: Battery thermal management systems, electric vehicles, phase change materials

3.1 Introduction

Battery thermal management is crucial for optimizing the performance of lithium-ion batteries in electric vehicles (EVs) and renewable energy storage systems [2, 7, 8, 10]. One promising approach involves integrating phase change materials (PCMs) with liquid cooling systems to efficiently regulate battery temperatures and tackle various challenges. This research delves into the modeling techniques and strategies used to develop and enhance battery thermal management systems utilizing PCMs and liquid cooling.

The current landscape of batteries used in EVs is detailed in Table 3.1. Overheating of batteries in EVs can lead to reduced performance, accelerated aging, and safety hazards. Effective thermal management systems are essential to mitigate these issues and maintain optimal operating

Table 3.1 Current different type of batteries performance [11].

Type	Specific energy (Wh/kg)	Specific power (kW/kg)	Nominal cell voltage (V)	Life cycle	Charge time (h)	Self-discharge rate (%)
Lead-acid	25–40	150–250	2	200–700	8	5
Nickel-cadmium	45–80	200	1.2	500–2000	1	20
Nickel-metal hydride	60–120	200	1.2	500–1000	1	30
Nickel-iron	50	100	1.2	2000	-	20–40
Nickel-zinc	70	150	1.7	300	-	20
Sodium-sulphur	100	150–200	2	≤1000	8	Quite low
Sodium-metal chloride	100	150	2	>1000	8	-
Li-ion	110–180	300	3.6	>1000	2–3	10
Li-polymer	100–130	300	3.6	300–500	2–4	10

temperatures. Strategies like pulsing heat pipes, heat pipe-based systems, and integrated-BTMS (IBTMs) have been proposed to address battery overheating, aiming to improve battery lifespan, ensure temperature consistency, and facilitate heat removal. Implementing advanced cooling solutions not only optimizes energy consumption efficiency in EVs but also ensures battery safety and performance [44–50]. Continuous research and development are necessary to meet the evolving demands of electric vehicle technologies.

PCMs are increasingly recognized as a viable solution for addressing thermal management challenges in batteries. These materials have the ability to keep and release enough amounts of latent heat during phase changes, effectively controlling temperature fluctuations within battery modules. PCMs, such as paraffin wax and composite materials, offer high latent heat capacity and minimal parasitic power usage, enhancing heat dissipation. In parallel, liquid cooling systems demonstrate efficient heat transfer capabilities, maintaining battery temperature within safe limits. Hybrid systems integrating PCMs with liquid cooling show potential for further improving thermal efficiency by leveraging the advantages of both approaches. These innovative methodologies address critical issues like thermal runaway and temperature uniformity in batteries, contributing to enhanced safety, durability, and performance in EVs and energy storage applications.

This study provides insights into the real-world applications and recent advancements in developing battery thermal control systems employing PCMs and liquid cooling. Through a comprehensive review of literature, it aims to advance our understanding and progress in developing state-of-the-art thermal management solutions for battery technologies.

3.2 Overview of Battery Thermal Management System

The constituents of a Battery Thermal Management System play a vital role in regulating the ideal temperature of a battery pack. Temperature sensors are strategically positioned within the pack to continuously observe the temperature of individual cells or modules, providing crucial data for the control system to make necessary adjustments in cooling or heating. This system comprises cooling/heating components such as a coolant circulation system with a pump, tubing, and heat exchanger, heat exchangers for heat transfer, fans, or blowers to enhance airflow, and thermal interface materials for efficient heat conduction. The control unit, equipped with a microcontroller or PLC, orchestrates the functioning of these components based on predetermined algorithms. Furthermore, a heat sink may be employed to disperse heat produced by the battery pack, ensuring optimal functionality and durability [51–62]. Table 3.1 presents the different type of batteries performance.

A Battery Thermal Management System (BTMS) is typically comprised of elements such as phase change materials (PCMs), heat pipes, liquid cooling tubes, and cooling fluids. PCMs, such as composite phase change materials (CPCM) [1], enhance heat absorption owing to their significant latent heat capacity. Heat pipes, such as pulsating heat pipes (PHP) [5], provide effective cooling through the dissipation of heat from batteries. Liquid cooling tubes, when combined with PCMs, offer secondary heat dissipation and prevent heat saturation [3]. Cooling fluids such as N-Pentane and N-Hexane can regulate battery temperatures within optimal ranges, thereby reducing energy consumption and emissions [12]. In summary, these components collaborate harmoniously to regulate battery heat, enhance performance, prolong lifespan, and ensure secure operation in electric vehicles.

3.2.1 Exploration of Prevalent Thermal Management Strategies

Various thermal management approaches, such as air cooling, liquid cooling, and phase change materials (PCMs), play crucial roles in diverse applications like electronic cooling. Air cooling is commonly used in electric vehicle air conditioning systems [13, 61–70], while liquid cooling is effective in BTMSs [4, 6]. PCMs, like n-eicosane with nanoparticles, have shown enhanced heat transfer capabilities in electronic cooling systems [14, 71–80]. Additionally, the integration of phase change processes in thermal management systems is recommended to dissipate heat efficiently and ensure battery safety in electric vehicles [9, 91–94]. Combining PCM with heat pipes in BTMSs can provide high heat transfer performance and uniform temperature distribution, addressing the limitations of individual cooling methods. These approaches collectively contribute to optimizing energy utilization, enhancing system efficiency, and ensuring thermal safety in various applications [81–90].

3.2.2 Pros and Cons of Various Thermal Management Approaches

Figure 3.1. shows the various thermal management strategies offer benefits and face limitations in different applications. For electric vehicles, integrating AC, battery, and motor thermal management systems maximizes energy

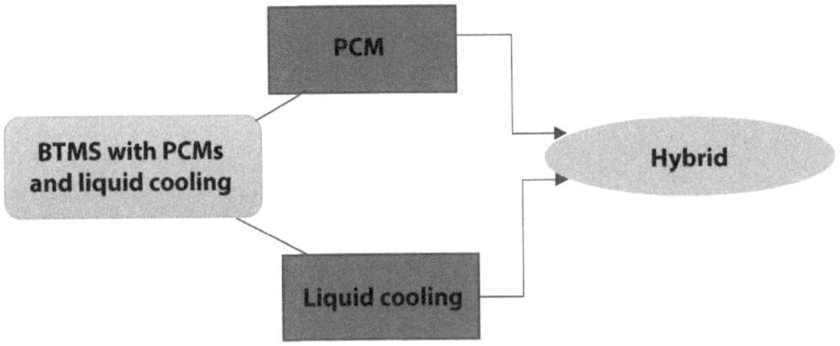

Figure 3.1 Battery thermal management system.

efficiency [4, 95–100]. Wearable thermoregulatory devices employ passive and active modes to regulate body temperature effectively [15, 101–105]. In smart electronic devices, forced air cooling with different airflow configurations significantly reduces temperatures, enhancing reliability [16]. Power lithium-ion batteries in EVs benefit from phase change material-based cooling for temperature control, but face challenges due to poor thermal conductivity [17]. Predictive thermal management strategies for electric vehicle powertrains aim to minimize energy consumption and enhance motor efficiency while ensuring safety and durability [18, 106]. These strategies demonstrate the importance of tailored thermal management approaches in various fields, each with its unique advantages and constraints.

3.3 Role of Phase Change Materials (PCMs) in Battery Thermal Management

3.3.1 Enhancement of Phase Change Material (PCM) Properties

The proposition of applying Phase Change Material (PCM) to Battery Thermal Management Systems (BTMS) was initially introduced by Hallaj and Selman [19]. This research duo delved into an Electric Vehicle (EV) BTMS utilizing Paraffin as the PCM, conducting a thorough investigation through the utilization of finite element software. The analysis involved the simulation of temperature circulation within a 100 Ampere-hour (Ah) battery comprising thousands of cells (18,650). The outcomes indicated a notable reduction in temperature elevation for the battery module integrated with PCM. Hallaj and Selman justified their selection of paraffin due to its cost-effectiveness, stability against decomposition, and alignment with the requisite phase change temperature. Consequently, paraffin emerged as a favored

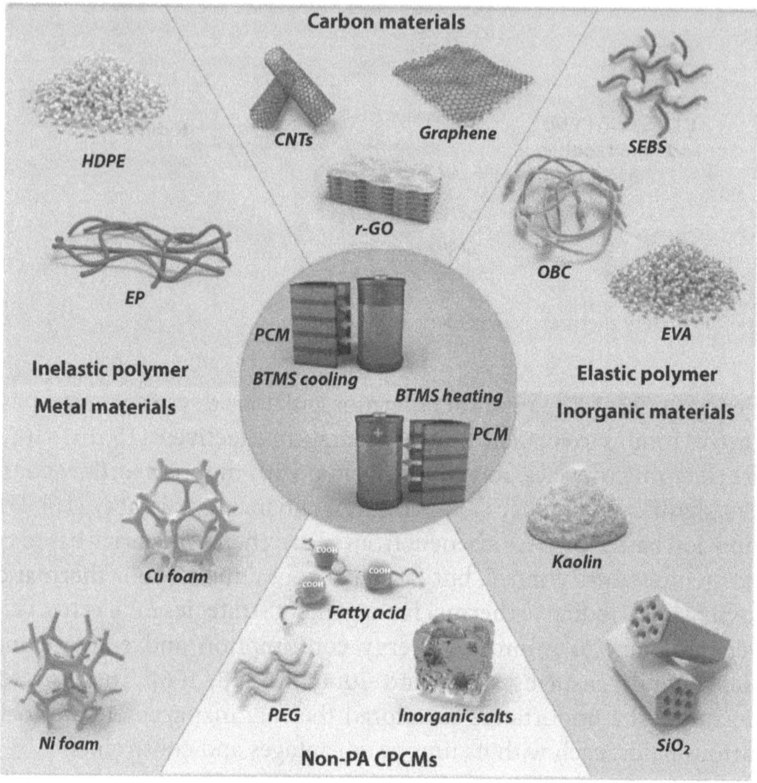

Figure 3.2 BTMS with PCMs [20].

PCM option among numerous researchers. Alongside Paraffin, scholarly interest has extended towards alternative materials like hydrated salts, fatty acids, and polyethylene glycols. An overarching observation reveals a common drawback among these PCM selections - poor thermal conductivity.

Figure 3.2 shows how PCM and BTMS work with each other. This deficiency significantly hampers heat dissipation, particularly when subjected to repeated charge/discharge cycles or elevated current rates. Hence, it becomes imperative to explore and implement robust enhancement techniques to ameliorate heat transfer efficiency.

3.3.1.1 Enhancement of Thermal Conductivity of Phase Change Material

Improving the thermal conductivity of phase change materials (PCMs) is imperative for maximizing their efficacy in battery thermal management systems. Various strategies are routinely utilized to achieve this enhancement. To improve the thermal conductivity of PCMs, various methods have been explored. One approach involves *in situ* reduction deposition of highly thermally conductive metals like copper onto PCM carriers, resulting in composite PCMs with significantly improved thermal conductivity [21]. Another method includes incorporating thermal conductivity enhancers (TCEs) such as fins, nanoparticles, and metal foam into PCMs to enhance their performance as heat sinks for electronics [22]. Additionally, impregnating PCM with TiO2 foam has been shown to significantly increase thermal conductivity, with a composite achieving a thermal conductivity 3.73 times higher than pure paraffin [23]. Furthermore, dispersing graphene nanotubes in paraffin has demonstrated a 22% increase in thermal conductivity in the solid state, showcasing the potential for enhancing PCM thermal properties [24].

Manipulating the size and dispersion of particles in phase change material (PCM) composites, creating innovative PCM architectures, and utilizing cutting-edge manufacturing methods such as 3D printing play a crucial role in enhancing thermal conductivity. By incorporating these techniques, scientists are able to customize PCMs with improved thermal characteristics, facilitating more effective thermal regulation in battery and energy storage systems. Each method presents distinct benefits and considerations based on the specific needs and limitations of the application at hand.

3.3.1.2 Enhancement of Flame Resistance of Phase Change Material

It is imperative to enhance the flame retardancy of phase change materials (PCMs) in order to guarantee the safety of PCM-based thermal management systems, especially in scenarios where fire risks are a significant consideration. Various methodologies can be utilized to bolster the flame resistance of PCMs. Surface coatings with flame retardants have been shown to significantly enhance the flame resistance of phase change materials (PCMs) [25]. Additionally, the introduction of phytic acid (PA) into PCM composites has been found to minimize the crest heat release rate and total heat release, indicating improved flame retardancy [26]. Furthermore, the use of three-dimensional (3D) melamine foam (MF)/phosphorous cellulose (PC)/graphite nanoplatelets (GNPs) network in PCM composites has led to a decrease in the crest heat release rate, addressing fire hazards [27]. It is crucial to consider the flammability of PCMs, as their burning behavior needs improvement for commercial applications, especially in fire fighters' protective clothing [28]. By incorporating these strategies, such as surface coatings and innovative 3D networks, the flame resistance of PCMs can be effectively enhanced for various applications.

Moreover, enhancing the thermal stability and increasing the decomposition temperature of phase change materials (PCMs), while also ensuring adherence to fire safety regulations, plays a crucial role in improving overall flame resistance. By incorporating these measures, thermal management systems utilizing PCMs can be rendered safer and more appropriate for applications necessitating fire-resistant materials. Each method presents distinct benefits and factors to consider, tailored to meet specific application needs and environmental conditions.

3.3.1.3 Enhancement of Mechanical Properties and Leakage

Improving the mechanical properties and mitigating leakage concerns are essential considerations in the optimization of phase change materials (PCMs) for thermal management applications. Strategies such as the integration of reinforcing materials like fibers or fillers into PCM formulations can bolster mechanical integrity and diminish the likelihood of leakage by offering structural support. Employing microencapsulation methods, wherein PCMs are enclosed within polymer shells, serves to prevent leakage and enhance mechanical resilience by confining the PCM within protective capsules. Additionally, the crosslinking of PCM matrices with polymers and the utilization of shape-stable PCMs embedded in supportive matrices further contribute to enhanced mechanical characteristics and reduced leakage risks.

The enhancement of mechanical properties and reduction of leakage in various materials have been achieved through different methods. For instance, the mechanical strength of polyurea coatings can be improved by modifying them with Ce/La-doped TiO2, leading to increased chemical durability and reduced microbial attachment [29]. Similarly, the mechanical properties of poly (arylene sulfide sulfone) (PASS) electrospun membranes can be significantly enhanced by co-electrospinning with semi-aromatic nylon poly (m-xylene adipamide) (MXD6), resulting in improved tensile strength and stability up to 180°C without decomposition [30]. Moreover, the addition of graphene oxide (GO) to microencapsulated phase change materials (MEPCMs) with calcium carbonate shells has shown to enhance thermal stability, reduce leakage rates, and improve mechanical properties significantly [31]. These modifications demonstrate effective strategies for enhancing mechanical properties and reducing leakage in various materials. Each strategy offers unique advantages and considerations based on specific application requirements and environmental conditions.

3.3.2 BTMS Based on Phase Change Material (PCM)

A PCM-utilizing Battery Thermal Management System (BTMS), functioning as a passive mechanism, offers numerous benefits such as low supplementary energy usage, a simple configuration, a lightweight construction, and the promotion of consistent cell temperature. PCM offers advantages like high heat storage capacity, but challenges such as low thermal conductivity exist. Studies propose innovative solutions like integrating metal fins with PCM systems for enhanced cooling performance [3]. Combining PCM with heat pipes in BTMS can effectively manage battery temperature, offering a balance between high heat transfer performance and uniform temperature distribution [32]. Research emphasizes the importance of PCM and heat pipes in maintaining battery temperature within safe limits, enhancing energy density, and improving thermal performance. Implementing PCM in BTMS through active or passive cooling methods significantly improves thermal management, ensuring better battery performance and safety. When examining the most suitable configuration of PCM-based BTMS, it is essential to consider the influence of heightened PCM mass on the energy consumption during operation. Furthermore, the expansion of PCM volume plays a role in diminishing the total specific energy. Figure 3.3 represents the Li-ion battery working.

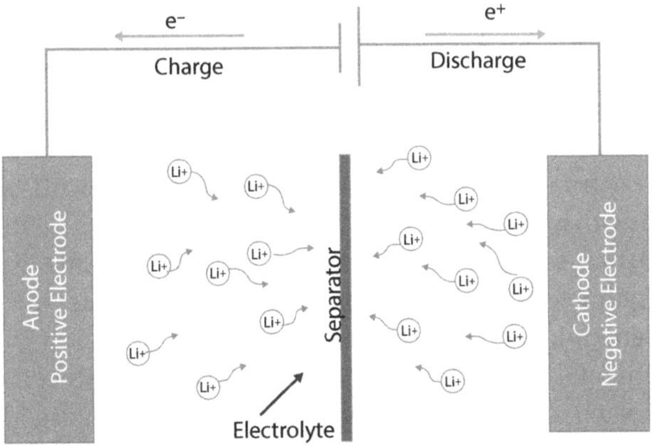

Figure 3.3 The representational of Li-ion battery [43].

3.4 Liquid Cooling Systems for Battery Thermal Management

3.4.1 Water

Water is commonly employed in liquid cooling systems for battery thermal management due to its elevated specific heat capacity and thermal conductivity, which facilitate efficient heat dissipation from batteries. It is a cost-effective, readily available, and environmentally friendly solution. However, the low viscosity of water may impede its effectiveness under high thermal loads, and its susceptibility to freezing in cold temperatures poses a risk of damage to cooling system components. To address these challenges, coolant additives such as ethylene glycol are often utilized to lower the freezing point and enhance overall performance. In conclusion, water-based liquid cooling systems offer effective heat management for batteries, particularly in electric vehicles, but proper system design and maintenance are crucial for achieving optimal performance and longevity. Additionally, water will in general deep-freeze when the encompassing temperature decreases under 0°C, provoking the expansion of liquid catalyst materials. $(CH_2OH)_2$ Ethylene glycol arises as the most normally used radiator fluid compound. The coolant stays thawed at temperature as minimum as – 45°C [33]. Since 1950s, the dominating glycol-to-water coolant proportion utilized has been 50–50% [34].

3.4.2 Oil

Contrasted with elective cooling-system, oil presents unmistakable benefits as it is non-conductive and non-combustible, permitting direct contact with the batteries. Oil is a commonly utilized medium in liquid cooling systems for battery thermal management, owing to its exceptional thermal stability and non-flammable properties, thereby enhancing safety in battery applications. With a higher viscosity compared to water, oil facilitates improved heat transfer efficiency and thermal contact with battery cells. Its ability to withstand a wide range of temperatures without freezing or boiling renders it suitable for use in extreme environments. Nevertheless, oil-based cooling systems may entail higher initial costs and necessitate specialized equipment. Moreover, considerations pertaining to disposal and environmental impact must be carefully evaluated. In certain applications where water-based systems may not be viable, oil-based liquid cooling systems present distinct advantages. However, the implementation of proper design and maintenance practices is imperative to ensure optimal performance and safety [35], after an inside and out investigation, proposed utilizing a minimum stream rate in oil cooling-system frameworks to ease the pressure drop. Furthermore, [36] blended a gatherer capacitor heat move model with a stream network way to deal with survey the execution of a business scale lithium-particle batteries pack under particular warm administration framework plans and shifting circumstances. Their outcomes underlined the significant effect of the coolant gulf and way-out structure on cell execution. Notwithstanding the high siphoning power prerequisite for oil, setting a proper stream rate keeps up with temperatures inside safe limits.

3.4.3 BTMS Based on Liquid Cooling

Battery Thermal Management Systems (BTMS) that utilize liquid cooling employ liquid-based techniques to effectively regulate battery temperature in various applications, such as electric vehicles (EVs). These systems can be implemented through direct liquid cooling, where coolant directly comes into contact with battery cells to absorb heat, or indirect cooling, which involves the use of separate heat exchangers to transfer heat from batteries to coolant. Additionally, immersion cooling is another viable approach, involving the submersion of batteries in a dielectric fluid like oil to facilitate efficient heat dissipation. Liquid cooling systems provide precise temperature control, effective heat dissipation, and the capability to manage high thermal loads, ultimately enhancing battery performance, extending lifespan, and improving safety. However, the successful implementation of liquid cooling systems necessitates meticulous design and maintenance to ensure optimal performance and reliability in maintaining battery temperatures within safe operating limits. To tackle the issue of direct liquid cooling, [37] proposed a drenching integrate direct cooling technique in which the battery is submerged in a proper liquid and embedded into an immediate cooling pipe (displayed in Figure 3.4) and examined the intensity move attributes of direct cooling and its affecting variables for modules of the batteries at 2C release rate and 25°C climate. The discoveries showed that direct cooling came about in altogether broadened ideal working time (temperature of cell underneath 35°C), accomplishing a 150 percent and 45 percent longer working time looked at to regular convection and inundation cooling, separately. In indirect cooling strategies, for example, cold plate cooling and channel cooling, albeit somewhat less effective in compare to direct cooling, offer more prominent security and

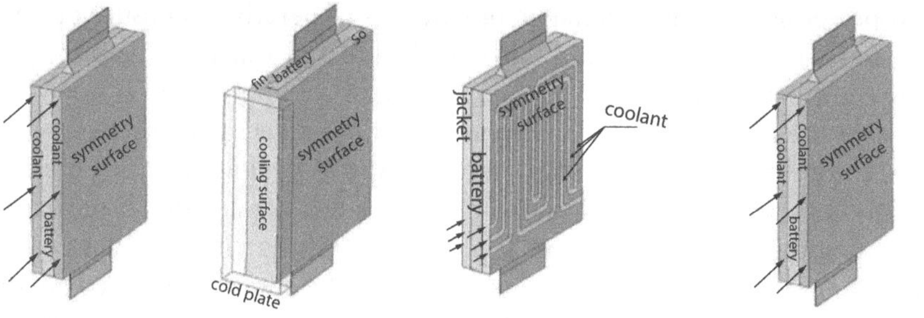

(**a**) Configuration of air cooling (**b**) Configuration of fin cooling (**c**) Configuration of indirect liquid cooling (**d**) Configuration of direct liquid cooling

Figure 3.4 Configuration of cooling with same gap between cells [38].

cost benefits. These strategies are as of now the essential center for exploration and business applications BTMs.

3.5 Modeling Approaches for Battery Thermal Management Systems

Modeling methodologies for Battery Thermal Management Systems (BTMS) are indispensable tools utilized to simulate and enhance the thermal dynamics of batteries across diverse operational scenarios [44, 107–119] These modeling strategies are pivotal in the development of effective cooling solutions and in guaranteeing the safety and efficacy of battery systems. There are some different approaches for BTMS, PCM modeling like CFD, FEA and some other, with the help of governing equation we can simulate the design using CFD for our BTMS and PCM system.

Hasan *et al.* [38] with the use of CFD model they done the parametric investigation on BTMS with PCM, fins and metal foam to minimize the battery temperature during 3C dis-charging in different conditions. During the process of they found that the temperature difference in the battery minimize by 75% to 66% in PCM. Additionally, copper metal form and copper fins give the best result in minimizing temperature of battery. Xing *et al.* [39] used an cylindrical battery which contain bionic spiral fin with PCM and liquid cooling plate and check performance of thermal management system. They observed that in 600 sec. duration and 0.09g/s of flow rate, the fraction of liquid PCM is reduced to 57% to 26%. Also, the temperature difference of BTMS is minimize by 0.5°C in compare to other 4.1°C at 0.5 g/s flow rate.

Hosanna *et al.* [40] study that thermal runaway propagation in Li-Ion batteries cell pack with nail penetration in CFD. 3D integrated model with experimental data and depth of impact of nail penetration (20 mm, 15 mm and 12.5 mm) which effects on battery thermal during the process they observed that the inception of thermal runaway for ISCs in middle of the bottom edge be delayed by 10 sec in compare to top, bottom and middle of battery lateral border. Additionally, this study found effect of convective heat transfer coefficient of 500 W/ m^2*K. Lebrouhi *et al.* [41] They used modeling of Li-ion battery pack thermal management system with PCM and liquid cooling this process carried 3C dis-charging or 0.5C charging rate and result were compared with 3D CFD. In conclusion they found that for BTMS combining both liquid cooling and PCM decrease the temperature of battery by 38°C to 4°C compared to natural convection and PCM thermal management. Also, with the increase in the number of cooling pipes that improve system performance and the temperature of battery cells decrease from 31°C to 20°C and inlet temperature of liquid coolant is decrease from 25°C to 10°C.

Guiqi *et al.* [42] study the battery thermal network model for thermal management system based on PCMs and liquid cooling they compared and found that the liquid cooling energy saved by 39.5%, reduced BTMS maximum temperature 0.38°C and maximum temperature reduced by 19.81%.

After concluding all the reference and there result we found that after combining PCMs and liquid cooling in BTMS we can reduce the heating temperature and decrease the charging time also.

3.6 Conclusions and Future Scope

Based on literature reviews, the following findings have been emerged:

- Critical for secure and efficient lithium-ion battery operation in EVs and renewable energy storage systems.
- Validated for maintaining battery temperatures within ideal ranges, enhancing performance, durability, and safety. PCMs absorb and release latent heat during phase transitions, controlling battery module temperatures. Ongoing research aims to improve PCM characteristics like thermal conductivity and flame resistance.
- Commonly use water and oil as cooling mediums due to their high heat capacity and effectiveness. Various methods

(direct, indirect) explored for suitability in battery thermal management.
- Enhances thermal efficiency by combining heat storage capacity of PCMs with heat dissipation capabilities of liquid cooling. Computational Fluid Dynamics (CFD) and
- Crucial for addressing battery overheating issues, ensuring operational dependability, and promoting electric transportation and sustainable energy solutions.
- Research conducted by Hasan *et al.*, Xing *et al.*, Hosanna *et al.*, Lebrouhi *et al.*, and Guiqi *et al.* has revealed substantial decreases in battery temperatures and improved thermal efficiency through in-depth parametric studies, novel cooling system designs, and analysis of thermal runaway utilizing computational fluid dynamics (CFD) modeling.

Future goals identified through literature review include: enhancing PCM development for improved thermal conductivity and flame resistance in battery thermal management; integrating PCM-liquid cooling hybrid systems for enhanced heat dissipation and energy efficiency in battery packs; refining modeling and simulation techniques to accurately predict thermal behavior and optimize BTMS designs; conducting practical implementation and validation through field trials to ensure scalability, reliability, and cost-effectiveness; optimizing BTMS designs for different applications based on battery chemistry and environmental factors; developing cost-effective solutions to promote widespread adoption of PCM-based BTMS; and prioritizing safety, sustainability, and reliability in PCM material selection and system design.

References

1. Lv, Y., Luo, W., Li, C., Zhang, G., Huang, S., Rao, Z., Experimental and numerical simulation study on the integrated thermal management system for electric vehicles. *J. Energy Storage*, 70, 107895, Oct. 2023, doi: 10.1016/j.est.2023.107895.
2. Kumar, R. and Goel, V., A study on thermal management system of lithium-ion batteries for electrical vehicles: A critical review. *J. Energy Storage*, 71, Nov. 01, 108025, 2023, Elsevier Ltd, doi: 10.1016/j.est.2023.108025.
3. Afzal, A., Abdul Razak, R.K., Mohammed Samee, A.D., Kumar, R., Ağbulut, U., Park, S.G., A critical review on renewable battery thermal management system using heat pipes. *J. Therm. Anal. Calorim.*, 148, 16, 8403–8442,

Aug. 01, 2023, Springer Science and Business Media B.V., doi: 10.1007/s10973-023-12100-9.
4. Dan, D., Zhao, Y., Wei, M., Wang, X., Review of Thermal Management Technology for Electric Vehicles. *Energies*, 16, 12, 4693, Jun. 01, 2023, MDPI, doi: 10.3390/en16124693.
5. Cattani, L., Malavasi, M., Bozzoli, F., D'Alessandro, V., Giammichele, L., Experimental Analysis of an Innovative Electrical Battery Thermal Management System. *Energies (Basel)*, 16, 13, 5071, Jul. 2023, doi: 10.3390/en16135071.
6. Xin, Q., Yang, T., Zhang, H., Yang, J., Zeng, J., Xiao, J., Experimental and numerical study of lithium-ion battery thermal management system using composite phase change material and liquid cooling. *J. Energy Storage*, 71, 127410, Nov. 2023, doi: 10.1016/j.est.2023.108003.
7. Souayfane, F., Fardoun, F., Biwole, P.H., Phase change materials (PCM) for cooling applications in buildings: A review. *Energy Build.*, 129, 396–431, Oct. 01, 2016, Elsevier Ltd, doi: 10.1016/j.enbuild.2016.04.006.
8. Wang, Y., Du, C., Wang, Z., Research on Thermal Management Coupling by CPCM and Liquid Cooling for Vehicle Lithium-Ion Batteries. *Energies (Basel)*, 16, 14, 5260, Jul. 2023, doi: 10.3390/en16145260.
9. Yu, Z., Zhang, J., Pan, W., A review of battery thermal management systems about heat pipe and phase change materials. *J. Energy Storage*, 62, 106827, Jun. 01, 2023, Elsevier Ltd, doi: 10.1016/j.est.2023.106827.
10. Sun, Z., *et al.*, Experimental study of battery passive thermal management system using copper foam-based phase change materials. *Int. J. Thermofluids*, 17, 100255, Feb. 2023, doi: 10.1016/j.ijft.2022.100255.
11. Liu, H., Wei, Z., He, W., Zhao, J., Thermal issues about Li-ion batteries and recent progress in battery thermal management systems: A review. *Energy Convers. Manage.*, 150, 304–330, 2017, Elsevier Ltd,, doi: 10.1016/j.enconman.2017.08.016.
12. Mo, C., Zhang, G., Yang, X., Wu, X., Li, X., A Battery Thermal Management System Coupling High-Stable Phase Change Material Module with Internal Liquid Cooling. *Energies (Basel)*, 15, 16, 5863, Aug. 2022, doi: 10.3390/en15165863.
13. Faraji, H., Yıldız, C., Arshad, A., Arıcı, M., Choukairy, K., El Alami, M., Passive thermal management strategy for cooling multiple portable electronic components: Hybrid nanoparticles enhanced phase change materials as an innovative solution. *J. Energy Storage*, 70, 108087, Oct. 2023, doi: 10.1016/j.est.2023.108087.
14. Li, W., Zhou, Y., Zhang, H., Tang, X., A Review on Battery Thermal Management for New Energy Vehicles. *Energies*, 16, 13, 4845, Jul. 01, 2023, Multidisciplinary Digital Publishing Institute (MDPI), doi: 10.3390/en16134845.
15. Jung, Y., Kim, M., Kim, T., Ahn, J., Lee, J., Ko, S.H., Functional Materials and Innovative Strategies for Wearable Thermal Management Applications. *Nano*

Micro Lett., 15, 1, 160, Dec. 01, 2023, Springer Science and Business Media B.V., doi: 10.1007/s40820-023-01126-1.
16. Saeed, S., Hussain, A., Ali, I., Shahid, H., Ali, H.M., Thermal Management of Electronics to Avoid Fire Using Different Air Flow Strategies. *Fire*, 6, 3, 87, Mar. 2023, doi: 10.3390/fire6030087.
17. Zhang, J., *et al.*, Advanced thermal management system driven by phase change materials for power lithium-ion batteries: A review. *Renew. Sustain. Energy Rev.*, 159, 112207, May 2022, doi: 10.1016/J.RSER.2022.112207.
18. Schaut, S., Arnold, E., Sawodny, O., Predictive Thermal Management for an Electric Vehicle Powertrain. *IEEE Trans. Intell. Veh.*, 8, 2, 1957–1970, Feb. 2023, doi: 10.1109/TIV.2021.3131944.
19. de Jaeger, T., *et al.*, A Novel Thermal Management System for Electric Vehicle Batteries Using Phase-Change Material. *J. Electrochem. Soc.*, 147, 9, 3231, 2000.
20. Li, Z., Zhang, Y., Zhang, S., Tang, B., Phase change materials for lithium-ion battery thermal management systems: A review. *J. Energy Storage*, 80, 110259, Mar. 01, 2024, Elsevier Ltd, doi: 10.1016/j.est.2023.110259.
21. Wu, D., *et al.*, Thermal conductivity enhancement of diatomite-based composite phase change materials by interfacial reduction deposition of Cu nanoparticles. *J. Energy Storage*, 61, May 2023, doi: 10.1016/j.est.2023.106861.
22. Maqbool, Z., Hanief, M., Parveez, M., Review on performance enhancement of phase change material based heat sinks in conjugation with thermal conductivity enhancers for electronic cooling. *J. Energy Storage*, 60, 106591, Apr. 01, 2023, Elsevier Ltd, doi: 10.1016/j.est.2022.106591.
23. Bora, N. and Joshi, D.P., Enhancement in Thermal Properties of Organic Phase Change Material (Paraffin) *via* TiO2 Foam Doping. *Prabha Mater. Sci. Lett.*, 2, 1, 1–15, Mar. 2023, doi: 10.33889/pmsl.2023.2.1.001.
24. Ivanovich, N.M., Nikolaevich, L.V., Nikolaevich, S.A., Increasing the thermal conductivity of phase-variable material by graphene nanotubes. *Proceedings of the Academy of Sciences of Higher Education of the Russian Federation*, vol. 4, pp. 50–65, 2022.
25. Yue, H., Wang, J., Wang, H., Du, Z., Cheng, X., Du, X., Flame-Retardant and Form-Stable Phase-Change Composites Based on Phytic Acid/ZnO-Decorated Surface-Carbonized Delignified Wood with Superior Solar-Thermal Conversion Efficiency and Improved Thermal Conductivity. *ACS Appl. Mater. Interfaces*, 15, 6, 8093–8104, 2023.
26. Xue, F., *et al.*, Largely improved thermal conductivity and flame resistance of phase change materials based on three-dimensional melamine foam/phosphorous cellulose/graphite nanoplatelets network with multiple energy transition abilities. *Compos Part A Appl. Sci. Manuf.*, 156, 106898, May 2022, doi: 10.1016/J.COMPOSITESA.2022.106898.
27. Xu, L., Liu, X., Yang, R., Flame retardant paraffin-based shape-stabilized phase change material *via* expandable graphite-based flame-retardant coating. *Molecules*, 25, 10, 2408, May 2020, doi: 10.3390/molecules25102408.

28. Angel, M. and Lazcano, G., Thermal performance and flammability of phase change material for medium and elevated temperatures for textile application. *J. Therm. Anal. Calorim.*, 117, 1, 9–18, 2014.
29. Li, Y., Fang, C., Zhuang, W.Q., Wang, H., Wang, X., Antimicrobial enhancement *via* Cerium (II)/Lanthanum (III)-doped TiO2 for emergency leak sealing polyurea coating system. *npj Mater. Degrad.*, 6, 1, 41, Dec. 2022, doi: 10.1038/s41529-022-00249-x.
30. Xu, M., Li, L., Liu, Y., Cai, X., Chen, Q., Chu, P.K., Mechanism of mechanical property enhancement in nitrogen and titanium implanted 321 stainless steel. *Mater. Sci. Engineering A*, 425, 1–2, 1–6, Jun. 2006, doi: 10.1016/J.MSEA.2006.03.040.
31. Chen, L., *et al.*, Enhanced Mechanical Properties of Poly(arylene sulfide sulfone) Membrane by Co-electrospinning with Poly(m-xylene adipamide). *Chin. J. Polym. Sci.*, 38, 63–71, 2019.
32. Nazar, M.W., Iqbal, N., Ali, M., Nazir, H., Bin Amjad, M.Z., Thermal management of Li-ion battery by using active and passive cooling method. *J. Energy Storage*, 61, 106800, May 2023, doi: 10.1016/j.est.2023.106800.
33. Azmi, W.H., Hamid, K.A., Usri, N.A., Mamat, R., Sharma, K.V., Heat transfer augmentation of ethylene glycol: Water nanofluids and applications - A review. *Int. Commun. Heat Mass Transfer*, 75, 13–23, Jul. 01, 2016, Elsevier Ltd, doi: 10.1016/j.icheatmasstransfer.2016.03.018.
34. Woodward, S. and Gershun, A., *Characterization of Used Engine Coolant by Statistical Analysis*, 1993.
35. Deng, Y., *et al.*, Effects of different coolants and cooling strategies on the cooling performance of the power lithium ion battery system: A review. *Appl. Therm. Eng.*, 142, 10–29, Sep. 01, 2018, Elsevier Ltd.
36. Karimi, G., Thermal analysis of high-power lithium-ion battery packs using flow network approach. *Energy Res.*, 38, 14, 1793–1811, 2014.
37. Wang, Z., Zhao, R., Wang, S., Huang, D., Heat transfer characteristics and influencing factors of immersion coupled direct cooling for battery thermal management. *J. Energy Storage*, 62, 106821, Jun. 2023, doi: 10.1016/j.est.2023.106821.
38. Hasan, K.N., Farzad, J., Abdul, A.A., Satyam, P., Amiratabak, A., Parametric investigation of battery thermal management system with phase change material, metal foam, and fins; utilizing CFD and ANN models. *Appl. Therm. Eng.*, 247, 123080, 2024.
39. Xing, C., Junjie, S., Xiaobin, X., Xiaolin, W., Yanghan, S., Jianguo, Q., Fei, Z., Performance of thermal management system for cylindrical battery containing bionic spiral fin wrapped with phase change material and embedded in liquid cooling plate. *Renew. Energy*, 223, 120087, 2024.
40. Hosanna, U., Aleksey, N., Vijay, N.M., Heehyang, K., Hankwon, L., CFD study of nail penetration induced thermal runaway propagation in Lithium-Ion battery cell pack. *Appl. Therm. Eng.*, 243, 122649, 2024.

41. Lebrouhi, B.E., Lamrani, B., Ouassaid, M., Abd-Lefdil, M., Maarouf, M., Kousksou, T., Low-cost numerical lumped modelling of lithium-ion battery pack with phase change material and liquid cooling thermal management system. *J. Energy Storage*, 54, 105293, 2022.
42. Guiqi, H., Lisheng, Y., Changhong, W., Xianqing, L., Wenxuan, H., Xiaoxing, Z., Optimization of the thermal management system of battery thermal network model based on coupled liquid cooling of phase change materials. *J. Energy Storage*, 73, 108936, 2023.
43. Changcheng, L., Dengji, X., Jingwen, W., Shujia, Z., Wenjuan, L., Yongqing, W., Shuaijun, J., Dechuang, Z., Jian, W., Que, H., Phase Change Materials Application in Battery Thermal Management System: A Review. *Materials*, 13, 4622, 2020.
44. Ahmed, A. and Yadav, A.K., Parametric Analysis of Wastewater Electrolysis for Green Hydrogen Production: A Combined RSM, Genetic Algorithm, and Particle Swarm Optimization Approach. *Int. J. Hydrogen Energy*, 59, 51–62, 2024.
45. Ahmad, A., Yadav, A.K., Hasan, S., Enhanced production of methane enriched biogas through intensified co-digestion process and its effective utilization in a biodiesel/biohydrogen fueled engine with duel injection strategies: ML-RSM based an efficient optimization approach. *Int. J. Hydrogen Energy*, 65, 671–686, 2024.
46. Ahmed, A., Yadav, A.K., Singh, A., Singh, D.K., A machine learning–genetic algorithm based models for optimization of intensification process through microwave reactor: A new approach for rapid and sustainable synthesis of biodiesel from novel Hiptage benghalensis seed oil. *Fuel*, 374, 132470, 2024.
47. Khan, O., Khan, M.P., Yadav, A.K., A Comparative Analysis of Various Nanocomposites for the Remediation of Heavy Metals from Biomass Liquid Digestate using Multi-Criteria Decision Methods. *Biomass Bioenergy*, 186, 107281, 2024.
48. Khan, O., Alsaduni, I., Equbal, A., Parvez, M., Yadav, A.K., Performance and Emission Analysis of Biodiesel Blends Enriched with Biohydrogen and Biogas in Internal Combustion Engines. *Process Saf. Environ. Prot.*, 183, 1013–1037, 2024.
49. Ahmed, A., Yadav, A.K., Singh, A., Singh, D.K., A machine learning-response surface optimization approach to enhance the performance of diesel engine using novel blends of aloe vera biodiesel with MWCNT nanoparticles and hydrogen. *Process Saf. Environ. Prot.*, 186, 738–755, 2024, doi: 10.1016/j.psep.2024.04.013.
50. Ahmed, A., Yadav, A.K., Singh, D.K., Singh, A., A Comprehensive Machine Learning-Coupled Response Surface Methodology Approach for Predictive Modeling and Optimization of Biogas Potential in Anaerobic Co-Digestion of Organic Waste. *Biomass Bioenergy*, 180, 106995, 2024.
51. Ahmad, A., Yadav, A.K., Singh, A., Singh, D.K., Ağbulut, U., A hybrid RSM-GA-PSO approach on optimization of process intensification of linseed

biodiesel synthesis using an ultrasonic reactor: Enhancing biodiesel properties and engine characteristics with ternary fuel blends. *Energy*, 288, 129077, 2024.

52. Ahmed, A., Yadav, A.K., Hasan, S., Enhancing Efficiency and Environmental Impacts of a Nano-Enhanced Oleander Biodiesel-Biohydrogen Dual Fuel Engine Equipped with EGR, through Operational Parameter Optimization using RSM-MOGA Technique. *Int. J. Hydrogen Energy*, 78, 1157–1172, 2024.
53. Khan, O., Parvez, M., Yahya, Z., Alhodaib, A., Yadav, A.K., Hoang, A.T., Ağbulut, U., Waste-to-energy power plants: Multi-objective analysis and optimization of landfill heat and methane gas by recirculation of leachate. *Process Saf. Environ. Prot.*, 186, 957–968, 2024.
54. Khanam, S., Khan, O., Ahmad, S., Sherwani, A.F., Khan, Z.A., Yadav, A.K., Ağbulut, U., A Taguchi-based hybrid multi-criteria decision-making approach for optimization of performance characteristics of diesel engine fuelled with blends of biodiesel-diesel and cerium oxide nano-additive. *J. Therm. Anal. Calorim.*, 149, 8, 3657–3676, 2024.
55. Ahmed, A., Yadav, A.K., Singh, A., Enhancing waste cooking oil biodiesel yield and characteristics through machine learning , response surface methodology , and genetic algorithm for optimal utilization in CI Engines. *Int. J. Green Energy*, 21, 1345–1365, 2024.
56. Singh, A.A., Singh, R.K., Yadav, A.K., Saha, A., Dewangan, A., Singh, B., Experimental investigations on the electrochemical machining of D3 die steel material and multi-objective parameter optimization. *Surf. Rev. Lett.*, 31, 2450074, 2024.
57. Ahmed, A., Yadav, A.K., Hasan, S., Biogas as a Sustainable and Viable Alternative Fuel for Diesel Engines: A Comprehensive Review of Production, Purification, Economic Analysis, and Performance Evaluation. *J. Process Mech. Engg. Part E.* 09544089241255930, https://doi.org/10.1177/09544089241255930.
58. Khan, F., Karimi, M.N., Khan, O., Yadav, A.K., Alhodaib, A., Gürel, A.E., Ağbulut, U., A hybrid MCDM optimization for utilization of novel set of biosynthesized nanofluids on thermal performance for solar thermal collectors. *Int. J. Thermofluids*, 22, 100686, 2024.
59. Alhodaib, A., Yahya, Z., Khan, O., Equbal, A., Equbal, M.S., Parvez, M., Idrisi, M.J., Sustainable coatings for green solar photovoltaic cells: performance and environmental impact of recyclable biomass digestate polymers. *Sci. Rep.*, 14, 1, 11221, 2024.
60. Singh, B., Yadav, A.K., Saha, A., Fabrication and Characterization of AA6082-T6 Produced by Friction Stir Additive Manufacturing. *Surf. Rev. Lett.*, DOI:10.1142/S0218625X25500222.
61. Equbal, A., Equbal, A., Ahmad, R., Kumar, M., Khan, O., Yadav, A.K., RSM based statistical approach to enhance the compressive performance of ABS P400 parts fabricated *via* FDM Technology. *J. Surf. Rev. Lett.*, DOI:10.1142/S0218625X25500210.

62. Dewangan, A., Ahmad, A., Yadav, A.K., Innovative research on waste tire recycling for sustainable biofuel production: Assessment of its usability on multi-cylinder diesel engine employing constant injection of oxyhydrogen and biogas through a premixing device. *Int. J. Hydrogen Energy*, 84, 155–163, 2024.
63. Dewangan, A., Mallick, A., Yadav, A.K., Islam, S., Saleel, C.A., Shaik, S., Ağbulut, U., Production of oxy-hydrogen gas and the impact of its usability on CI engine combustion, performance, and emission behaviors. *Energy*, 278, 127937, 2023.
64. Khan, O., Parvez, M., Kumari, P., Yadav, A.K., Akram, W., Ahmad, S., Idrisi, M.J., Modelling of compression ignition engine by soft computing techniques (ANFIS-NSGA-II and RSM) to enhance the performance characteristics for leachate blends with nano-additives. *Sci. Rep.*, 13, 1, 15429, 2023.
65. Ahmad, A., Yadav, A.K., Singh, A., Singh, D.K., Multi-response optimization of a microalgae-spirulina-fueled VCR diesel engine: a comprehensive RSM-GA approach. *Environ. Dev. Sustain.*, 27, 1–29, 2023.
66. Ahmed, A., Yadav, A.K., Singh, A., Process optimization of spirulina microalgae biodiesel synthesis using RSM coupled GA technique: a performance study of a biogas powered dual fuel engine. *Int. J. Environ Sci. Tech.*, 21, 169–188, 2023.
67. Khan, O., Ali, V., Parvez, M., Alhodaib, A., Yahya, Z., Yadav, A.K., Ağbulut, U., Exploring the performance of biodiesel-hydrogen blends with diverse nanoparticles in diesel engine: A hybrid machine learning K-means clustering approach with weighted performance metrics. *Int. J. Hydrogen Energy*, 78, 547–563, 2024.
68. Ahmed, A., Yadav, A.K., Singh, A., Application of machine learning and genetic algorithm for prediction and optimization of Biodiesel Yield from Waste Cooking Oil. *Korean J. Chem. Engg.*, 40, 2941–2956, 2023.
69. Ahmed, A., Yadav, A.K., Singh, A., Enhancement of Biogas Yield from Dual Organic Waste Using Hybrid Statistical Approach and Its Effects on Ternary Fuel Blend (Biodiesel/n-butanol /Diesel) Powered Diesel Engine Environmental Progress & Sust. *Energy*, 42, 5, e14163, 2023. doi.org/10.1002/ep.14163.
70. Ahmed, A., Yadav, A.K., Singh, A., Optimization of Cavitation-Assisted Biodiesel Production and Fuel Properties from Neochloris Oleoabundans Microalgae oil using Genetic Algorithm and Response Surface Methodology. *J. Process Mech. Engineering Part E.*, 238, 4, 1821–1835, 2023. doi.org/10.1177/09544089231159832.
71. Yadav, M., Karimi, M.N., Yadav, A.K., Effect of Graphene Oxide Nanoparticles Dispersed Biodiesel on Combustion, Performance, and Emission Characteristics of a Diesel Engine. *J. Process Mech. Engineering Part E.*, 2023. https://doi.org/10.1177/09544089231209139.

72. Dwivedi, S.P., Yadav, A.K., Saxena, A., Dwivedi, V.K., Tribo-mechanical, physical and thermal behaviour of Al/Si3N4 composite with and without the addition of Cu, Ni and Cr entropy elements. *Proc. Inst. Mech. Eng., Part E: J. Process Mech. Eng.*, 2023. 09544089231189663.
73. Das, D., Yadav, A.K., Ahmed, A., Numerical Simulation of the Combustion Chamber and Performance Analysis of Microalgae-Based Biodiesel/Diesel-Powered CI Engine: An Axisymmetric Flow Approach. *J. Process Mech. Engineering Part E*, 2023. https://doi.org/10.1177/09544089231218121.
74. Dewangan, A., Mallick, A., Yadav, A.K., Ahmad, A., Alqahtani, D., Islam., S., Combined Effect of Operating Parameters and Nano Particles on Performance of a Diesel Engine: Response Surface Methodology coupled Genetic algorithm Approach. *ACS Omega*, 8, 27, 24586–24600, 2023.
75. Singh, D., Singh, S., Yadav, A.K., Khan, O., Dewangan, A., Alkahtani, M.Q., Islam, S., From Theory to Practice: A Sustainable Solution to Water Scarcity by Using a Hybrid Solar Distiller with a Heat Exchanger and Aluminum Oxide Nanoparticles. *ACS Omega*, 37, 8, 33543–33553, 2023.
76. Ahmed, A., Yadav, A.K., Singh, A., Optimization of Biogas Yield from Anaerobic Co-digestion of Dual waste for environmental sustainability: ANN, RSM, and GA, Approach. *Int. J. Oil Gas Coal Tech.*, 33, 1, 75, 2023.
77. Ahmed, A., Yadav, A.K., Singh, A., An environmental impact assessment and optimization study of biodiesel production from microalgae. *Int. J. Global Warming*, 31, 3, 294–313, 2023.
78. Ahmed, A., Yadav, A.K., Singh, A., Biodiesel yield optimization from a third-generation feedstock (microalgae spirulina) using a hybrid statistical approach. *Int. J. Ambient Energy*, 44, 1, 1202–1213, 2023.
79. Ahmed, A., Yadav, A.K., Singh, A., Modelling and optimisation of VCR diesel engine parameters fuelled with microalgae methyl-ester: a RSM-coupled Taguchi approach. *Int. J. Ambient Energy*, 44, 1, 2498–2506, 2023.
80. Ahmed, A., Yadav, A.K., Singh, A., Biodiesel Production from Mahua Oil: Characterization, Optimization, and Modeling with a Hybrid Statistical Approach. *Int. J. Ambient Energy*, 44, 1, 2600–2609, 2023.
81. Khan, T.A., Khan, T.A., Yadav, A.K., Khan, M.E., Experimental Studies on Utilization of Neochloris Oleoabundans Microalgae Biodiesel as an Alternative Fuel for Diesel Engine. *Int. J. Ambient Energy*, 44, 1, 115–123, 2023.
82. Singh, D., Yadav, A.K., Kumar, A., Samsher, Energy matrices and life cycle conversion analysis of N-identical hybrid double slope solar distiller unit using Al2O3 nanoparticle. *J. Water Environ. Nanotechnol.*, 8, 3, 267–284, 2023.
83. Khan, T.A., Yadav, A.K., Khan, T.A., A hydrodynamic cavitation-assisted system for optimization of biodiesel production from green microalgae oil using a genetic algorithm and response surface methodology approach. *Environ. Sci. Pollut. Res*, 29, 49465–49477, 2022.

84. Dewangan, A., Yadav, A.K., Mallick, A., Singh, S., Comparative study of Manilkara Zapota and Karanja based Biodiesel Properties and Its Effect on Multi-Cylinder Diesel Engine Characteristics. *Energy Sources Part A*, 44, 2, 5143–5153, 2022.
85. Yadav, A.K., Khan, T.A., Khan, T.A., Kumar, S., Methyl ester of Gmelina arborea oil as a substitute for petroleum diesel: an experimental study on its performance and emissions in a diesel engine. *Energy Sources Part A*, 43, 11, 1307–1314, 2021.
86. Khan, O., Yadav, A.K., Khan, M.E., Parvez, M., Part, A., Characterization of Bio-Ethanol Obtained from Eichhornia Crassipes Plant; Its Emission and Performance Analysis on CI Engine. *Energy Sources Part A*, 43, 14, 1793–1803, 2021.
87. Khan, I.A., Singh, S.K., Pal, A., Yadav, A.K., Efficient Production of Biodiesel from Cannabis Sativa oil using Intensified Transesterication (Hydrodynamic Cavitation) Method. *Energy Sources Part A*, 42, 20, 2461–2470, 2020.
88. Dewangan, A., Yadav, A.K., Mallick, A., Tyagi, A., Optimisation of biodiesel production and engine performance from Simarouba oil in compression ignition engine. *Int. J. Oil Gas Coal Technol.*, 25, 3, 357–376, 2020.
89. Khan, M.E., Khan, T.A., Yadav, A.K., Pal, A., A comprehensive review on performance, emission and combustion characteristics of diesel engines fuelled with algae oil and methyl ester. *WEENTECH Proc. Energy*, 6, 2, 160–168, 2020.
90. Yadav, A.K., Khan, M.E., Pal, A., Custard Apple Seed (Annona Squamosa) Oil as a Promising Feedstock for Biodiesel Production using Advanced Techniques and Its Experimental Investigations on an Agricultural Diesel Engine. *Int. J. Oil Gas Coal Tech.*, 20, 4, 473–492, 2019.
91. Yadav, A.K., Pal, A., Ghosh, U., Gupta, S.K., Comparative Study of Biodiesel Production Methods from Yellow Oleander Oil and its Performance Analysis on an Agricultural Diesel Engine. *Int. J. Ambient Energy*, 40, 2, 152–157, 2019.
92. Khan, I.A., Singh, S.K., Yadav, A.K., Sharma, D., Enhancement in the Performance of a Diesel Engine Fueled with Pongamia Methyl Ester and n-Butanol as Oxygenated Additive. *Int. J. Ambient Energy*, 40, 8, 842–846, 2019.
93. Yadav, A.K., Khan, O., Khan, M.E., Utilization of High FFA Landfill Waste (Leachates) as a Feedstock for Sustainable Biodiesel Production: Its Characterization and Engine Performance Evaluation. *Environ. Sci. Pollut. Res.*, 25, 32, 32312–32320, 2018.
94. Yadav, A.K., Dewangan, A., Mallick, A., Effect of n-Butanol and Diethyl Ether Additives on Performance and Emission Characteristics of a Diesel Engine Fuelled with Diesel-Pongamia Biodiesel Blends. *J. Energy Eng.*, 144, 6, 04018062, 2018.
95. Vinay, Singh, B. and Yadav, A.K., Optimization of Performance and Emission Characteristics of CI Engine Fuelled with Mahua Oil Methyl Ester-Diesel

Blend using Response Surface Methodology. *Int. J. Ambient Energy*, 44, 6, 674–85, 2020.

96. Yadav, A.K., Dewangan, A., Mallick, A., Synthesis and Stability Study of Biodiesel from Kachnar Seed Oil. *J. Energy Eng.*, 144, 5, 04018053, 2018.
97. Yadav, A.K., Pal, A., Dubey, A.M., Experimental Studies on Utilization of Prunus Armeniaca L. (Wild Apricot) Biodiesel as an Alternative Fuel for CI Engine. *Waste Biomass Valorization*, 9, 10, 1961–1969, 2018.
98. Yadav, A.K., Khan, M.E., Pal, A., Ultrasonic Assisted Production of Biodiesel from Karabi Oil using Heterogeneous Catalyst. *Biofuels*, 9, 1, 101–112, 2018.
99. Yadav, A.K., Vinay, Singh, B., Optimization of Biodiesel Production from Annona Squamosa Seeds Oil Using Response Surface Methodology and Its Characterization. *Energy Sources Part A*, 40, 9, 1051–1059, 2018.
100. Dewangan, A., Yadav, A.K., Mallick, A., Current Scenario of Biodiesel Development in India: Prospects and Challenges. *Energy Sources Part A*, 40, 20, 2494–2501, 2018.
101. Yadav, A.K., Khan, M.E., Pal, A., Performance and Emission Characteristics of a Stationary Diesel Engine fuelled by Schleichera Oleosa oil Methyl Ester (SOME) Produced through Hydrodynamic Cavitation Process. *Egypt. J. Pet.*, 27, 1, 89–93, 2018.
102. Yadav, A.K., Khan, M.E., Pal, A., Performance, Emission and Combustion Characteristics of an Indica Diesel Engine Operated with Yellow Oleander (Thevetia Peruviana) Oil Biodiesel Produced through Hydrodynamic Cavitation Method. *Int. J. Ambient Energy*, 39, 4, 365–371, 2018.
103. Dewangan, A., Mallick, A., Yadav, A.K., Effect of Metal Oxide Nanoparticles and Engine Parameters on the Performance of a Diesel Engine: A Review. *Mater. Today Proc.*, 21, 1722–27, 2020.
104. Yadav, A.K., Khan, M.E., Pal, A., Dubey, A.M., Experimental Investigations of Performance and Emissions Characteristics of Kusum (Schleichera Oleosa) Biodiesel in a Multi-Cylinder Transportation Diesel Engine. *Waste Biomass Valor*, 8, 4, 1331, 2017.
105. Yadav, A.K., Khan, M.E., Pal, A., Biodiesel Production from Oleander (Thevetia Peruviana) Oil and Its Performance Testing on a Diesel Engine. *Korean J. Chem. Eng.*, 34, 2, 340–345, 2017.
106. Dubey, A.M., Kumar, A., Yadav, A.K., Wear behaviour of friction stir weld joint of cast Al (4–10%) Cu alloy welded at different operating parameters. *J. Mater. Process. Tech.*, 240, 87–97, 2017.
107. Khan, O., Khan, M.E., Yadav, A.K., Sharma, D., The Ultrasonic Assisted Optimization of Biodiesel Production from Eucalyptus Oil. *Energy Sources Part A*, 39, 13, 1323–1331, 2017.
108. Dewangan, A. and Yadav, A.K., Wax Deposition during Production of Waxy Crude Oil and Its Remediation. *Pet. Sci. Techn*, 35, 18, 1831–1838, 2017.
109. Yadav, A.K., Khan, M.E., Pal, A., Kaner biodiesel production through hybrid reactor and its performance testing on a CI engine at different compression ratios. *Egypt. J. Pet.*, 26, 2, 525–532, 2017.

110. Yadav, A.K., Khan, M.E., Pal, A., Dubey, A.M., Performance and emission characteristics of a transportation diesel engine operated with non-edible vegetable oils biodiesel. *Case Stud. Therm. Eng.*, 8, 236–244, 2016.
111. Yadav, A.K., Khan, M.E., Pal, A., Biodiesel Production from Nerium Oleander (Thevetia Peruviana) Oil through Conventional and Ultrasonic Irradiation Methods. *Energy Sources Part A*, 38, 23, 3447–3452, 2016.
112. Yadav, A.K., Khan, M.E., Pal A., A Comparative Study on Ultrasonic Cavitation and Mechanical Stirring Method towards Efficient Production of Biodiesel from Non-Edible Oils and Performance Testing on a C.I. Engine. *Int. J. Envt. Waste Managt.*, 18, 4, 349–367, 2016.
113. Dewangan, A., Mallick, A., Yadav, A.K., Kumar, R., Combustion-generated pollutions and strategy for its control in CI engines: A review. *Mater. Today Proc.*, 21, 1728–1733, 2020.
114. Yadav, A.K., Khan, M.E., Pal, A., Sharma, D., Optimization of biodiesel production from bitter groundnut oil using Taguchi method and its performance and emissions characteristics on a 4-cylinder Tata Indica engine. *Int. J. Sust. Agri. Mgt. Inf.*, 1, 285–300, 2015.
115. Khan, F., Khan, O., Parvez, M., Ahmad, S., Yahya, Z., Alhodaib, A., Yadav, A.K., Ağbulut, U., K-means clustering optimization of various quantum dots and nanoparticles-added biofuels for engine performance, emission, vibration, and noise characteristics. *Therm. Sci. Eng. Prog.*, 54, 102815, 2024.
116. Khan, O., Parvez, M., Yadav, A.K., *et al.*, Investigation of the Thermal Properties of Lauric Acid Melting as a Phase Change Material Using Priority based K-Means Clustering Method. *Energy Storage*, 6, 6, e70026, 2024, doi: 10.1002/est2.70026.
117. Sharma, G., Dewangan, A.K., Yadav, A.K., Ahmed, A., Current status of research on hydrogen generation, storage and transportation technologies: A state-of-the-art reviewtowards sustainable energy. *Process Saf. Environ. Prot.*, 191, Part A, 1445–1460, 2024.
118. Singh, S., Singh, B., Kumar, S., Yadav, A.K., Temperature-dependent dynamic hysteresis scaling of ferroelectric hysteresis parameters of lead free [(Ba 0.825 + xCa0.175-x) (Ti1- x Snx)O3]ceramics. *Ferroelectrics*, 551, 1, 133–142, 2019.
119. Dubey, A.M., Ram, R., Yadav, A.K., Ion cell performance using single particle representation of battery electrode. *J. Electro Analytical Chem.*, 768, 11–17, 2016.

4

Green Nano Fluids: Applications in Thermal Management

Sunil Baloda[1], Naveen Sharma[2*] and Mukesh Kumar[1]

[1]Department of Mechanical Engineering, The Technological Institute of Textile and Sciences, Bhiwani, India
[2]Department of Mechanical Engineering, Netaji Subhas University of Technology, New Delhi, India

Abstract

The present chapter discovers the state-of-the art and sustainable realm of green nanofluids and their growing applications in thermal management (TM). Since the need for efficient thermal control and heat dissipation increases continuously in various industries, the incorporation of nanotechnology into fluid systems offers a promising solution. Green nanofluids, attributed by environmental friendly composition and superior thermal properties, offer an exciting opportunity for addressing the challenges faced by traditional heat transfer fluids.

This chapter begins with an overview of basic principles of nanofluids, highlighting the integration of environmental friendly nanoparticles into base fluids. It probes into the preparation methods and considerations that add to the green facet of the nanofluids, confirming their compatibility with sustainable practices. The chapter also puts light on the thermo-physical properties (TPs) of green nanofluids. A considerable portion of the chapter is dedicated to the discussion of several applications of green nanofluids in thermal management. From food processing to electronics cooling of automotive systems, and renewable energy technologies to industrial cooling, the exceptional thermal features of the green nanofluids play an important role in improving heat transfer efficiency and system performance. The chapter concludes by highlighting the challenges of green nanofluids providing bibliophiles with valuable intuition into the emerging field of green nanofluids and their key role in shaping the future of thermal management.

**Corresponding author*: naveen.sharma@nsut.ac.in; ORCID: 0000-0002-3326-1800
Sunil Baloda: ORCID: 0000-0001-9260-8248
Mukesh Kumar: ORCID: 0000-0002-7975-9196

Ashwani Kumar, Mukesh Kumar Awasthi, Nitesh Dutt, Yogesh Kumar Singla and Sivasakthivel Thangavel (eds.) Thermal Battery Management System for Hybrid and Electric Vehicles, (73–90) © 2025 Scrivener Publishing LLC

Keywords: Green nanofluids, nanoparticles, thermal management systems, thermo-physical properties, thermal battery management

4.1 Introduction

Over the years, there has been growing emphasis within engineering and industrial sectors on improving heat transfer efficiency. Researchers persistently attempt to augment heat transfer rates and thermal conductivity while simultaneously aiming at minimizing frictional losses, pressure drop, and pumping power related with heat transfer fluids used in thermal systems. Therefore, a novel heat transfer fluid (HTF) has been engineered well-known as 'nanofluid' proposing excellent enhanced heat transfer characteristics. Choi and Eastman coined the term 'nanofluid' in 1993. A nanofluid is synthesized by dispersing small quantities of nanoparticles such as diamond, Ag, Al, Au, Cu, Si, SiC, CuO, Al_2O_3, MgO, TiO_2, ZnO, multi walled carbon nanotube etc. into the conventional base fluids (water, toluene, engine oils, ethylene glycol etc.). Nanoparticles (NPs) average size is typically below 100 nm [1]. A perfect example to clearly understand the comparison of nanometer, micrometer, and millimeter was suggested by Serrano *et al.* [2] as revealed in Figure 4.1.

The use of NPs into the base fluid (BF), exceptionally improves TPs of conventional fluids which further contribute to augment heat transfer (HT) coefficient. The extent to which heat dissipation rate increases relies on the concentration of NPs dispersed within the conventional fluid. Peyghambarzadeh *et al.* [3] performed a study on car engine radiators replacing pure water with Al_2O_3/water nanofluid. The study found that the heat dissipation rate was enhanced by 45% due to Al_2O_3/water nanofluid. Another experimental investigation on radiator type heat exchanger determined that using Al_2O_3/water nanofluid with 6.8% volume nanoparticle

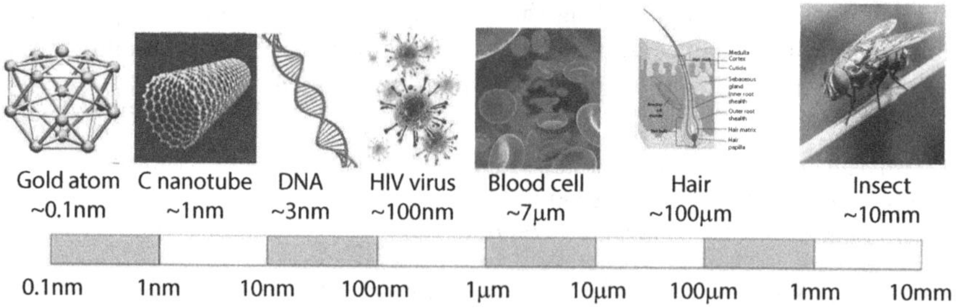

Figure 4.1 A comparison of length scale with related examples [2].

concentration could achieve 40% enhanced heat transfer coefficient [4]. Hwang *et al.* [5] carried out a study using 3% concentration of Al_2O_3/water nanofluid and discovered 8% enhancement in heat transfer coefficient. Figure 4.2 shows structures of nanofluids [6].

The ever-increasing energy demand is driving an untiring escalation in global temperatures beyond pre-industrial levels, simultaneously releasing harmful gases and radiation which lead to austere environmental conditions. Consequently, to address this pressing issue there is an imperious need to develop robust and highly effective thermal systems. The recent scientific investigations in nanofluids for improved performance of thermal systems have garnered substantial attention because of the exceptional properties of nanofluids. However, this heightened interest in use of conventional nanofluids also fetches safety and health concerns due to reactive nature of nanofluids in the solvents. Moreover, the presence of strong chemicals in the nanofluids can lead to detrimental effects like

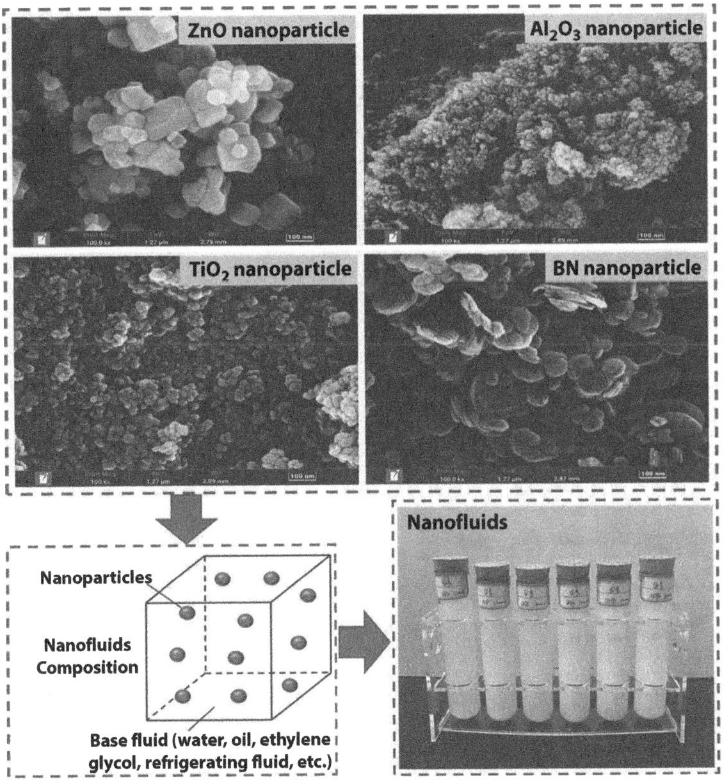

Figure 4.2 Schematic diagram of structure of nanofluids [6].

equipment corrosion and the release of non-biodegradable byproducts. Subsequently, this compels us to discover an economical and environmental friendly nanofluid, the answer to this issue is Green nanofluids. Green nanofluids offer a great choice to improve efficiency of the heat exchange systems due to their exceptional properties. Natural extracts of plants, such as gallic acid, biochar, neem, and rice bran etc., are used to produce green nanoparticles that have specific qualities like optical features, stability, distinct shape, increased total area, least toxic and enhanced thermo-physical properties (TPs). These green NPs are further employed to create green nanofluids with excellent thermo-physical properties [7]. These innovative nanofluids synthesized from green nanoparticles using natural extracts offer numerous benefits. The advantages of green synthesis of nanofluids over conventional physio-chemical methods are summarized in Figure 4.3.

Thanks to the excellent TPs of green nano fluids which open up a new paradigm for HTFs to find applications in diverse categories requiring high heat-dissipation rates. Undoubtedly, green nanofluids have appeared as a captivating area of research in thermal management (TM) systems with wide range of applications like coolants in processed food industry, solar industry, oil recovery, atomic reactors, microelectronics, fuel cells, hybrid powered engines, heat exchangers and boilers etc. Finally, green nanofluids offer a great choice for efficient TM and address the challenges posed by rise in global temperature and increased energy consumption. The distinctive characteristics of green nanofluids make them valuable across several industries and applications [7, 9].

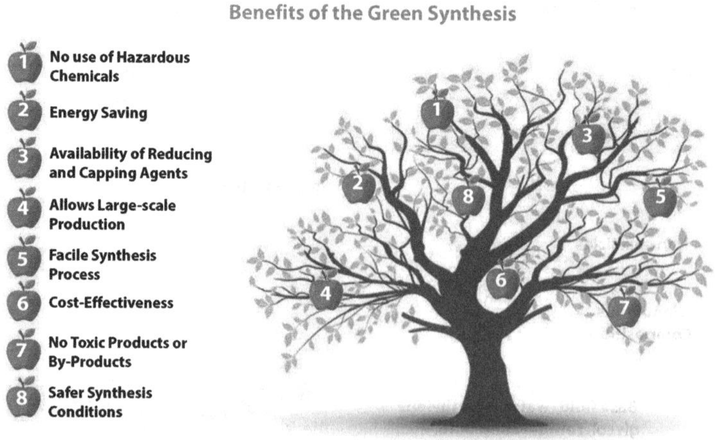

Figure 4.3 Advantages of the green production of nanoparticles [8].

Section 4.2 of the chapter covers various methods for preparing nanofluids, with a particular emphasis on green nanofluids. Section 4.3 explores the key TPs of these green nanofluids. In Section 4.4, the potential applications of green nanofluids are highlighted. The chapter concludes with a summary in Section 4.5.

4.2 Nanofluids Preparation Steps

NPs are suspended or dispersed in the BF to prepare nanofluids. The surface-active agents or surfactants are added to the nanofluids to evade NPs settling and to enhance the nanofluid stability. According to colloidal theory the sedimentation in suspensions stops below a critical radius of the particle size because the Brownian forces counterweigh gravity forces. The smaller size of nanoparticles results in agglomerates creation among them since it has a surface [10, 11]. Hence, researchers find it as a huge challenge to prepare a stable nanofluid having suitable particle diameter and concentration. Various methods like one-step, two-step, and others have worked up some innovative methods to create nanofluids which are discussed below.

4.2.1 One-Step Method

In this method, the NPs are directly incorporated into the BF sans intermediary stages like drying, storage, and dispersion. The nanofluids prepared through one-step process have good stability; but because of their high cost

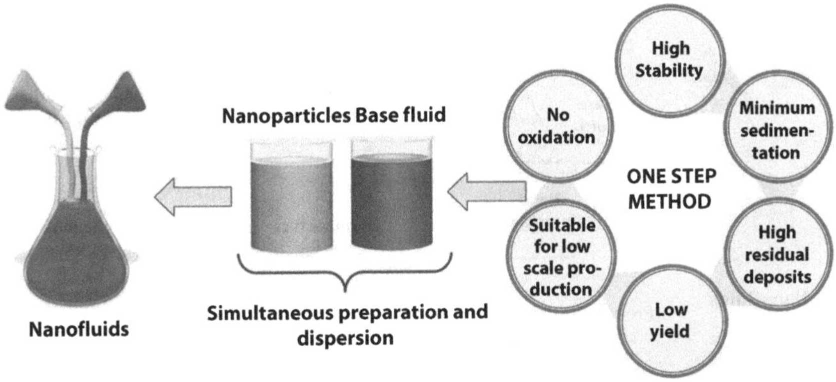

Figure 4.4 Nanofluids production using single-step process [14, 15].

this technique is only appropriate for limited manufacture of nanofluids [12, 13]. The benefits of one step process, as shown in Figure 4.4, include less agglomeration, good stability, no oxidation, and no requirements of transport, storing, drying, and re-dispersal. The disadvantages associated with single step method are large heaps of left over reactants, not suited for large-scale manufacturing, suitable only for low pressure BFs as well as expensive synthesis approach [14, 15].

4.2.2 Two-Step Method

The two-step approach is regarded as the best cost-effective technique for mass production of nanofluids. In this approach, first nanoparticles are synthesized using different routes such as biological route (plant extracts, bio-templates and microorganisms aided biogenesis), chemical route (chemical vapor-, sol-gel-, and hydrothermal synthesis and micro mixture technique etc.), physical route (pulse vapor deposition, melt mixing, laser pyrolysis and high energy milling etc.), and physiochemical route (sono-chemical, or sono-electrochemical). Then these nanoparticles in powdered form are dispersed in base fluids with the aid of ultrasonic agitators, magnetic steering, shear mixing, and ball millers etc. Figure 4.5 reveals the advantages and disadvantages of the two-step approach [14, 15]. NPs agglomeration because of high surface activity is the main demerit of two-step technique which leads to reduced stability. An additive (surfactant/dispersant) is used to maintain good stability [16]. Several researchers preferably utilized this commercial approach to prepare nanofluids [17–20].

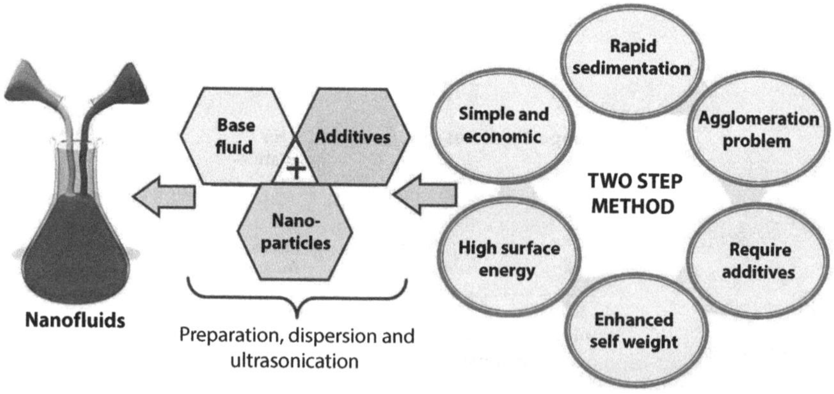

Figure 4.5 Nanofluids production: two-step method [14, 15].

4.2.3 Other Methods

Some other techniques have been used by researchers to prepare nanofluids with comparatively improved characteristics and increased stability. Feng *et al.* [21] prepared NPs of Au, Ag, and Pt soluble in water using aqueous-organic phase-transfer approach. Zhu *et al.* [22] and Wei and Wang [23] developed a new technique using an innovative precursor modification to produce copper nanofluids through ultrasonic and microwave irradiation. Chen and Wang [24] and Yu *et al.* [25] used phase-transfer technique to prepare mono-disperse noble metal colloids and stable kerosene-based Fe_3O_4 nanofluids respectively. Research also showed that nanofluids prepared using chemical solution process have superior conductivity with more stability [26].

4.2.4 Summary of Green Nanofluids Preparation Methods

Green nanofluids are produced using green nanoparticles synthesized by the following techniques discussed here from natural extracts. Some common preparation approaches for green nanofluids are:

Single-Step Physical Methods:

- **Submerged Arc Method**: It includes direct synthesis of NPs in the BF by means of an electric arc.
- **Precipitation (Ion Exchange) Method**: In this approach NPs are precipitated from a solution into the base fluid.
- **Chemical Reduction Method**: It involves reduction of metal ions in a solution to form nanoparticles.
- **Emulsion Polymerization**: In this approach nanoparticles are formed within an emulsion.
- **Sol-Gel (Hydrolysis) Method**: It includes gelation of a precursor solution to form nanoparticles.
- **Microwave-Assisted Reaction**: Synthesis of nanoparticles using microwave irradiation.

Double-Step Physical Methods:

- **Adding Acid or Base**: To stabilize the nanofluid the pH value of the suspension is adjusted.
- **Surface Active Agents and Dispersants**: Surfactants/dispersants are added to disperse nanoparticles into the fluid.
- **Ultrasonic Vibration**: Ultrasonic waves are used to disperse and stabilize nanoparticles.

Chemical Process:

- Involving chemical reactions to prepare nanoparticles and disperse them in the base fluid is an emerging technique for preparing nanofluids.

4.3 Nanofluids Thermo-Physical Properties (TPs)

The TPs of the nanofluids play a key role in augmenting heat transfer rate in various thermal processes. The most widely studied nanofluids TPs are density (ρ), specific heat (C_p), viscosity (μ), and heat conductivity (k). Incorporating NPs into the BFs improves the TPs of nanofluids significantly which have an influence on convective HT. The heat exchange rate of a TM scheme mostly relies upon the TPs of nanofluids. Enhanced TPs

Table 4.1 TPs of different nanoparticles.

Nanoparticles	Thermal conductivity (W/m.K)	Specific heat (J/kg.K)	Density (kg/m^3)
Al_2O_3	40.0	773	4000
ZnO	29.0	544	5610
TiO_2	8.4	692	4260
BN	260	900	2270
Diamond	3300	509	3500
MWCNT	2000–3000	710	1300
MgO	54.9	1050	3580
CuO	32.9	551	6315
SiC	490	700	3210
Ag	429	235	10490
Au	315	129	19320
Al	237	904	2700
Cu	398	385	8960
Si	148	714	2330

of nanofluids enable to develop highly energy efficient, durable, portable, and compact industrial systems along with benefits of reduced operational expenses and green environment. The important factors affecting TPs of nanofluids include temperature, particle loading, size, shape, material, sonication, aggregation, additives, and interfacial layer formation. The representative values of thermal attributes of commonly used nanoparticles are tabulated in Table 4.1.

4.3.1 Density

The density of substance is the ratio of mass to volume. The density of the nanofluid affects the thermal performance of the system. The variation in density values directly affects the flow parameters, friction factor, pumping power, pressure loss, and Nusselt number. The density of nanofluids is higher than most of the BFs because of extra mass of nanoparticles. The commonly used relation to estimate nanofluid density according to mixture law is given in Equation 4.1 [27].

$$\rho_{nf} = \rho_{np} \times \varphi + (1-\varphi) \times \rho_{bf} \tag{4.1}$$

Here, $\rho_{nf}, \rho_{bf}, \rho_{np}$ and φ represent density of nanofluid, BF, NP and nanoparticle volume concentration respectively.

4.3.2 Specific Heat Capacity

The specific heat of a substance is measured using a "differential scanning calorimeter", which measures how much energy it takes to raise its temperature by one degree. The heat dissipation rate of nanofluids is largely assessed by its specific thermal capacity. The specific thermal capacity of nanofluids is affected by various factors like magnitude, profile, concentration of nanoparticles, and the attributes of BFs. A large specific heat is advantageous in car engine cooling applications where it permits the system to engross a huge quantity of heat sans corresponding increase in working temperature. On the other hand, in applications where a quick change of temperature is desired, the heat transfer fluids having small specific heat can be helpful to such systems. The most important finding from literature showed that the specific heat increases with temperature and decreases with density. Pak and Cho [27], introduced the prediction method to guess the specific heat of nanofluids and the developed correlation is given in Equation 4.2.

$$Cp_{nf} = Cp_{np} \times \varphi + (1-\varphi) \times Cp_{bf} \tag{4.2}$$

Here, $Cp_{nf}, Cp_{bf}, Cp_{np}$ and φ represent specific heat of nanofluid, BF, NP and nanoparticle volume concentration respectively.

4.3.3 Viscosity

The resistance to fluid flow between adjacent layers of fluid is known as dynamic viscosity. It represents the ratio of shear stress to shear strain rate. The viscometer and rheometer can be used to measure dynamic viscosity. The higher viscosity of nanofluids results in increased pumping power, friction, and energy demands. In nanofluids, the viscosity diminishes with rise in temperature but surges with rising NP concentration. The causes influencing nanofluids viscosity comprise size, type, concentration of nanoparticles and type of BFs. The most prominent model for estimating the dynamic viscosity for low volume concentration was suggested by Einstein [28] as shown in Equation 4.3.

$$\frac{\mu_{nf}}{\mu_{bf}} = 1 + 2.5\varphi \tag{4.3}$$

Here, μ_{nf}, μ_{bf}, and φ represent nanofluid dynamic viscosity, BF dynamic viscosity and volume concentration respectively.

4.3.4 Thermal Conductivity

Thermal conductivity stands out as the most significant TP of nanofluids which can be experimentally measured through transient hot-wire and thermal constant analyzers. The investigations on augmenting heat conductivity of nanofluids for higher heat dissipation are very noteworthy. The key factors influencing the heat conductivity of nanofluids are shape, dimensions, materials, NPs concentration, temperature, sonication time, and type of BF as displayed in Figure 4.6. Elevating the concentration of NPs leads to a rise in thermal conductivity of nanofluids. The primary heat conductivity model applicable for spherical sized particles of solid-liquid mixture was suggested by Maxwell in 1873 based on EMT (Effective Medium Theory) theory [1].

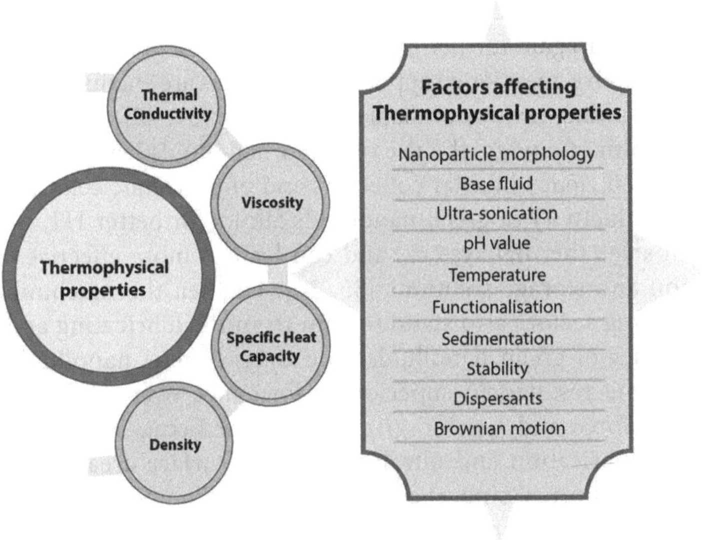

Figure 4.6 Graphical representation of TPs of nanofluids and factors influencing them [30].

The generalized Maxwell model considering empirical shape factor (n) for nanofluids heat conductivity was proposed by Hamilton-Crosser [29] in 1962 which is expressed in Equation 4.4.

$$\frac{k_{nf}}{k_{bf}} = \frac{k_p + (n-1) \times k_{bf} + (n-1) \times (k_p - k_{bf}) \times \varphi}{k_p + (n-1) \times k_{bf} - (k_p - k_{bf}) \times \varphi} \tag{4.4}$$

Here, k_{nf}, k_{bf}, k_p, represent heat conductivity of nanofluid, BF and NP respectively and φ stand for volume faction of nanofluid.

4.4 Applications from Thermal Management Perspective

The interest in green nanofluids, which are a subset of nanofluids, is increasing continuously due to their enhanced heat dissipation capacity, making them a promising option for TM in diverse applications, such as energy generation and utilization, machining and manufacturing sectors, food processing, heat exchangers, cooling of electronics, automotive industry, nuclear reactors, biomedical and environmental remediation, across various industries. To gain further insight into the use of green nanofluid utilized for the above-mentioned applications, readers are encouraged to refer several outstanding reviews available in literature [31–45].

In the realm of energy, the use of green nanofluids leads to enhancement in the efficiency of solar collectors and photovoltaic cells. The high thermal conductivity of green nanofluids allows for better HT, which is crucial in solar thermal systems and can lead to more effective energy conversion and storage solutions [35, 36]. Further, the machining and manufacturing sectors also stand to gain from the lubricating and cooling properties of green nanofluids. The hybrid green nanofluid added with graphene resulted in superior performance over Al_2O_3 in terms of grinding force and specific grinding energy because of their interlayer shear lubrication and ultrahigh specific surface area [37]. In the domain of food processing, the utilization of nanofluids has the potential to diminish the duration required for thermal processing, thereby preserving and potentially enhancing the nutritional value of the food products [38, 39]. In industrial processes, green nanofluids are employed in heat exchangers and cooling systems to improve TM. Their superior heat transfer capabilities allow for smaller, more efficient systems that can operate at lower energy costs [40, 41]. In electronics, where overheating is a significant challenge, green nanofluids serve as advanced coolants that can dissipate heat more effectively than traditional coolants, thereby protecting delicate electronic components and extending their lifespan [31, 32]. The automotive industry also benefits from green nanofluids, particularly in TM systems for electric and hybrid vehicles, where maintaining optimal battery temperatures is essential for performance and safety [42]. Green nanofluids can aid in the removal of pollutants from water and air, offering a novel approach to tackling environmental pollution [43]. Further, in aerospace and defense, green nanofluids can be used in advanced cooling systems for high-performance equipment, where reliability and efficiency are paramount [44]. Table 4.2 summarized some of the latest research articles related to the use of green nanofluids in diverse applications from TM perspective.

A hybrid cooling system that uses green nanofluids for battery TM systems reduces temperature rises and enhances temperature uniformity. This results in improved cooling performance, with nanofluids achieving about 10% to 15% greater efficiency compared to conventional fluids [52].

Table 4.2 Summary of the latest research on applications of green nano fluids in thermal management.

Applications	Nanofluids configurations	Purpose
Energy	Cu, CuO, TiO_2, and Al_2O_3 in the oil heat transfer fluid	For performance improvement of organic Rankine cycle using a new two-stage solar collector [36]
	Polyaniline nanofibers-palm oil nanofluids	For thermal energy storage application [45]
	Photothermal phase change hydrogel	For solar thermal storage and human TM [46]
Machining and manufacturing sectors	Hybrid nanofluids (Added graphene and Al_2O_3 nano-additives into oleic acid-based nanofluids)	For grinding hard-to-cut materials [37]
Heat exchangers	Water based PANI (Polyaniline) [47]	For performance improvement of helical coiled heat exchanger
	Silver nano particles using Neem leaf extracts [48]	
Cooling of electronic components	Single-phase covalently functionalized Graphene Nanoplatelets (CGNPs)/ H_2O	For designing heat sink to cool CPU [49]
Automotive industry	Nanofluid and hybrid nanoparticle-enhanced phase change material	For TM of Li-ion battery [50]
	Loop Heat Pipe filled with copper oxide–water and alumina–water nanofluids	Battery TM in electric vehicles [51]
	Al_2O_3, CuO, and SiO_2 - suspended in base fluids such as water and ethylene glycol	Effectiveness in cooling Li-ion battery packs [52]

4.5 Conclusion

Green nanofluids represent a frontier in nanotechnology with vast potential for sustainable development. Their unique properties and the green synthesis methods make them an attractive option for industries looking to improve efficiency and reduce environmental impact. However, the real-world application and progress of nanofluids are hindered by various challenges, such as their enduring stability, additional energy required for their circulation, an increase in pressure penalty, and the high production cost. As research continues to advance, the scope of applications for green nanofluids is expected to expand, paving the way for innovative solutions to some of the most pressing challenges in energy, automotive, environment, and healthcare.

References

1. Ali, A.R.I. and Salam, B., A review on nanofluid: preparation, stability, thermophysical properties, heat transfer characteristics and application. *SN Appl. Sci.*, 2, 1636, 2020, https://doi.org/10.1007/s42452-020-03427-1.
2. Serrano, E., Rus, G., Martínez, J.G., Nanotechnology for sustainable energy. *Renew. Sust. Energy Rev.*, 9, 2373–84, 13, December, 2009.
3. Peyghambarzadeh, S.M., Hashemabadi, S.H., Jamnani, M.S., Hoseini, S.M., Improving the cooling performance of automobile radiator with Al_2O_3/water nanofluid. *J. Appl. Therm. Eng.*, 31, 1833–1838, 2011.
4. Nguyen, C.T., Roy, G., Gauthier, C., Galanis, N., Heat transfer enhancement using Al_2O_3–water nanofluid for an electronic liquid cooling system. *Appl. Therm. Eng.*, 27, 1501–1506, 2007.
5. Hwang, K.S., Jang, S.P., Choi, S.U.S., Flow and convective heat transfer characteristics of water-based Al_2O_3 nanofluids in fully developed laminar flow regime. *Int. J. Heat Mass Transf.*, 52, 193–199, 2009.
6. Tao, Q., Zhong, F., Deng, Y., Wang, Y., Su, C., A Review of Nanofluids as Coolants for Thermal Management Systems in Fuel Cell Vehicles. *Nanomaterials*, 13, 2861-1-31, 2023, https://doi.org/10.3390/nano13212861.
7. Kalaiyarasan, A., Sankar, K., Sundaram, S., Priya, S., Applications of Green Nanofluids in Thermal Systems: An Innovative Study, in: *Sustainable Utilization of Nanoparticles and Nanofluids in Engineering Applications*, Boopathi, S. and Davim, J. (Eds.), pp. 235–264, 2023.
8. Pereira, J.E., Moita, A.S., Moreira, A.L.N., The pressing need for green nanofluids: A review. *J. Environ. Chem. Eng.*, 10, 3, 107940, 2022. ISSN 2213-3437, https://doi.org/10.1016/j.jece.2022.107940.

9. Salem Ahmed, M., Nanofluid: New Fluids by Nanotechnology, in: *Thermophysical Properties of Complex Materials*, IntechOpen, London, UK, 2020, doi:10.5772/intechopen.86784.
10. Witharana, S., Palabiyik, I., Musina, Z., Ding, Y., Stability of glycol nanofluids—The theory and experiment. *Powder Technol.*, 239, 72–77, 2013.
11. Jailani, S., Franks, G.V., Healy, T.W., The potential of nanoparticle suspensions: Effect of electrolyte concentration, particle size and volume fraction. *J. Am. Ceram. Soc.*, 91, 1141–1147, 2008.
12. Zhu, H.T., Lin, Y.S., Yin, Y.S., A novel one-step chemical method for preparation of copper nano-fluids. *J. Colloid Interface Sci.*, 277, 100–103, 2004.
13. Yu, W., Xie, H., Wang, X., Wang, X., Highly efficient method for preparing homogeneous and stable colloids containing graphene oxide. *Nanoscale Res. Lett.*, 6, 1–7, 2011.
14. Bakthavatchalam, B., Habib, K., Saidur, R., Saha, B.B., Irshad, K., Comprehensive study on nanofluid and ionanofluid for heat transfer enhancement: a review on current and future perspective. *J. Mol. Liq.*, 305, 112787-1-33, 2020, https://doi.org/10.1016/j.molliq.2020.112787.
15. Kalsi, S., Kumar, S., Kumar, A., Alam, T., Dobrota, D., Thermophysical Properties of Nanofluids and Their Potential Applications in Heat Transfer Enhancement: A Review. *Arabian J. Chem.*, 16, 105272, 2023. 10.1016/j.arabjc.2023.105272.
16. Chakraborty, S. and Panigrahi, P.K., Stability of nanofluid: A review. *Appl. Therm. Eng.*, *174*, 115259, 2020.
17. Xuan, Y. and Li, Q., Heat transfer enhancement of nanofluids. *Int. J. Heat Fluid Flow*, 21, 58–64, 2000.
18. Kole, M. and Dey, T.K., Thermal conductivity and viscosity of AIN-ethanol nanofluids based on car engine coolant. *J. Phys. D: Appl. Phys.*, 43, 10, 2010.
19. Nerella, S., Sudheer, N.V.V.S., Bhramara, P., Enhancement of heat transfer by nanofluids in solar collectors. *Int. J. Innov. Eng. Technol.*, 3, 115–20, 2014.
20. Zhu, D., Li, X., Wang, N., *et al.*, Dispersion behavior and thermal conductivity characteristics of Al_2O_3–H_2O nanofluids. *Curr. Appl Phys.*, 9, 131–139, 2009.
21. Feng, X., Ma, H., Huang, S., *et al.*, Aqueous-organic phase transfer of highly stable gold, silver, and platinum nanoparticles and new route for fabrication of gold nanofilms at the oil/water interface and on solid supports. *J. Phys. Chem. B*, 110, 12311–12317, 2006.
22. Zhu, H.T., Zhang, C.Y., Tang, Y.M., Wang, J.X., Novel synthesis and thermal conductivity of CuO nano-fluid. *J. Phys. Chem. C*, 111, 1646–1650, 2007.
23. Wei, X. and Wang, L., Synthesis and thermal conductivity of microfluidic copper nano-fluids. *Particuology*, 8, 262–271, 2010.
24. Chen, Y. and Wang, X., Novel phase-transfer preparation of monodisperse silver and gold nanoparticles at room temperature. *Mater. Lett.*, 62, 2215–2218, 2008.

25. Yu, W., Xie, H., Chen, L., Li, Y., Enhancement of thermal conductivity of kerosene-based Fe_3O_4 nano-fluids prepared *via* phase-transfer method. *Colloids Surf., A*, 355, 109–113, 2010.
26. Wang, L. and Fan, J., Nano-fluids research: Key issues. *Nanoscale Res. Lett.*, 5, 1241–1252, 2010.
27. Pak, B.C. and Cho, Y.I., Hydrodynamic and heat transfer study of dispersed fluids with submicron metallic oxide particles. *Exp. Heat. Transf.m*, 11, 151–170, 1998, https://doi.org/10.1080/08916159808946559.
28. Einstein, A., A new determination of molecular dimensions. *Ann. Phys.*, 19, 289–306, 1906.
29. Hamilton, R.L. and Crosser, O.K., Thermal conductivity of heterogeneous two-component systems. *IEC Fundam.*, 1, 187–191, 1962.
30. Harish Kumar, L., Kazi, S.N., Masjuki, H.H., Zubir, M.N.M., A review of recent advances in green nanofluids and their application in thermal systems. *Chem. Eng. J.*, 429, 132321, 2022. ISSN 1385-8947, https://doi.org/10.1016/j.cej.2021.132321.
31. Souza, R.R., Gonçalves, I.M., Rodrigues, R.O., Minas, G., Miranda, J.M., Moreira, A.L., Lima, R., Coutinho, G., Pereira, J.E., Moita, A.S., Recent advances on the thermal properties and applications of nanofluids: From nanomedicine to renewable energies. *Appl. Therm. Eng.*, *201*, 117725, 2022, https://doi.org/10.1016/j.applthermaleng.2021.117725.
32. Singh, B. and Sood, S., Hybrid nanofluids preparation, thermo-physical properties, and applications: A Review. *Hybrid Adv.*, 6, 100192, 2024, https://doi.org/10.1016/j.hybadv.2024.100192.
33. Soudagar, M.E.M., Shelare, S., Marghade, D., Belkhode, P., Nur-E-Alam, M., Kiong, T.S., Fattah, I.M.R., Optimizing IC engine efficiency: A comprehensive review on biodiesel, nanofluid, and the role of artificial intelligence and machine learning. *Energy Convers. Manage.*, *307*, 118337, 2024, https://doi.org/10.1016/j.enconman.2024.118337.
34. Bacha, H.B., Ullah, N., Hamid, A., Shah, N.A., A comprehensive review on nanofluids: Synthesis, cutting-edge applications, and future prospects. *Int. J. Thermofluids*, 22, 100595, 2024, https://doi.org/10.1016/j.ijft.2024.100595.
35. Liu, C., Geng, L., Xiao, T., Liu, Q., Zhang, S., Ali, H.M., Sharifpur, M., Zhao, J., Recent advances of plasmonic nanofluids in solar harvesting and energy storage. *J. Energy Storage*, *72*, 108329, 2023, https://doi.org/10.1016/j.est.2023.108329.
36. Ge, Z., Xiang, Q., Li, J., Zhang, S., Wang, Z., Xie, J., Xie, Z., Yang, F., Thermo-economic analysis of organic Rankine cycle using a new two-stage solar collector with nanofluids. *Int. J. Heat Fluid Flow*, *107*, 109393, 2024, https://doi.org/10.1016/j.ijheatfluidflow.2024.109393.
37. Li, W., Zeng, Z., Le, S., Zhu, K., Huang, X., Hegab, H., Ibrahim, A.M.M., Investigation of a green nanofluid added with graphene and Al_2O_3 nano-additives for grinding hard-to-cut materials. *Tribol. Int.*, *195*, 109580, 2024, https://doi.org/10.1016/j.triboint.2024.109580.

38. Salari, S. and Jafari, S.M., Application of nanofluids for thermal processing of food products. *Trends Food Sci. Technol.*, *97*, 100–113, 2020, https://doi.org/10.1016/j.tifs.2020.01.004.
39. Tarafdar, A., Sirohi, R., Negi, T., Singh, S., Badgujar, P.C., Shahi, N.C., Kumar, S., Sim, S.J., Pandey, A., Nanofluid research advances: Preparation, characteristics and applications in food processing. *Food Res. Int.*, *150*, 110751, 2021, https://doi.org/10.1016/j.foodres.2021.110751.
40. Sundar, L.S., Synthesis and characterization of hybrid nanofluids and their usage in different heat exchangers for an improved heat transfer rates: A critical review. *Eng. Sci. Technol. Int. J.*, *44*, 101468, 2023, https://doi.org/10.1016/j.jestch.2023.101468.
41. Babar, H., Wu, H., Zhang, W., Shah, T.R., McCluskey, D., Zhou, C., The promise of nanofluids: A bibliometric journey through advanced heat transfer fluids in heat exchanger tubes. *Adv. Colloid Interface Sci.*, 325, 103112, 2024, https://doi.org/10.1016/j.cis.2024.103112.
42. Olabi, A.G., Maghrabie, H.M., Adhari, O.H.K., Sayed, E.T., Yousef, B.A., Salameh, T., Abdelkareem, M.A., Battery thermal management systems: Recent progress and challenges. *Int. J. Thermofluids*, *15*, 100171, 2022, https://doi.org/10.1016/j.ijft.2022.100171.
43. Elsaid, K., Olabi, A.G., Wilberforce, T., Abdelkareem, M.A., Sayed, E.T., Environmental impacts of nanofluids: A review. *Sci. Total Environ.*, *763*, 144202, 2021, https://doi.org/10.1016/j.scitotenv.2020.144202.
44. Giwa, S.O., Adegoke, K.A., Sharifpur, M., Meyer, J.P., Research trends in nanofluid and its applications: A bibliometric analysis. *J. Nanopart. Res.*, *24*, 3, 63, 2022, https://doi.org/10.1007/s11051-022-05453-z.
45. Sofiah, A.G.N., Pasupuleti, J., Samykano, M., Sulaiman, N.F., Ramli, Z.A.C., Kumar, R.R., Shahabuddin, S., Pandey, A.K., Tiong, S.K., Koh, S.P., Multi-objective optimization and price performance factor evaluation of polyaniline nanofibers-palm oil nanofluids for thermal energy storage application. *Case Stud. Therm. Eng.*, *52*, 103673, 2023, https://doi.org/10.1016/j.csite.2023.103673.
46. Song, M., Wang, L., Shao, F., Xie, H., Xu, H., Yu, W., Thermally induced flexible phase change hydrogels for solar thermal storage and human thermal management. *Chem. Eng. J.*, 464, 142682, 2023, https://doi.org/10.1016/j.cej.2023.142682.
47. Bhanvase, B.A., Sayankar, S.D., Kapre, A., Fule, P.J., Sonawane, S.H., Experimental investigation on intensified convective heat transfer coefficient of water based PANI nanofluid in vertical helical coiled heat exchanger. *Appl. Therm. Eng.*, *128*, 134–140, 2018, https://doi.org/10.1016/j.applthermaleng.2017.09.009.
48. Kulkarni, H.R., Dhanasekaran, C., Rathnakumar, P., Sivaganesan, S., Experimental study on thermal analysis of helical coil heat exchanger using Green synthesis silver nanofluid. *Mater. Today Proc.*, *42*, 1037–1042, 2021, https://doi.org/10.1016/j.matpr.2020.12.087.

49. Ghadikolaei, S.S., Siahchehrehghadikolaei, S., Gholinia, M., Rahimi, M., A CFD modeling of heat transfer between CGNPs/H2O Eco-friendly nanofluid and the novel nature-based designs heat sink: Hybrid passive techniques for CPU cooling. *Therm. Sci. Eng. Prog.*, *37*, 101604, 2023, https://doi.org/10.1016/j.tsep.2022.101604.
50. Jilte, R., Afzal, A., Ağbulut, U., Shaik, S., Khan, S.A., Linul, E., Asif, M., Battery thermal management of a novel helical channeled cylindrical Li-ion battery with nanofluid and hybrid nanoparticle-enhanced phase change material. *Int. J. Heat Mass Transfer*, *216*, 124547, 2023, https://doi.org/10.1016/j.ijheatmasstransfer.2023.124547.
51. Gabsi, I., Saad, I., Maalej, S., Zaghdoudi, M.C., Thermal management of Li-ion batteries in electric vehicles by nanofluid-filled loop heat pipes. *Sci. Technol. Energy Transition*, 79, 1–17, 2024, https://doi.org/10.2516/stet/2024019.
52. Thorat, P., Sanap, S., Gawade, S., A comparative study of thermo-physical properties of different nanofluids for effective heat transfer leading to Li-ion battery pack cooling. *Energy Storage*, 6, 1, 566, 2024, https://doi.org/10.1002/est2.566.

5

Modeling and Simulation of Batteries Thermal Management System

Milad Heidari[1]*, Sivasakthivel Thangavel[1], Morteza Khashehchi[2], Pooyan Rahmanivahid[1], Ashwani Kumar[3] and Yatika Gori[4]

[1]*Mechanical Engineering Department, Global College of Engineering and Technology (GCET), Muscat, Sultanate of Oman*
[2]*Systems Engineering Department, Militry Technological College, Muscat, Sultanate of Oman*
[3]*Department of Mechanical Engineering, Technical Education Department Uttar Pradesh, Kanpur, India*
[4]*Department of Mechanical Engineering, Graphic Era (Deemed to be University) Dehradun, Uttarakhand, India*

Abstract

Batteries Thermal Management Systems (BTMS) are essential for maximizing battery technologies' efficiency, robustness, and safety in a range of applications, including portable electronics, renewable energy storage, and electric cars. This manuscript offers a thorough examination of BTMS, covering all the important details, including principles, simulation techniques, and validation procedures. The importance of BTMS in preventing negative consequences like degradation and thermal runaway is explained in the introduction. The mechanisms of heat generation, thermal management strategies, and the impact of temperature on battery performance and safety are all covered in the Fundamentals of BTMS. The modeling of complex thermal dynamics within battery packs using software tools such as ANSYS Fluent and COMSOL Multiphysics is the main topic of discussion when it comes to the simulation of BTMS. To accurately model thermal behavior, BTMS simulations must be set up with geometry creation, material property assignment, and boundary conditions

**Corresponding author*: milad@gcet.edu.om; ORCID: 0000-0001-7577-4789
Sivasakthivel Thangavel: ORCID: 0000-0001-6802-4015
Morteza Khashehchi: ORCID: 0000-0001-8636-4718
Pooyan Rahmanivahid: ORCID: 0000-0002-1350-2259
Ashwani Kumar: ORCID: 0000-0003-4099-935X

Ashwani Kumar, Mukesh Kumar Awasthi, Nitesh Dutt, Yogesh Kumar Singla and Sivasakthivel Thangavel (eds.) Thermal Battery Management System for Hybrid and Electric Vehicles, (91–116) © 2025 Scrivener Publishing LLC

specification. The article highlights the importance of sensitivity analysis, material property characterization, and geometry simplification to create effective and accurate simulations. Experimental methods such as thermocouples, calorimetry, and thermal imaging are used to validate BTMS models and show how important they are for validating simulation results and improving models. The final section of the manuscript discusses potential developments in multi-physics modeling, data-driven methodologies, and sophisticated thermal management techniques, as well as potential trends and challenges in BTMS modeling. Overall, this manuscript offers a thorough overview of BTMS, highlighting the critical role that modeling and simulation techniques play in improving battery efficiency, reliability, and safety.

Keywords: BTMS, batteries, electric vehicle, ANSYS, COMSOL, simulation

5.1 Introduction

The purpose of the Batteries Thermal Management System (BTMS), a vital part of contemporary battery technology, is to control battery cell temperature to maximize performance, longevity, and safety. This system is critical to many applications, including renewable energy storage systems, portable electronics, and electric and hybrid electric vehicles (HEVs). The main goal of BTMS is to keep batteries in the optimal operating temperature range, which is normally between 20°C and 45°C, to maximize efficiency and minimize the negative effects of extremely high or low temperatures. Temperature is a critical factor in battery performance and longevity because high temperatures can hasten the breakdown of electrode materials and electrolyte, reducing capacity and raising the risk of thermal runaway [1–3]. BTMS employs different thermal management techniques, including active and passive cooling methods:

Active cooling: This technique uses heat-absorbing and heat-dissipating cooling fluids (such as water or refrigerants) that are pumped through battery pack channels. Passive cooling: Uses phase-change materials or natural convection to regulate heat without requiring extra energy. Furthermore, to ensure effective heat dissipation, advanced thermal interface materials (TIMs) are used to improve heat transfer between cooling systems and battery cells. Improving battery efficiency, preserving steady operation under changing circumstances, and boosting overall system reliability and safety all depend on the successful deployment of BTMS. Continued research aims to improve the performance and sustainability of battery technologies across industries by creating novel BTMS solutions to handle the particular thermal challenges presented by high-power applications and emerging battery chemistries. Because it directly affects battery performance, safety, and longevity, effective thermal management in batteries is crucial.

Because batteries are temperature-sensitive, it's essential to maintain ideal operating temperatures to maximize their longevity and efficiency.

The growing need for safer, more dependable, and efficient battery technologies across a range of industries, including electric vehicles (EVs), renewable energy storage, and portable electronics, is the driving force behind the modeling and simulation of battery technology management systems (BTMS). Extensive and expensive experimental testing is not necessary when understanding the complex thermal dynamics within battery systems can be accomplished through modeling and simulation. Engineers and researchers can optimize BTMS designs and strategies before physical implementation through modeling, which provides accurate predictions of temperature distributions, heat transfer mechanisms, and thermal behavior under various operating conditions. Moreover, modeling and simulation are essential due to the inherent difficulties with Battery-Testing Systems (BTMS), including controlling heat generation, reducing thermal gradients, and guaranteeing consistent temperature distribution throughout battery cells. With the use of these tools, one can investigate various cooling techniques, thermal interface materials, and system configurations in a virtual setting, making it easier to find the best solutions that improve performance while maintaining dependability and safety. The ability of modeling and simulation to promote innovation, enhance design efficiency, and reduce risks related to thermal management in progressively complex battery systems is ultimately what drives its use in BTMS [1, 4, 5].

5.2 Fundamentals of BTMS

An important component of battery design and operation is thermal management, especially in light of the heat that is naturally produced during the charge and discharge cycles. The electrochemical reactions and internal resistance of the battery cells are the main sources of heat production. Overheating can shorten battery life and capacity and speed up the deterioration of electrode materials. It can also increase internal resistance. Furthermore, there are safety concerns associated with high temperatures, such as the possibility of thermal runaway and catastrophic failure. Therefore, effective thermal management strategies are essential to maintain batteries within optimal temperature ranges, typically between 20°C to 45°C depending on the battery chemistry, to maximize performance and longevity while ensuring safety [2, 3, 6].

In battery systems, a variety of thermal management techniques are used to efficiently manage heat. Cooling fluids (like water or refrigerants) are pumped through battery pack channels as part of active cooling techniques to absorb and release heat. On the other hand, passive cooling techniques use thermal interface materials (TIMs), phase change materials, or natural convection to dissipate heat without requiring additional energy input. Thermal modeling and control algorithms are integrated by advanced battery management systems (BMS) to monitor battery temperature and adjust cooling systems in real-time. In order to precisely control cooling mechanisms and maintain ideal operating conditions and guarantee battery safety and reliability across a variety of applications, including electric vehicles, grid energy storage, and portable electronics, these systems use temperature sensors embedded within the battery pack to provide feedback to the BMS.

5.2.1 Role of Temperature in Battery Performance and Safety

One of the most important areas for research and development in contemporary battery technology is the effect of temperature on battery performance and safety. Temperature affects the electrochemical reactions that take place in battery cells, which has an effect on cycle life, power output, capacity, and other performance indicators. Higher temperatures often quicken chemical reactions in the battery, boosting ion mobility and enabling quicker rates of charge and discharge. On the other hand, extended exposure to high temperatures can hasten the deterioration of battery components, such as electrolytes and electrode particles. Reduced capacity and shortened cycle life are the ultimate results of this degradation, which can also show up as elevated impedance, active material loss, and morphological changes in the electrode. Conversely, low temperatures can inhibit ion mobility and decrease battery performance, particularly in terms of power output and efficiency [7, 8].

Furthermore, temperature is a major factor in battery safety. High temperatures can encourage thermal runaway, an exothermic reaction that can sustain itself and cause a sharp rise in temperature as well as possible battery failure. Internal short circuits, overcharging, and exposure to high outside temperatures can all cause thermal runaway. Battery management systems (BMS) use temperature sensors and control algorithms to monitor and manage battery temperature to reduce safety hazards. Good thermal management techniques are used to keep batteries within ideal operating temperature ranges and avoid overheating-related safety incidents. These techniques include active cooling with liquid cooling systems and passive cooling with phase change materials. The goal of ongoing research is to improve our understanding of how temperature, battery chemistry, and performance interact to increase battery systems' safety and efficiency for a variety of uses.

5.2.2 Components of BTMS

The main elements of BTMS are listed in Table 5.1. Together, these elements support efficient thermal management in battery systems, guaranteeing ideal operating temperatures to enhance efficiency, durability,

Table 5.1 Components of batteries thermal management system [45–51].

Component	Description
Cooling Systems	
Liquid Cooling	Circulation of coolant (e.g., water or refrigerant) through channels within the battery pack for heat dissipation.
Air Cooling	Utilization of fans or blowers to direct air over the battery pack to remove heat through natural convection.
Phase Change Materials	Materials that absorb and release latent heat during phase transitions, effectively managing temperature variations.
Heat Pipes	Conductive heat transfer devices that efficiently transport heat away from hot spots to cooler regions.
Thermal Interface Materials	
Thermal Greases	Silicone-based compounds applied between surfaces to enhance thermal conductivity and fill air gaps.
Thermal Pads	Soft, compressible pads made of thermally conductive materials used for heat transfer and gap filling.
Thermal Paste	High-performance paste applied between surfaces to improve thermal contact and heat transfer efficiency.
Battery Management System (BMS)	Control system incorporating temperature sensors and algorithms to monitor and regulate battery temperature.
Heat Exchangers	Devices that transfer heat between two or more fluids to control battery temperature, often used in liquid cooling systems.

and security. To create effective BTMS solutions that are suited to particular battery chemistries and application requirements, it is essential to choose and integrate the right cooling systems and thermal interface materials.

5.2.3 Challenges in BTMS Design and Operation

Some of the major difficulties in the development and use of BTMS are listed in Table 5.2. A multidisciplinary strategy combining knowledge in

Table 5.2 Challenges in the design and operation of BTMS [52–54].

Challenge	Description
Heat Dissipation	Efficiently removing heat generated during battery operation to maintain optimal temperature and prevent thermal runaway.
Temperature Uniformity	Achieving uniform temperature distribution across battery cells to avoid hotspots and ensure consistent performance.
Energy Efficiency	Balancing the energy consumption of cooling systems with the overall energy efficiency of the battery system.
Size and Weight Constraints	Integrating thermal management components within limited space and weight budgets, especially in portable applications.
Thermal Resistance	Minimizing thermal resistance between battery cells and cooling components to enhance heat transfer efficiency.
Dynamic Operating Conditions	Adapting to varying environmental conditions and operating loads while maintaining stable battery temperature profiles.
Thermal Management Integration	Integrating thermal management solutions seamlessly with battery packs and surrounding systems without compromising design.
Cost-Effectiveness	Designing BTMS solutions that are cost-effective and scalable, considering manufacturing, installation, and maintenance costs.
Reliability and Safety	Ensuring the reliability and safety of BTMS components and systems to prevent potential hazards and ensure long-term operation.

thermal engineering, materials science, control systems, and battery technology is needed to address these issues. To overcome these challenges and improve the effectiveness and performance of battery thermal management systems, novel solutions are constantly being developed. These include enhanced cooling methods, optimized thermal interface materials, and intelligent battery management algorithms.

5.2.4 Mathematical Modeling of BTMS

The fundamental equations for conduction, convection, and radiation govern heat transfer in BTMS components. The basic formulas for each mode of heat transfer are as follows:

1 - Conduction (Fourier's Law):
Conduction refers to heat transfer through a solid medium without bulk motion of the material. Fourier's Law describes conduction mathematically:

$$q=-k\nabla T$$

where:

q is the heat flux (heat transfer per unit area per unit time),

k is the thermal conductivity of the material,

∇T is the temperature gradient (rate of change of temperature concerning distance).

This equation indicates that heat flux q is proportional to the negative of the temperature gradient

∇T and the thermal conductivity k of the material.

2 - Convection (Newton's Law of Cooling):
Convection involves heat transfer through a fluid (liquid or gas) due to the motion of the fluid itself. Newton's Law of Cooling describes convective heat transfer:

$$q=hA(T_s-T_\infty)$$

where:

q is the heat transfer rate,

h is the convective heat transfer coefficient,

A is the surface area through which heat is transferred,

T_s is the surface temperature,

T_∞ is the fluid (ambient) temperature.

This equation shows that the heat transfer rate q is proportional to the temperature difference between the surface T_s and the fluid T_∞, scaled by the convective heat transfer coefficient h and the surface area *A*.

3 - Radiation (Stefan-Boltzmann Law):
Radiation is heat transfer through electromagnetic waves emitted by a surface. The Stefan-Boltzmann Law describes the power radiated per unit area by a black body:

$$q=\sigma\epsilon A(T_s^4 - T_\infty^4)$$

where:

q is the radiative heat transfer rate,
σ is the Stefan-Boltzmann constant ($\sigma \approx 5.67\times10^{-8}\,W/m^2K^4$
ϵ is the emissivity of the surface,
A is the surface area,
T_s is the surface temperature,
T_∞ is the ambient temperature.

This equation shows that the radiative heat transfer rate q is proportional to the fourth power of the absolute temperatures T_s and T_∞, scaled by the surface emissivity ϵ, Stefan-Boltzmann constant σ, and surface area A.

The design and optimization of efficient thermal management techniques for battery systems are made possible by these fundamental equations, which are essential for comprehending and simulating heat transfer phenomena inside BTMS components.

5.3 Simulation of BTMS

Advanced software tools that can model complex thermal dynamics within battery packs are often used in BTMS simulations. COMSOL Multiphysics and ANSYS Fluent are two popular simulation software programs for BTMS modeling; they each have special features and capabilities suited to various facets of thermal management analysis. Engineers and researchers can simulate coupled physical phenomena, such as heat transfer, fluid flow, and electrochemical processes related to BTMS, using the flexible multiphysics simulation program COMSOL Multiphysics. The temperature distribution, heat fluxes, and fluid flow characteristics within the battery thermodynamic mass separator (BTMS) can be thoroughly analyzed by users through the creation of intricate 3D models of battery cells, cooling systems, and thermal interface materials using COMSOL. The software is a great option for researching the thermal behavior of intricate battery systems under various operating conditions because of its user-friendly interface and large library of predefined physics modules [9, 10].

On the other hand, ANSYS Fluent is a computational fluid dynamics (CFD) program that is frequently used to examine fluid flow and heat transfer phenomena in BTMS parts like heat exchangers and cooling systems. To solve the Navier-Stokes equations and energy equations, ANSYS Fluent uses complex numerical algorithms. This allows it to provide insights into the fluid behavior and convective heat transfer processes inside the battery pack. This software is especially helpful for forecasting thermal performance under various flow regimes and configurations and for optimizing the design of cooling systems. Advanced post-processing features are provided by ANSYS Fluent and COMSOL Multiphysics, enabling users to see simulation results through precise heat transfer profiles, velocity vectors, and contour plots. These software tools are essential for speeding up the development and optimization of BTMS designs because they offer insightful information about thermal management techniques, boost system effectiveness, and guarantee the dependability and safety of battery systems in a variety of applications [11–14].

5.3.1 Setting Up BTMS Simulations

Creating geometry, assigning material properties, and defining boundary conditions are some of the crucial steps in setting up simulations for BTMS. To optimize thermal management strategies and guarantee battery performance and safety, these elements are essential for accurately modeling and simulating the thermal behavior of battery systems. We go into more detail about each of these elements below [4, 9, 15, 16].

Geometry Creation
Usually, this entails specifying the battery pack's geometrical configuration, which includes the placement of the battery cells, cooling channels, thermal interface materials, and any other pertinent parts. Robust 3D modeling capabilities are offered by sophisticated simulation software tools like ANSYS Fluent or COMSOL Multiphysics, which let users construct geometry from scratch or import CAD (Computer-Aided Design) files. The simulation results are highly dependent on the geometric model's accuracy and fidelity. For accurate and trustworthy simulations, therefore,

close attention to detail and adherence to the actual physical layout of the BTMS components are essential.

Important things to think about when creating geometry are:

- Cell Arrangement: Describe the arrangement and spatial distribution of the battery cells in the pack, taking into consideration dimensions, spacing, and stacking direction.
- Cooling System Layout: Make sure that fluid flow paths and heat transfer surfaces are accurately represented by including the design of cooling channels or heat exchangers.
- The location and characteristics of thermal interface materials (TIMs), which are used to improve heat transfer between battery components, are included.

Material Properties Assignment

The next stage is to give the different parts of the BTMS the proper material properties after the geometric model has been established. Heat transfer, thermal conductivity, and other physical phenomena that are simulated in the model are largely determined by the properties of the materials. Among the important attributes that must be specified are:

- Thermal conductivity: Defines a material's capacity to conduct heat and affects how hot or cold a battery pack is inside.
- Specific Heat Capacity: Indicates how much heat energy is needed to raise a material's temperature by one degree Celsius.
- Density: This factor affects a material's thermal inertia and reaction to temperature changes by determining its mass per unit volume.
- Emissivity: A surface's ability to emit thermal radiation efficiently is indicated by its relevance to radiation heat transfer.

The material properties of batteries, cooling fluids, and thermal interface materials used in battery thermodynamic storage systems (BTMS) are typically derived from experimental data or literature sources. To ensure that simulation results are realistic and accurately represent the behavior of the physical system, accurate characterization of material properties is necessary.

Boundary Conditions Specification

Boundary conditions define the external constraints and environmental factors that influence the behavior of the BTMS. In order to accurately model the interactions between the components of the BTMS and the surrounding environment, boundary conditions must be specified. Important boundaries to think about are as follows:

- Ambient Temperature: The temperature outside the battery system that influences heat transfer to the surrounding air.
- Heat Generation: Internal heat production from electrochemical reactions in battery cells affects how heat is distributed throughout the cell.
- Fluid Flow Conditions: Give specifics regarding fluid properties, flow velocities, and inlet/outlet flow rates in the case of cooling systems.
- Thermal Interfaces: To accurately simulate heat transfer across interfaces, define the contact conductance and boundary resistance at interfaces between different materials.

Simulation Setup and Solving

Setting up the simulation solver and solving the mathematical model numerically comes last, following the creation of the geometry, assignment of material properties, and specification of boundary conditions. Users of sophisticated simulation software tools can specify simulation parameters like time steps, convergence criteria, and solution techniques in addition to choosing relevant physics modules (such as heat transfer, fluid flow, and electrochemistry).

During simulation setup:

- Physics Settings: Based on the BTMS components and phenomena of interest, configure the simulation to include pertinent physics modules (such as heat transfer and fluid flow).
- Mesh Generation: To discretize the geometric model, generate a finite element mesh with a resolution and quality appropriate for precisely capturing fluid dynamics and temperature gradients.
- Solver Settings: Establish the parameters for the solver, such as time integration schemes, convergence criteria, and numerical techniques (such as finite element and finite volume).

- Post-processing: To analyze simulation results, such as temperature distributions, heat fluxes, and fluid velocities, specify output variables and visualization settings.

After the simulation is set up, the solver uses the boundary conditions and defined physics to calculate the transient or steady-state solutions. Through the use of post-processing tools in simulation software, users can examine and interpret simulation results to learn more about the BTMS's thermal behavior and assess the efficacy of various thermal management techniques.

5.3.2 Considerations for Efficient and Accurate Simulation of BTMS

Table 5.3 lists the most important factors to take into account to simulate BTMS accurately and efficiently. These factors are crucial for streamlining the simulation process and guaranteeing that battery system performance and thermal behavior are appropriately modeled in BTMS simulations. Engineers and researchers can effectively use simulation tools to design and optimize effective thermal management strategies for a variety of applications, such as electric vehicles, renewable energy storage, and portable electronics, by methodically addressing these factors.

5.4 Validation of BTMS Models

To evaluate the precision and dependability of simulated thermal behaviors, experimental methods are vital for validating and verifying BTMS simulations. These experimental approaches cover a wide range of techniques, such as calorimetry, thermocouples, and thermal imaging, each of which provides a different perspective on the thermal performance and heat transfer of battery systems [17–20].

Thermal Imaging

A non-contact technique called thermal imaging is used to see and measure temperature distributions inside the BTMS and among battery components. Researchers can identify temperature variations, hotspots, and patterns of heat dissipation in real-time by using infrared cameras, which record thermal images based on emitted infrared radiation. Through the visualization of heat dissipation pathways and the identification of areas of excessive heat accumulation, thermal imaging is especially useful for

Table 5.3 How to achieve efficient and accurate simulation of BTMS [55–58].

Consideration	Description
Geometry Simplification and Meshing	Simplify complex geometries while maintaining essential features for accurate representation. Generate high-quality meshes for precise resolution of thermal gradients and fluid flow dynamics.
Material Property Characterization	Obtain accurate material properties from experimental data or literature sources to ensure realistic simulation outcomes.
Model Validation and Verification	Validate simulation models against experimental data to verify accuracy and reliability. Adjust parameters based on validation results.
Time Step and Convergence Criteria	Choose appropriate time steps and convergence criteria to ensure numerical stability and convergence of simulation solutions.
Physics Module Selection	Select relevant physics modules (e.g., heat transfer, fluid flow) based on BTMS components and phenomena of interest.
Boundary Condition Specification	Define realistic boundary conditions (e.g., ambient temperature, heat generation rates) to capture external influences on the BTMS.
Sensitivity Analysis	Perform sensitivity analysis to identify influential parameters and assess the robustness of simulation results.
Computational Resources	Utilize sufficient computational resources (CPU, memory) to handle complex simulations efficiently and reduce computational time.
Post-Processing and Visualization	Employ advanced post-processing tools to analyze and visualize simulation results, facilitating insights into thermal behavior.
Iterative Model Refinement	Continuously refine simulation models based on feedback and insights gained from simulation results and experimental observations.

evaluating the efficacy of cooling systems and thermal management strategies. Researchers can verify simulation results and improve models to precisely capture complex thermal dynamics within battery systems by comparing thermal images with simulated temperature profiles.

Thermocouples

Thermocouples are temperature sensors that are frequently used in the BTMS to measure local temperatures at particular locations. The voltage produced by these sensors, which are made up of two different metal wires connected at one end, is proportional to the temperature differential between the measurement point and the reference junction. Researchers can monitor temperature gradients, transient behaviors, and thermal response under a variety of operating conditions thanks to thermocouples' accurate and localized temperature measurements. The performance of thermal management strategies in maintaining ideal operating temperatures can be evaluated and simulated temperature profiles verified by researchers by carefully placing thermocouples at key locations within the battery pack and cooling systems.

Calorimetry

A quantitative method for calculating heat transfer and thermal energy variations inside the BTMS is calorimetry. Differential scanning calorimeters (DSC) and thermal conductivity meters are commonly employed to characterize the thermal properties of battery materials and assess heat generation rates during charge/discharge cycles. In order to validate heat transfer models and the electrochemical-thermal coupling that occurs within battery systems, calorimetry experiments yield important data on heat capacity, thermal conductivity, and heat generation. Through the comparison of simulated heat generation rates and thermal responses with experimental calorimetry data, researchers can optimize BTMS designs for improved safety and performance by fine-tuning simulation parameters.

Combined Approaches and Data Integration

In order to thoroughly validate BTMS simulations and improve thermal management strategies, researchers frequently use a combination of experimental techniques in practice. To obtain a comprehensive understanding of thermal behavior, thermal imaging can be utilized in tandem with thermocouple measurements to establish a correlation between spatial temperature distributions and localized temperature readings. Furthermore, the integration of calorimetry data into simulation models can enhance the precision of heat generation forecasts and optimize the designs of thermal management systems.

Importance of Experimental Validation

To make BTMS simulations more credible and useful in real-world applications, experimental validation is necessary. Researchers can determine areas for optimization or improvement and gain confidence in the predictive power of simulation models by using experimental techniques to validate simulation results. Verified simulation models are also useful for predicting battery system thermal performance under various operating conditions, investigating alternative designs, and optimizing thermal management techniques.

5.4.1 Comparison of Simulation Results with Experimental Data

A crucial stage in verifying and improving BTMS model performance is comparing simulation outcomes with experimental data. Researchers can use this comparison to find differences, hone simulation parameters, and enhance the predictive power of BTMS models by learning more about the precision and dependability of simulation predictions.

Building confidence in the predictive power of BTMS models requires validating simulation results against experimental data. Researchers can determine whether simulation models are accurate and pinpoint areas that need improvement by contrasting measured data from physical experiments with simulated temperature distributions, heat fluxes, and thermal behaviors. Additionally, validation makes sure that simulation predictions accurately represent real-world behaviors, which helps engineers design and optimize BTMS strategies and components with knowledge [21–23]. Several important factors need to be taken into account to compare simulation results and experimental data in a meaningful way:

1. Consistency of the Experimental Setup: Make sure that the parameters and assumptions utilized in the simulation models closely match the experimental conditions (such as battery configuration, operating conditions, and measurement methods). Regular experimental designs make it easier to compare directly and increase the validity of validation efforts.
2. Data Acquisition and Processing: To obtain high-quality experimental data, use dependable data acquisition techniques

(such as precise measurement tools and accurate sensors). To obtain significant insights and compare simulation and experimental results quantitatively, proper data processing and analysis are essential.

3. Quantitative Metrics: Establish suitable quantitative metrics for evaluating the degree of agreement between simulation and experimental data, such as coefficient of determination and root mean square error. In addition to guiding iterative improvements in simulation fidelity, quantitative comparisons offer objective measures of model accuracy.
4. Sensitivity Analysis: Use sensitivity analysis in both simulation and experimental setups to find key parameters and sources of uncertainty. By addressing sensitivity issues, simulation models become more resilient and maintain consistent performance under varying operating conditions.

The integration of simulation and experimental approaches offers synergistic benefits for advancing BTMS research and development:

- Model Calibration and Optimization: To make sure that simulation results closely resemble observed behaviors and phenomena, use experimental data to calibrate simulation parameters and optimize model settings.
- Understanding of Complex Phenomena: To obtain a thorough grasp of the intricate thermal dynamics within BTMS components, integrate simulation insights with experimental observations. The identification of underlying mechanisms is made easier by this integrated approach, which also provides targeted improvements in thermal management strategies.
- Iterative Design and Validation: Use a cycle of iterative design and validation in which simulation predictions and vice versa inform experimental designs. This iterative process promotes continuous improvement, sparks innovation, and eventually results in the creation of BTMS solutions that are optimized.

5.4.2 Sensitivity Analysis to Identify Influential Parameters and Validate Model Robustness

Sensitivity analysis is a potent modeling and simulation technique that is used to determine critical parameters and confirm the accuracy of mathematical models, such as the ones in BTMS. The thorough examination entails methodically altering input parameters within designated intervals to evaluate their influence on simulation results and model forecasts. We examine the fundamentals of sensitivity analysis, how it is used in BTMS modeling, important techniques for carrying out sensitivity analysis, and the importance of using this method to validate the robustness of the model. Table 5.4 outlines the sensitivity analysis procedure and how it is used to verify the robustness of the model.

Table 5.4 Process of sensitivity analysis and its application in validating model robustness [59, 60].

Step	Task	Description
1	Parameter Selection	Identify key input parameters (e.g., material properties, boundary conditions, operating parameters) in the BTMS model.
2	Range Definition	Define plausible ranges or variations for each selected parameter to explore their impact on simulation outcomes.
3	Simulation Execution	Perform multiple simulations by systematically varying individual parameters within their defined ranges.
4	Output Analysis	Analyze simulation results (e.g., temperature profiles, heat fluxes) to assess changes and trends induced by parameter variations.
5	Sensitivity Indices	Calculate sensitivity indices (e.g., sensitivity coefficients, variance-based indices) to quantify the influence of each parameter on model outputs.
6	Ranking Parameters	Rank parameters based on their sensitivity indices to identify the most influential factors affecting simulation outcomes.
7	Model Validation	Validate model robustness by comparing sensitivity analysis results with experimental data or known physical behaviors.
8	Refinement and Optimization	Use sensitivity analysis insights to refine model parameters, optimize designs, and enhance the predictive accuracy of the BTMS model.

Table 5.5 Example of sensitivity analysis.

Parameter	Range	Simulation output	Sensitivity index	Rank
Thermal Conductivity	0.1 - 1.0 W/mK	Temperature Profiles	0.75	High
Heat Generation Rate	10 - 50 W/m^3	Heat Flux Distribution	0.60	Medium
Fluid Flow Rate	0.1 - 1.0 m/s	Cooling Effectiveness	0.45	Low
Battery Cell Arrangement	Various layouts	Temperature Uniformity	0.55	Medium

Table 5.5 presents an example of a sensitivity analysis table that highlights significant parameters and their effect on the results of BTMS simulations:

In this example:

- Thermal conductivity: Temperature profiles within the BTMS are greatly influenced by variations in thermal conductivity, with a high sensitivity index indicating a strong influence on simulation results.
- Heat Generation Rate: Variations in this rate have a medium sensitivity and moderate impact on simulation results in terms of heat flux distribution.
- Fluid Flow Rate: While changes in this rate have an effect on cooling efficiency, they are less sensitive than changes in thermal conductivity and heat generation rate.
- Battery Cell Arrangement: Various cell configurations affect temperature uniformity, demonstrating a moderate degree of sensitivity and significance in BTMS design optimization [39–44].

5.5 Future Trends and Challenges in BTMS Modeling and Simulation

Emerging trends and challenges are driving major advancements in the field of Batteries Thermal Management Systems (BTMS) modeling and simulation. The development and use of BTMS technologies are anticipated to be shaped by several significant future trends and challenges.

Integration of Multi-Physics and Multi-Scale Modeling
Integrating multi-physics and multi-scale modeling techniques is one popular trend in BTMS modeling. It is crucial to take into consideration coupled phenomena like electrochemical reactions, heat transfer, fluid flow, and structural interactions as battery systems get more intricate and networked. To accurately simulate these interdependencies, future BTMS models are anticipated to integrate sophisticated computational techniques. Furthermore, the integration of multi-scale modeling will make it possible to comprehend thermal behavior and performance across a wider range of length and time scales, from atomistic to system-level simulations [24–26].

Data-Driven Modeling and Machine Learning
The acceptance of data-driven modeling and machine learning approaches in BTMS simulation will be fueled by the development of data acquisition technologies and the growing accessibility of massive datasets from experimental investigations. Researchers can create predictive models that learn from actual operational data by utilizing artificial intelligence and data analytics. This will allow for the development of more precise and flexible thermal management strategies. To improve battery safety and efficiency, machine learning algorithms can predict thermal runaway incidents, optimize cooling control strategies, and support real-time decision-making.

Advanced Thermal Management Strategies
To handle the rising power densities and energy capacities of next-generation batteries, the development of sophisticated thermal management techniques will be the main focus of future BTMS technologies. This covers cutting-edge liquid cooling systems, phase change materials, and sophisticated thermal interface materials that improve heat dissipation and sustain ideal operating temperatures. The amalgamation of intelligent sensors and predictive control algorithms will facilitate adaptive thermal management systems that can maximize efficiency while reducing energy usage and thermal strains [8, 15, 27–29].

Safety and Reliability Enhancement
In BTMS modeling, ensuring the security and dependability of battery systems continues to be a major obstacle. The creation of simulation tools and predictive models to evaluate thermally induced risks, forecast thermal

runaway events and enhance safety features will be a key focus of future trends. This includes the incorporation of predictive analytics for early fault detection, thermal management strategies for thermal runaway mitigation, and advanced battery design considerations to minimize safety hazards under extreme conditions. Comprehensive standards and laws will also keep changing to solve safety issues and guarantee the use of trustworthy BTMS technologies in a variety of applications [30–38].

Conclusion

This manuscript highlights the crucial significance and intricacy of BTMS in modern battery technology. A thorough examination of BTMS concepts, simulation approaches, and validation strategies yields several important insights that influence both scholarly study and business operations. Designing successful battery test and management systems (BTMS) that maximize battery efficiency and guarantee long-term reliability requires a solid understanding of heat generation mechanisms and thermal management strategies. The investigation of battery pack thermal dynamics can be accomplished economically and analytically through the simulation of BTMS using sophisticated software tools such as ANSYS Fluent and COMSOL Multiphysics. To produce realistic models that mimic real-world behaviors, careful simulation setup—including geometry creation, material property assignment, and boundary condition specification—is essential. Enhancing model credibility and reliability requires the validation of BTMS models using experimental techniques like calorimetry, thermocouples, and thermal imaging. Through the correlation of simulation outcomes with empirical data, scholars can enhance models and optimize thermal management approaches for a variety of uses. Future developments in BTMS modeling are expected to incorporate data-driven techniques and sophisticated thermal management strategies, as well as multi-physics and multi-scale modeling approaches. The need to handle new issues brought on by next-generation batteries' growing power densities and energy capacities is what is driving this evolution. This manuscript adds to the scholarly investigation of BTMS by emphasizing its crucial role in the advancement of battery technologies. This research contributes to ongoing efforts to optimize battery efficiency, reliability, and safety through advanced modeling and simulation techniques by bridging theoretical insights with real-world applications. It will take sustained interdisciplinary innovation and collaboration to solve changing problems and move the field of BTMS toward long-term, game-changing solutions.

References

1. Cattani, L., Malavasi, M., Bozzoli, F., D'Alessandro, V., Giammichele, L., Experimental Analysis of an Innovative Electrical Battery Thermal Management System. *Energies*, 16, 5071, 2023, 10.3390/en16135071.
2. Altuntop, E.S., Erdemir, D., Kaplan, Y., Özceyhan, V., A comprehensive review on battery thermal management system for better guidance and operation. *Energy Storage*, 5, 8, e501, 2023, Available from: 10.1002/est2.501.
3. Li, W., Zhou, Y., Zhang, H., Tang, X., A Review on Battery Thermal Management for New Energy Vehicles. *Energies*, 16, 13, 4845, 2023, Available from: 10.3390/en16134845.
4. Guo, Z., Wang, Y., Zhao, S., Zhao, T., Ni, M., Modeling and optimization of liquid-based battery thermal management system considering battery electrochemical characteristics. *J. Energy Storage*, 70, 108028, 2023, ISSN 2352-152X, https://doi.org/10.1016/j.est.2023.108028.
5. Lv, Y., Luo, W., Li, C., Zhang, G., Rao, Z., Experimental and numerical simulation study on the integrated thermal management system for electric vehicles. *J. Energy Storage*, 70, 107895, 2023, ISSN 2352-152X, https://doi.org/10.1016/j.est.2023.107895
6. Waqas Nazar, M., Iqbal, N., Ali, M., Nazir, H., Zain, M., Amjad, B., Thermal management of Li-ion battery by using active and passive cooling method. *J. Energy Storage*, 61, 106800, 2023, ISSN 2352-152X, https://doi.org/10.1016/j.est.2023.106800.
7. Gupta, R., Bahga, S.S., Gupta, A., Effect of Ambient Temperature and Discharge Current on Thermo-Electrochemical Behaviour of Lithium-Ion Cells Using Surrogate Modelling and Analysis. *J. Electrochem. Soc.*, 170, 6, 060526, 2023, 10.1149/1945-7111/acd965.
8. Zhao, Z., Li, L., Ou, Y., Wang, Y.M., Wang, S., Yu, J.-C., Feng, R., A Comparative Study on the Energy Flow of Electric Vehicle Batteries among Different Environmental Temperatures. *Energies*, 16, 14, 5253, https://doi.org/10.3390/en16145253.
9. Zhu, X., Xu, X., Kong, B., Wang, J., Shi, H.-J., Jiang, Y., Coupling simulation of the cooling air duct and the battery pack in battery energy storage systems. *Phys. Scr.*, 98, 075906, 2023, 10.1088/1402-4896/acd824.
10. Carello, M., Ricci, F., Rizzello, A., CFD Simulation and Modelling of a Battery Thermal Management System: Comparison between Indirect and Immersion Cooling. *SAE Tech. Pap. Ser.*, 6, 1, 545–558, 2024, https://doi.org/10.4271/2023-01-0514.
11. Afaq, M., and Ahmad, R., Comparison Between ANSYS Fluent and Solidworks Internal Flow Simulation for Analysis of A Fuzzy Logic Controller-Based Heating/Cooling System in A Mobile Robot Design, 2023, 10.1109/ICRAI57502.2023.10089571.

12. Kurkin, A.S., Dobrov, A.G., Doronkov, D.V., Pronin, A.N., Solntsev, D.N., CFD Modeling of Heat Exchanger with Small Bent Radius Coils Using Porous Media Model. *Fluids*, 8, 5, 141, 2023, 10.3390/fluids8050141.
13. Papkov, V.V., Shadymov, N., Pashchenko, D., CFD-modeling of fluid flow in Ansys Fluent using Python-based code for automation of repeating calculations. *Int. J. Mod. Phys. C*, 34, 09, 2350114, 2023, 10.1142/s0129183123501140.
14. Iakovlev, L., and Morozov, I., Evaluation of liquid non-freezing conditions in large vessels with local heating scheme using the Ansys Fluent software. *Energy Syst.*, 4, 002, 2022, 10.34031/es.2022.4.002.
15. Ahmad, S., Liu, Y., Khan, S.A., Hao, M.-C., Huang, X., Hybrid battery thermal management by coupling fin intensified phase change material with air cooling. *J. Energy Storage*, 64, 107167, 2023, ISSN 2352-152X, https://doi.org/10.1016/j.est.2023.107167.
16. Kang, C., Yang, J., Yuan, X., Qiu, C.-H., Cai, Y., A novel multilayer composite structure based battery thermal management system. *Front. Energy Res.*, 11, 1187904, 2023, 10.3389/fenrg.2023.1187904.
17. Sarmadian, A., Widanage, W.D., Shollock, B,.A., Restuccia, F., Experimentally-verified thermal-electrochemical simulations of a cylindrical battery using physics-based, simplified and generalised lumped models. *J. Energy Storage*, 70, 107910, 2023, 10.1016/j.est.2023.107910.
18. Patil, K.R., Raul, A., Gopal Barve, M., Performance Evaluation of BTMS for Electric Vehicles using Heat Sink - A Numerical Study. *Int. J. Veh. Struct. Syst.*, 15, 1, 37–45, 2023, 10.4273/ijvss.15.1.07.
19. Cattani, L., Malavasi, M., Bozzoli, F., D'Alessandro, V., L,.G., Experimental Analysis of an Innovative Electrical Battery Thermal Management System. *Energies*, 16, 13, 5071, 2023, Available from: 10.3390/en16135071.
20. Xu, Y., Wang, Z., Ke, Z., Lai, B., Zhang, Y., Huang, X., Experimental and Simulation Research on Heat Pipe Thermal Management System Coupled with Battery Thermo-Electric Model. *Processes*, 11, 4, 1204, 2023, Available from: 10.3390/pr11041204.
21. Paquette-Rufiangea, A., Prudhommea, S., Laforesta, M., Optimal Design of Validation Experiments for the Prediction of Quantities of Interest, 2023, 10.48550/arxiv.2303.06114.
22. Naveen, S.N., Ratnam Gunti, S., Naveen Kumar, M.A., Louis, S., Pawan Sharma, U., Optimizing BTM of HV Battery Pack for Automotive Application Using Electro-Thermal Simulation, 2022, 10.1109/iSPEC54162.2022.10032980.
23. Jin, Y.-D., Bajorek, S.M., Cheung, F.B., Validation and Uncertainty Quantification of Transient Reflood Models Using COBRA-TF and Machine Learning Techniques Based on the NRC/PSU RBHT Benchmark. *Nucl. Sci. Eng.*, 197, 5, 967–986, 2023, Available from: 10.1080/00295639.2022.2087834.
24. Peissner, S., and Weigand, B., Enhancing Simulation Efficiency and Quality of Transient Conjugate Thermal Problems by Using an Advanced Meta-modeling Approach. *SAE Int. J. Passenger Veh. Syst.*, 16(15-16-03-0016), 255–273, 2023, 10.4271/15-16-03-0016.

25. Wolf, A., Baudrin, E., Nirschl, H., A Multi-scale Flow Battery Modeling Approach Using Mass Transfer Coefficients. *Energy Technol.*, 11, 7, 2300175, 2023, 10.1002/ente.202300175.
26. Włodarczyk, J.K., Mourouga, G., Schärer, R.P., Schumacher, J.O., Continuum Modelling and Simulation of Flow Batteries, 2023, 10.1002/9783527832767.ch17.
27. Xin, Q., Yang, T., Zhang, H., Yang, J., Zeng, J., Xiao, J.-K., Experimental and numerical study of lithium-ion battery thermal management system using composite phase change material and liquid cooling. *J. Energy Storage*, 71, 108003, 2023, 10.1016/j.est.2023.108003.
28. Garud, K.S., Hwang, S.-G., Lee, M.-Y., A Review of Advanced Cooling Strategies for Battery Thermal Management Systems in Electric Vehicles. *Symmetry*, 15, 7, 1322, 2023, 10.3390/sym15071322.
29. Ranjbaran, Y.S., Shojaeefard, M.H., Molaeimanesh, G.R., Thermal performance enhancement of a passive battery thermal management system based on phase change material using cold air passageways for lithium batteries. *J. Energy Storage*, 68, 107744, 2023, Available from: 10.1016/j.est.2023.107744.
30. Verma, V., Thangavel, S., Dutt, N., Kumar, A., Weerasinghe, R. (Eds.), *Highly Efficient Thermal Renewable Energy Systems: Design, Optimization and Applications*, CRC Press, Boca Raton, FL USA, 2024, https://doi.org/10.1201/9781003472629 ISBN 9781032595641.
31. Kundu, A., Kumar, A., Dutt, N., Meena, C.S., Singh, V.P., Introduction to Thermal Energy Resources and Its Smart Application, in: *Thermal Energy Systems: Design, Computational Techniques and Applications*, vol. Chapter 1, A. Kumar, V.P. Singh, C.S. Meena, N. Dutt, (Eds.), pp. 1–15, CRC Press, Boca Raton, FL, USA, 2023, http://dx.doi.org/10.1201/9781003395768-1.
32. Dewangan, A.K., Moinuddin, S.Q., Cheepu, M., Sajjan, S.K., Kumar, A., Thermal Energy Storage: Opportunities, Challenges and Future Scope, in: *Thermal Energy Systems: Design, Computational Techniques and Applications*, vol. Chapter 2, A. Kumar, V.P. Singh, C.S. Meena, N. Dutt, (Eds.), pp. 17–28, CRC Press, Boca Raton, FL, USA, 2023, http://dx.doi.org/10.1201/9781003395768-2.
33. Kundu, A., Kumar, A., Dutt, N., Singh, V.P., Meena, C.S., Modelling and Simulation of Thermal Energy System for Design Optimization, in: *Thermal Energy Systems: Design, Computational Techniques and Applications*, vol. Chapter 7, A. Kumar, V.P. Singh, C.S. Meena, N. Dutt, (Eds.), pp. 103–140, CRC Press, Boca Raton, FL, USA, 2023, http://dx.doi.org/10.1201/9781003395768-7.
34. Kumar, A., Gori, Y., Meena, C.S., Singh, V.P., Sharma, V.K., Design and analysis of heavy vehicle medium duty transmission gearbox system, in: *Additive Manufacturing in Industry 4.0: Methods, Techniques, Modeling and Nano Aspects*, vol. Chapter 4, V.K. Sharma, A. Kumar, M. Gupta, V. Kumar, D.K. Sharma, S.K. Sharma, (Eds.), pp. 67–86, CRC Press, Boca Raton, FL, USA, 2022, DOI: 10.1201/9781003360001-4.
35. Singh, V.P., Kumar, A., Meena, C.S., Dwivedi, G., Introduction to Energy-Efficient Vehicles for Sustainable Transportation: Transitions and Challenges,

in: *Energy Efficient Vehicles Technologies and Challenges*, vol. Chapter 01, V.P. Singh, A. Kumar, C.S. Meena, G. Dwivedi (Eds.), pp. 01–18, CRC Press, Boca Raton, FL, USA, 2024, DOI: https://doi.org/ 10.1201/9781003464556-1.

36. Kumar, A., Jaiswal, H., Patil, P.P., Parametric Optimization of Vibrating Heavy Vehicle Medium Duty Transmission Gearbox Housing Using Response Surface Method. *Int. J. Veh. Struct. Syst.*, 9, 3, 149–153, 2017. doi:10.4273/ijvss.9.3.04.
37. Singh, V.P. and Kumar., A., Techno-Economic and Future Aspects of the HEV-EV-FCV: Decarbonization, Digitalization and Sustainability, in: *Energy Efficient Vehicles Technologies and Challenges*, vol. Chapter 11, V.P. Singh, A. Kumar, C.S. Meena, G. Dwivedi (Eds.), pp. 212–231, CRC Press, Boca Raton, FL, USA, 2024, DOI: https://doi.org/ 10.1201/9781003464556-11.
38. Kumar, A. and Patil, P.P., Modal Analysis of Heavy Vehicle Truck Transmission Gearbox Housing Made From Different Materials. *J. Eng. Sci. Technol.*, 11, 2, 252–266, 2016.
39. Singh, V.P., Kumar, A., Meena, C.S., Dwivedi, G. (Eds.), *Energy Efficient Vehicles: Technologies and Challenges*, CRC Press, Boca Raton, FL USA, 2024, https://doi.org/10.1201/9781003464556 ISBN: 9781032548111.
40. Kumar, A., Jain, R., Patil, P.P., Dynamic Analysis of Heavy Vehicle Medium Duty Drive Shaft Using Conventional and Composite Material. *IOP Mater. Sci. Eng.*, 149, 1–11, 2016, DOI: 10.1088/1757-899X/149/1/012156.
41. Kumar, A., Singh, V.P., Meena, C.S., Dutt, N., *Thermal Energy Systems: Design, Computational Techniques and Applications*, Taylor & Francis (CRC Press, Boca Raton FL, USA), 2023, https://doi.org/10.1201/9781003395768.
42. Kumar, A. and Patil, P.P., Dynamic Vibration Analysis of Heavy Vehicle Truck Transmission Gearbox Housing Using FEA. *J. Eng. Sci. Technol. Rev.*, 7, 4, 66–72, 2014, ISSN: 1791-2377.
43. Upadhyay, R.K., Singh, V.P., Kumar, A., Sustainable Transportation: Policy, Planning and Implementation, in: *Energy Efficient Vehicles Technologies and Challenges*, vol. Chapter 05, V.P. Singh, A. Kumar, C.S. Meena, G. Dwivedi (Eds.), pp. 74–95, CRC Press, Boca Raton, FL, USA, 2024, DOI: https://doi.org/ 10.1201/9781003464556-5.
44. Kumar, A., Low to High Speed Transient Structural and Thermal Temperature Measurement of Oil-Lubricated Multi-Speed Heavy Vehicle Transmission Gearbox System Based on FEA, in: *Advanced Numerical Simulations in Mechanical Engineering*, pp. 1–21, IGI Global, Hershey, Pennsylvania, USA, 2018, DOI: 10.4018/978-1-5225-3722-9.ch001, ISBN13: 9781522537229.
45. Singh, A.R., Vishnuram, P., Alagarsamy, S., Bajaj, M., Blazek, V., Damaj, I., Othman, K.M., Electric vehicle charging technologies, infrastructure expansion, grid integration strategies, and their role in promoting sustainable e-mobility. *Alexandria Eng. J.*, 105, 300–330, 2024, 10.1016/j.aej.2024.06.093.
46. Ravindran, M.A., Nallathambi, K., Vishnuram, P., Rathore, R.S., Bajaj, M., Rida, I., Alkhayyat, A., A novel technological review on fast charging

infrastructure for electrical vehicles: Challenges, solutions, and future research directions. *Alexandria Eng. J.*, 82, 260–290, 2023, 10.1016/j.aej.2023.10.009.
47. Gupta, A., Singh, R.K., Paul, J., Sharma, M., Joshi, S., Battery-Operated Electric Vehicles (BOEVs) Adoption in India: Analysis of Barriers and Strategies. *IEEE Trans. Eng. Manage.*, 71, 7076–7087, 2023, http://dx.doi.org/10.1109/TEM.2023.3253621.
48. Jena, N.K., Sahoo, S., Sahu, B.K., Naik, A.K., Bajaj, M., Misak, S., Prokop, L., Impact of a redox flow battery on the frequency stability of a five-area system integrated with renewable sources. *Energies*, 16, 14, 5540, 2023, 10.3390/en16145540.
49. Hamed, S.B., Hamed, M.B., Sbita, L., Bajaj, M., Blazek, V., Prokop, L., Ghoneim, S.S., Robust optimization and power management of a triple junction photovoltaic electric vehicle with battery storage. *Sensors*, 22, 16, 6123, 2022, 10.3390/s22166123.
50. Kumar, A., Prasad, A., Kumar., G., *Sustainable Mobility: Policies, Challenges and Advancements*. John Wiley & Sons, Inc., USA, 2024, https://doi.org/10.1002/9781394166831, ISBN: 9781394166831.
51. Bansal, P., Joshi, J., Hans, S., Kumar, A., Autonomous Navigation of Unmanned Aerial Vehicle Using Reinforcement Learning, in: *Sustainable Mobility Policies, Challenges and Advancements*, A. Kumar, A. Prasad, G. Kumar (Editors), John Wiley & Sons, Inc. USA, Chapter 05, pp. 79–103, 2024, https://doi.org/10.1002/9781394166831.ch5.
52. Almutairi, A., Albagami, N., Almesned, S., Alrumayh, O., Malik, H., Electric vehicle load estimation at home and workplace in Saudi Arabia for grid planners and policy makers. *Sustainability*, 15, 22, 15878, 2023, http://dx.doi.org/10.3390/su152215878.
53. Badri, Y.I., Sudabattula, S.K., Hussain, I., Malik, H., Márquez, F.P.G., Ultracapacitor as selectable energy buffer in electric vehicle application. *J. Energy Storage*, 73, 109200, 2023, http://dx.doi.org/10.1016/j.est.2023.109200.
54. Kasri, A., Ouari, K., Belkhier, Y., Bajaj, M., Zaitsev, I., Optimizing electric vehicle powertrains peak performance with robust predictive direct torque control of induction motors: A practical approach and experimental validation. *Sci. Rep.*, 14, 1, 14977, 2024, 10.1038/s41598-024-65988-0.
55. Ramakrishnan, V., Vishnuram, P., Yang, T., Bajaj, M., Rathore, R.S., Zaitsev, I., Design and implementation of a high misalignment-tolerance wireless charger for an electric vehicle with control of the constant current/voltage charging. *Sci. Rep.*, 14, 1, 13165, 2024, 10.1038/s41598-024-63952-6.
56. Dimri, S.C., Indu, R., Bajaj, M., Rathore, R.S., Blazek, V., Dutta, A.K., Alsubai, S., Modeling of traffic at a road crossing and optimization of waiting time of the vehicles. *Alex. Eng. J.*, 98, 114–129, 2024, 10.1016/j.aej.2024.04.050.
57. Kumar, B.A., Jyothi, B., Singh, A.R., Bajaj, M., Rathore, R.S., Berhanu, M., A novel strategy towards efficient and reliable electric vehicle charging for

the realisation of a true sustainable transportation landscape. *Sci. Rep.*, 14, 1, 3261, 2024, 10.1038/s41598-024-53214-w.

58. Singh, A.R., Vishnuram, P., Alagarsamy, S., Bajaj, M., Blazek, V., Damaj, I., Othman, K.M., Electric vehicle charging technologies, infrastructure expansion, grid integration strategies, and their role in promoting sustainable e-mobility. *Alex. Eng. J.*, 105, 300–330, 2024, 10.1016/j.aej.2024.06.093.
59. Rekioua, D., Mokrani, Z., Kakouche, K., Rekioua, T., Oubelaid, A., Logerais, P.O., Ghoneim, S.S., Optimization and intelligent power management control for an autonomous hybrid wind turbine photovoltaic diesel generator with batteries. *Sci. Rep.*, 13, 1, 21830, 2023, 10.1038/s41598-023-49067-4.
60. Kanouni, B., Badoud, A.E., Mekhilef, S., Bajaj, M., Zaitsev, I., Advanced efficient energy management strategy based on state machine control for multi-sources PV-PEMFC-batteries system. *Sci. Rep.*, 14, 1, 7996, 2024, 10.1038/s41598-024-58785-2.

6
Analysis of BTMS for Electric Vehicles (EV)

Mira Chitt

Department of Mechanical Engineering, Global College of Engineering and Technology, Muscat, Oman

Abstract
The global transition towards electric vehicles (EVs) is driving significant advancements in battery technologies to meet the growing demands for sustainability, performance, and reliability. Central to the efficiency and longevity of EV batteries is the Battery Thermal Management System (BTMS), which regulates battery temperature to ensure safe and optimal operation under various conditions. This chapter presents a comprehensive analysis of BTMS, highlighting its critical role in managing thermal behavior during charging, discharging, and exposure to external temperature fluctuations. Key areas explored include the design principles, components, operational mechanisms, and performance indicators of BTMS. The chapter also discusses the implications of thermal management on overall vehicle performance and identifies emerging trends and opportunities for innovation. By offering an in-depth understanding of BTMS, this work aims to support the development of more efficient and reliable thermal management strategies, ultimately contributing to the advancement of electric vehicle technology.

***Keywords*:** Battery thermal management system, development, electric vehicles, performance, thermal

6.1 Introduction

The automotive industry is undergoing a transformative shift towards electric vehicles (EVs) to reduce carbon emissions and mitigate the impacts of

Email: mira.c@gcet.edu.om

Ashwani Kumar, Mukesh Kumar Awasthi, Nitesh Dutt, Yogesh Kumar Singla and Sivasakthivel Thangavel (eds.) Thermal Battery Management System for Hybrid and Electric Vehicles, (117–132) © 2025 Scrivener Publishing LLC

climate change. As the adoption of EVs continues to accelerate, the demand for robust, efficient, and reliable battery systems becomes increasingly critical. Among the key components ensuring the optimal performance and longevity of EV batteries is the Battery Thermal Management System (BTMS). The BTMS plays a crucial role in maintaining the temperature of the battery pack within a safe and efficient operating range. This system is designed to manage heat generated during charging and discharging cycles, as well as external temperature variations, to prevent overheating or overcooling of the battery cells. By ensuring the optimal temperature, the BTMS enhances the overall performance, efficiency, and longevity of the battery pack, thus maximizing the driving range and reliability of the EV [1, 2].

The analysis of BTMS for electric vehicles encompasses various aspects, including the design, components, operating principles, and performance metrics of the system. This analysis is essential for understanding the impact of BTMS on the overall performance and efficiency of EVs, as well as for identifying opportunities for improvement and innovation in this critical component of electric vehicle technology.

This chapter aims to provide a comprehensive analysis of BTMS for electric vehicles, focusing on its importance, functions, design considerations, and future trends. By examining the key aspects of BTMS technology, this analysis will contribute to a deeper understanding of its role in enhancing the performance and efficiency of electric vehicles, thereby driving further innovation and advancement in this rapidly evolving field [3, 4].

6.2 Literature Review

Analyzing Battery Thermal Management Systems (BTMS) in electric vehicles (EVs) involves a comprehensive review of existing literature, encompassing various aspects such as design methodologies, modeling techniques, control strategies, and experimental studies. In what follows is a literature review summarizing key findings and methodologies in this field.

Battery Thermal Management Systems (BTMS) are crucial components in electric vehicles (EVs) to maintain optimal operating temperatures of the battery pack. Effective BTMS ensures battery safety, longevity, and performance under various operating conditions.

Designing a BTMS involves considerations such as thermal management requirements, integration with the vehicle's cooling system, packaging constraints, and thermal insulation. Researchers employ computational fluid dynamics (CFD) simulations, thermal modeling, and experimental validation to optimize BTMS designs [5].

In their study, Lee *et al.* (2019) proposed a design methodology for a BTMS using a combination of CFD simulations and experimental validation. Their approach optimized the cooling channels and heat exchanger configurations to enhance thermal performance.

Modeling BTMS involves developing mathematical models to simulate thermal behavior, predict temperature distributions, and assess heat dissipation capabilities. These models range from simple empirical equations to complex multiphysics simulations considering heat generation, conduction, convection, and radiation [6].

Chen *et al.* (2018) developed a dynamic thermal model for an EV battery pack considering transient heat generation, thermal resistances, and cooling system dynamics. Their model accurately predicted battery temperatures under different driving cycles, facilitating the design of efficient BTMS.

Control strategies play a vital role in regulating battery temperatures within safe limits while optimizing energy efficiency. Proactive thermal management strategies anticipate temperature variations and adjust cooling/heating mechanisms accordingly to prevent overheating or overcooling [7].

Wang *et al.* (2020) proposed a model predictive control (MPC) strategy for BTMS in EVs, considering battery temperature constraints, ambient conditions, and energy consumption. Their MPC algorithm effectively balanced thermal management and energy efficiency, improving battery lifespan and performance.

Experimental validation is essential to verify the effectiveness and reliability of BTMS designs and control strategies under real-world conditions. Researchers conduct experiments using test benches, prototype vehicles, and instrumented battery packs to measure temperatures, thermal resistances, and heat transfer rates.

In a study by Zhang *et al.* (2017), experimental tests were conducted on a prototype EV equipped with a BTMS to evaluate its thermal performance under different driving scenarios. The experiments validated the effectiveness of the BTMS in maintaining battery temperatures within the desired range, ensuring safe and reliable operation.

Another research paper provides a comprehensive evaluation of energy management systems for electric vehicles, focusing on hybrid energy storage systems and their control algorithms. The study outlines the characteristics of electric vehicles, research methods, hybrid energy storage system architecture, converter topology, and energy management techniques. A systematic research framework is used to analyze the strength and co-occurrence of keywords over the past decade, revealing trends in the literature.

The hybrid energy storage system and converter circuit architecture are assessed, with a non-isolated DC-DC converter connected to a supercapacitor identified as a suitable configuration due to its simplicity, reliability, and cost-effectiveness. Control strategies like deterministic and fuzzy-rule-based techniques are successfully evaluated through real-scale vehicle experiments, while real-time optimization-based energy management techniques show promise in lab-scale tests for future real-scale vehicle applications. The study provides insights into the benefits, drawbacks, and suitability of different energy management systems for battery/supercapacitor electric vehicle applications [8–10].

In the research article, Design and Analysis of Control Strategy for an Efficient Battery Thermal Management System of an Electric Vehicle by K.R. Patil *et al*, the study focuses on battery thermal management systems (BTMS) for electric vehicles, crucial for battery safety and reliability in hybrid electric vehicles (HEVs) and electric vehicles (EVs). An air-cooling system with a V-type staggered arrangement of plate-fin heat sink is proposed to maintain lithium-ion battery pack temperatures within safe limits, preventing overheating and potential fire hazards. A comparison is made between straight rectangular fins and V-type staggered arrangement heat sinks to optimize battery thermal management. Numerical simulations are conducted to analyze the impact of the number of fins, fin spacing, and fin inclination angle on heat dissipation efficiency under varying air mass flow rates and heat inputs. The research aims to enhance battery performance and longevity in electric vehicles through effective thermal management strategies, combining computational modeling with traditional heat transfer principles.

The research paper done by S., N., Naveen in 2022 focuses on optimizing the Battery Thermal Management System (BTMS) of a 15-module battery pack for electric vehicles. BTMS is crucial for maintaining battery safety, performance, and longevity by regulating cell temperatures and preventing capacity loss. The study uses Ansys Fluent Battery Model to optimize the power requirement of BTMS. The battery pack consists of 15 modules in a 7S2P configuration with active cooling at flow rates of 10 to 30 lpm. The Equivalent Circuit Model (ECM) is employed to represent cell dynamics using Hybrid Power Pulse Characteristics (HPPC) data. Experimental validation is done on an 80Ah single cell for a real drive cycle, showing an 8% underprediction in mean cell temperature. The optimal flow rate of 10 to 20 lpm is determined to maintain the cell's maximum temperature below 35°C during the drive cycle. This study emphasizes the importance of efficient thermal management for electric vehicle battery packs to ensure optimal performance and longevity.

In summary, Battery Thermal Management Systems (BTMS) are critical components in electric vehicles (EVs) to ensure battery safety, longevity, and performance. Designing efficient BTMS involves a combination of computational simulations, thermal modeling, control strategies, and experimental validation. Advanced methodologies such as model predictive control (MPC) and dynamic thermal modeling have shown promise in optimizing BTMS performance while balancing energy efficiency. Continued research and development in this field are essential to address the evolving challenges of thermal management in next-generation electric vehicles [11–13].

6.3 Importance of BTMS in Electric Vehicles

The battery pack is one of the most expensive and critical components of an electric vehicle, accounting for a significant portion of the vehicle's cost and weight. The performance, efficiency, and longevity of the battery pack are directly influenced by its operating temperature, making thermal management a key consideration in EV design.

The BTMS plays a crucial role in ensuring that the battery pack operates within a safe and efficient temperature range. By actively managing the temperature of the battery cells, the BTMS helps to prevent overheating, which can reduce battery life and performance, and minimize overcooling, which can negatively impact efficiency and range [14, 15].

Furthermore, the BTMS helps to optimize the charging and discharging processes of the battery pack, ensuring faster charging times, longer battery life, and improved overall performance. By maintaining the battery pack at an optimal temperature, the BTMS helps to maximize the driving range of the EV, enhance its performance in extreme weather conditions, and improve its overall reliability.

In addition to its impact on battery performance and efficiency, the BTMS also plays a crucial role in ensuring the safety of the EV. Overheating of the battery pack can lead to thermal runaway, a dangerous condition that can result in fire or explosion. The BTMS helps to prevent such incidents by actively managing the temperature of the battery cells and preventing them from reaching critical levels.

6.4 BTMS for Battery Electric Vehicles (BEV)

The emergence of Battery Electric Vehicles (BEVs) has revolutionized the automotive industry, offering a sustainable and environmentally friendly alternative to traditional internal combustion engine vehicles. At the heart of every BEV lies the battery pack, a complex system of lithium-ion cells that stores and delivers electrical energy to power the vehicle. The performance, longevity, and safety of BEV batteries are heavily influenced by their operating temperature. Extreme temperatures, whether too hot or too cold, can degrade battery performance, reduce its lifespan, and even pose safety risks. In response to these challenges, Battery Thermal Management Systems (BTMS) have become essential components of modern BEVs, tasked with regulating battery temperature to ensure optimal performance and safety under varying conditions. This analysis will delve into the intricacies of BTMS, examining its components, functions, benefits, and the impact it has on the overall performance of Battery Electric Vehicles [16–19].

6.4.1 Components of BTMS

A comprehensive BTMS comprises several key components working in harmony to monitor, control, and regulate battery temperature:

- Temperature Sensors: Distributed strategically throughout the battery pack, temperature sensors continuously monitor the temperature of individual cells or modules. These sensors provide real-time data to the BTMS, allowing it to make precise adjustments to maintain optimal operating conditions.
- Cooling System: The cooling system is responsible for dissipating excess heat generated during charging, discharging, or high ambient temperatures. It typically consists of heat exchangers, pumps, and coolant fluids that circulate through the battery pack to absorb and remove heat efficiently.
- Heating System: In cold climates, maintaining battery temperature within the optimal range becomes challenging. A heating system, often integrated with the cooling system, helps warm up the battery pack to ensure efficient performance during cold starts and operation in low temperatures.
- Battery Management System (BMS): The BMS serves as the central control unit of the BTMS, receiving data from temperature sensors and other relevant parameters. It utilizes this information to make informed decisions regarding thermal management, controlling the operation of cooling

and heating components to maintain the battery within the desired temperature range [20–22].

6.4.2 Functions of BTMS

The primary functions of BTMS in BEVs are multifaceted and critical for ensuring optimal battery performance, longevity, and safety [23, 24]:

- Temperature Regulation: BTMS actively regulates battery temperature to prevent overheating during charging or discharging and to maintain optimal conditions for performance and longevity. By keeping the battery within the ideal temperature range, BTMS ensures efficient energy transfer and minimizes the risk of thermal degradation.
- Thermal Runaway Prevention: Overheating of lithium-ion batteries can lead to thermal runaway, a dangerous chain reaction resulting in fire or explosion. BTMS employs various cooling strategies to prevent such catastrophic events by rapidly dissipating heat and keeping the battery within safe temperature limits.
- Cold Weather Performance: In cold climates, BTMS ensures that the battery maintains sufficient temperature to deliver optimal performance and range. By preheating the battery pack before operation and providing continuous heating during operation, BTMS mitigates the adverse effects of low temperatures on battery efficiency and power output.
- Optimizing Charging Efficiency: BTMS plays a crucial role in optimizing charging efficiency by managing battery temperature during charging sessions. By maintaining the battery within the optimal temperature range, BTMS facilitates faster charging while minimizing degradation and ensuring long-term battery health.

6.4.3 Benefits of BTMS

The implementation of BTMS in BEVs offers numerous benefits that directly impact the overall performance and usability of electric vehicles [25, 26]:

- Extended Battery Life: By maintaining the battery within the optimal temperature range, BTMS helps mitigate the factors that contribute to battery degradation, such as high temperatures, overcharging, and thermal stress. This results in extended battery life and improved long-term reliability, reducing the need for premature battery replacements and lowering ownership costs for BEV owners.

- Enhanced Performance and Range: Optimal battery temperature is crucial for maximizing the performance and driving range of BEVs. BTMS ensures that the battery operates efficiently under diverse environmental conditions, delivering consistent power output and maintaining range even in challenging climates.
- Improved Safety: Overheating of lithium-ion batteries poses significant safety risks, including thermal runaway and battery fires. BTMS actively monitors and regulates battery temperature to prevent such hazards, enhancing the overall safety of BEVs for both occupants and surrounding environments.
- Enhanced Charging Efficiency: Efficient thermal management is essential for maximizing charging efficiency and reducing charging times for BEVs. BTMS optimizes battery temperature during charging sessions, allowing for faster charging rates while minimizing the risk of overheating or damage to the battery pack.

6.4.4 Impact of BTMS on BEV Performance

The integration of BTMS has a profound impact on the overall performance and usability of Battery Electric Vehicles [27–29]:

- Range and Efficiency: BTMS plays a crucial role in maximizing the driving range and efficiency of BEVs by ensuring optimal battery temperature. By minimizing energy losses due to thermal inefficiencies and maintaining consistent performance under varying conditions, BTMS enhances the overall range and efficiency of electric vehicles, making them more practical for daily use.
- Reliability and Durability: The implementation of BTMS enhances the reliability and durability of BEV batteries by mitigating the factors that contribute to premature degradation and failure. By maintaining the battery within the optimal temperature range, BTMS helps preserve battery health

and extend its lifespan, resulting in more reliable and durable electric vehicles.

- Safety and Peace of Mind: Safety is paramount in electric vehicles, and BTMS plays a crucial role in ensuring the safe operation of BEVs under diverse conditions. By actively monitoring and regulating battery temperature, BTMS reduces the risk of thermal hazards such as thermal runaway and battery fires, providing occupants with peace of mind and confidence in the safety of electric vehicles.
- User Experience: The integration of BTMS enhances the overall user experience of BEVs by providing consistent performance, extended range, and faster charging times. By mitigating the impact of external factors such as temperature on battery performance, BTMS improves the usability and practicality of electric vehicles, making them more appealing to a broader range of consumers.

6.4.5 Challenges and Future Developments

Despite the numerous benefits offered by BTMS, several challenges and opportunities for improvement remain [30]:

- Cost and Complexity: The implementation of comprehensive BTMS adds complexity and cost to BEVs, impacting their affordability and mass adoption. Future developments in thermal management technologies, such as advanced materials and innovative cooling strategies, may help mitigate these challenges and make BTMS more cost-effective and efficient.
- Integration with Vehicle Design: Integrating BTMS seamlessly into the overall vehicle design without compromising interior space or aerodynamics remains a challenge for automakers. Future advancements in packaging and integration techniques may enable more efficient and compact BTMS solutions that optimize space and enhance vehicle performance.
- Adaptability to Extreme Conditions: BEVs are subjected to a wide range of environmental conditions, from scorching heat to freezing cold. BTMS must be capable of adapting to these extremes and providing effective thermal management under all circumstances. Continued research and

development in thermal management technologies will be essential for addressing these challenges and improving the performance and reliability of BTMS in BEVs.
- Standardization and Regulation: As the adoption of electric vehicles continues to grow, standardization and regulation of BTMS technologies will become increasingly important. Establishing industry-wide standards and regulations for BTMS design, performance, and safety will ensure consistency and interoperability across different vehicle models and manufacturers, fostering consumer confidence and trust in electric vehicles.

Battery Thermal Management Systems (BTMS) are integral components of Battery Electric Vehicles (BEVs), playing a crucial role in regulating battery temperature to ensure optimal performance, longevity, and safety. By actively monitoring and controlling battery temperature, BTMS mitigates the factors that contribute to battery degradation, enhances driving range and efficiency, and improves the overall reliability and safety of electric vehicles. Despite the challenges and opportunities for improvement, the continued advancements in thermal management technologies promise to further enhance the performance and usability of BTMS in BEVs, accelerating the transition towards a sustainable and electrified transportation future.

6.5 BTMS for Hybrid Electric Vehicles (HEV)

In the pursuit of a more sustainable transportation future, Hybrid Electric Vehicles (HEVs) have emerged as a promising solution. By integrating both internal combustion engines and electric propulsion systems, HEVs offer improved fuel efficiency and reduced emissions compared to traditional gasoline-powered vehicles. However, as technology advances and environmental concerns grow, the need for more efficient and intelligent systems becomes increasingly evident. This is where Battery Thermal Management Systems (BTMS) play a crucial role in enhancing the performance and longevity of HEVs.

BTMS serves as the backbone of the electrical system in HEVs, ensuring optimal operating conditions for the vehicle's battery pack. The battery pack is the heart of an HEV, providing power for electric propulsion and energy storage. Efficient operation of the battery pack is essential for maximizing the vehicle's range, performance, and lifespan. BTMS manages the thermal conditions of the battery pack, maintaining it within the ideal temperature range for optimal performance.

One of the primary functions of BTMS is to regulate the temperature of the battery pack during charging and discharging cycles. Lithium-ion batteries, commonly used in HEVs, are sensitive to temperature fluctuations. High temperatures can accelerate degradation and reduce battery life, while low temperatures can impede performance and energy output. BTMS employs various cooling and heating mechanisms to maintain the battery pack within the optimal temperature range, ensuring consistent and reliable operation.

Furthermore, BTMS enhances the safety of HEVs by preventing overheating and thermal runaway events. Thermal runaway, a phenomenon where the temperature of the battery pack rapidly increases, can lead to catastrophic failure and even fire. BTMS monitors the temperature of individual battery cells and employs thermal management strategies such as active cooling or thermal insulation to prevent thermal runaway. By proactively managing thermal conditions, BTMS mitigates the risk of safety incidents and ensures the reliability of HEVs [31, 32].

In addition to thermal management, BTMS also plays a crucial role in optimizing the charging and discharging processes of the battery pack. By monitoring parameters such as state of charge (SOC) and state of health (SOH), BTMS can adapt charging rates and energy flow to maximize efficiency and extend battery life. Smart charging algorithms and predictive analytics enable BTMS to optimize energy usage based on driving conditions, user preferences, and environmental factors. This not only improves the performance of HEVs but also enhances overall energy efficiency and reduces environmental impact.

Moreover, BTMS contributes to the longevity and reliability of HEVs by implementing advanced battery diagnostics and prognostics. By continuously monitoring the health and performance of the battery pack, BTMS can detect early signs of degradation or malfunction and alert users or maintenance personnel before serious issues arise. Predictive maintenance based on real-time data analytics enables proactive servicing and replacement of battery components, minimizing downtime and reducing maintenance costs.

The integration of BTMS with advanced vehicle telematics and connectivity further enhances the capabilities of HEVs. By leveraging real-time data from sensors, GPS, and weather forecasts, BTMS can optimize thermal management strategies and energy usage in response to dynamic driving conditions. Smart grid integration allows HEVs to participate in demand response programs, enabling bidirectional energy flow between the vehicle and the grid. This not only provides additional revenue streams for vehicle owners but also contributes to grid stability and renewable energy integration.

Looking ahead, the evolution of BTMS will continue to drive innovation in the field of hybrid and electric vehicles. Advances in materials science, thermal management technologies, and artificial intelligence will enable the development of more efficient, reliable, and intelligent BTMS solutions. Integration with emerging technologies such as solid-state batteries and wireless charging will further enhance the performance and convenience of HEVs.

To summarize this section, Battery Thermal Management Systems (BTMS) play a critical role in enhancing the performance, safety, and longevity of Hybrid Electric Vehicles (HEVs). By optimizing thermal conditions, managing energy flow, and implementing advanced diagnostics, BTMS ensures the efficient operation of the battery pack and maximizes the benefits of electrified transportation. As we strive towards a sustainable future, continued investment and innovation in BTMS will be essential for realizing the full potential of HEVs and accelerating the transition to cleaner and greener mobility solutions.

6.6 BTMS for Plug-In Hybrid Electric Vehicles (PHEV)

In the landscape of sustainable transportation, Plug-In Hybrid Electric Vehicles (PHEVs) stand out as a versatile solution offering the benefits of both electric and conventional propulsion systems. With their ability to operate on electric power alone for shorter trips and seamlessly transition to hybrid mode for longer journeys, PHEVs represent a compelling option for reducing fuel consumption and emissions. Central to the efficient operation and longevity of PHEVs is the Battery Thermal Management System (BTMS), which plays a pivotal role in optimizing the performance of the vehicle's battery pack.

At the core of every PHEV lies its battery pack, which serves as the energy storage unit for electric propulsion. Unlike traditional hybrid vehicles, PHEVs feature larger battery packs capable of providing extended electric-only range. Efficient management of the thermal conditions within the battery pack is essential for maximizing its performance, lifespan, and safety. This is where BTMS comes into play, employing a range of strategies to regulate temperature and ensure optimal operating conditions.

One of the primary functions of BTMS in PHEVs is to manage the thermal dynamics of the battery pack during charging and discharging cycles. Lithium-ion batteries, commonly used in PHEVs, are sensitive to temperature fluctuations, with excessive heat or cold potentially degrading performance and shortening battery life. BTMS employs sophisticated cooling and heating mechanisms to maintain the battery pack within the optimal temperature range, thereby enhancing its efficiency and longevity.

Moreover, BTMS plays a crucial role in optimizing the charging process for PHEVs, ensuring fast and efficient replenishment of the battery pack. By monitoring parameters such as state of charge (SOC) and ambient temperature, BTMS can adjust charging rates and strategies to maximize efficiency and battery health. Fast charging capabilities integrated into BTMS enable PHEV owners to replenish their battery packs quickly, further enhancing the convenience and usability of electric propulsion.

In addition to thermal management and charging optimization, BTMS contributes to the safety and reliability of PHEVs by implementing advanced battery diagnostics and prognostics. Continuous monitoring of battery health parameters allows BTMS to detect early signs of degradation or malfunction, enabling proactive maintenance and minimizing the risk of safety incidents. Predictive analytics integrated into BTMS facilitate real-time monitoring and alerting, ensuring the reliability and safety of PHEVs in all driving conditions.

Furthermore, the integration of BTMS with vehicle telematics and connectivity enhances the capabilities of PHEVs, enabling intelligent energy management and optimization. By leveraging real-time data from sensors, GPS, and weather forecasts, BTMS can adapt thermal management strategies and energy usage to maximize efficiency and performance. Smart grid integration allows PHEVs to participate in demand response programs, enabling bidirectional energy flow between the vehicle and the grid, thereby enhancing grid stability and supporting renewable energy integration.

Looking ahead, the evolution of BTMS will continue to drive innovation in the realm of PHEVs, unlocking new levels of efficiency, performance, and sustainability. Advances in materials science, thermal management technologies, and artificial intelligence will enable the development of more intelligent and adaptive BTMS solutions. Integration with emerging technologies such as solid-state batteries and wireless charging will further enhance the capabilities and convenience of PHEVs, accelerating their adoption and mainstream acceptance.

To summarize this section, Battery Thermal Management Systems (BTMS) play a critical role in maximizing the efficiency, performance, and longevity of Plug-In Hybrid Electric Vehicles (PHEVs). By optimizing thermal conditions, managing charging processes, and implementing advanced diagnostics, BTMS ensures the reliable and safe operation of PHEVs in all driving conditions. As we strive towards a more sustainable transportation future, continued investment and innovation in BTMS will be essential for realizing the full potential of PHEVs and advancing the transition to cleaner and greener mobility solutions.

References

1. Hamed, S.B., Hamed, M.B., Sbita, L., Bajaj, M., Blazek, V., Prokop, L., Ghoneim, S.S., Robust optimization and power management of a triple junction photovoltaic electric vehicle with battery storage. *Sensors*, 22, 16, 6123, 2022, 10.3390/s22166123.
2. Rawat, T., Niazi, K.R., Gupta, N., Sharma, S., Multi-objective techno-economic operation of smart distribution network integrated with reactive power support of battery storage systems. *Sustain. Cities Soc.*, 75, 103359, 2021, http://dx.doi.org/10.1016/j.scs.2021.103359.
3. Gupta, A., Singh, R.K., Paul, J., Sharma, M., Joshi, S., Battery-Operated Electric Vehicles (BOEVs) Adoption in India: Analysis of Barriers and Strategies. *IEEE Trans. Eng. Manage.*, 71, 7076–7087, 2023, http://dx.doi.org/10.1109/TEM.2023.3253621.
4. Lee, S., *et al.*, Design and Optimization of Battery Thermal Management System for Electric Vehicles. *J. Power Sources*, 437, 226946, 2019.
5. Kanouni, B., Badoud, A.E., Mekhilef, S., Bajaj, M., Zaitsev, I., Advanced efficient energy management strategy based on state machine control for multi-sources PV-PEMFC-batteries system. *Sci. Rep.*, 14, 1, 7996, 2024, 10.1038/s41598-024-58785-2.
6. Bagga, P., Das, A.K., Wazid, M., Rodrigues, J.J., Choo, K.K.R., Park, Y., On the design of mutual authentication and key agreement protocol in internet of vehicles-enabled intelligent transportation system. *IEEE Trans. Veh. Technol.*, 70, 2, 1736–1751, 2021, 10.1109/TVT.2021.3050614.
7. Chen, Y., *et al.*, Dynamic Thermal Modeling of Lithium-Ion Battery Pack for Electric Vehicles Considering Thermal Runaway. *Appl. Energy*, 228, 245–256, 2018.
8. Thapliyal, S., Wazid, M., Singh, D.P., Das, A.K., Islam, S.H., Robust authenticated key agreement protocol for internet of vehicles-envisioned intelligent transportation system. *J. Syst. Archit.*, 142, 102937, 2023, 10.1016/j.sysarc.2023.102937.
9. Das, S., Thakur, P., Singh, A.K., Singh, S.N., Optimal management of vehicle-to-grid and grid-to-vehicle strategies for load profile improvement in distribution system. *J. Energy Storage*, 49, 104068, 2022, http://dx.doi.org/10.1016/j.est.2022.104068.

10. Wang, L., *et al.*, Model Predictive Control of Battery Thermal Management System for Electric Vehicles. *IEEE Trans. Veh. Technol.*, 69, 4, 3834–3844, 2020.
11. Oubelaid, A., Kakouche, K., Belkhier, Y., Khosravi, N., Taib, N., Rekioua, T., Tuka, M.B., New coordinated drive mode switching strategy for distributed drive electric vehicles with energy storage system. *Sci. Rep.*, 14, 1, 6448, 2024, 10.1038/s41598-024-56209-9.
12. Ravindran, M.A., Nallathambi, K., Vishnuram, P., Rathore, R.S., Bajaj, M., Rida, I., Alkhayyat, A., A novel technological review on fast charging infrastructure for electrical vehicles: Challenges, solutions, and future research directions. *Alexandria Eng. J.*, 82, 260–290, 2023, 10.1016/j.aej.2023.10.009.
13. Amoussou, I., Paddy, E.Y., Agajie, T.F., Ibrahim, F.S., Agajie, E.F., Nsanyuy, W.B., Mohammadi, S.A.D., Enhancing residential energy access with optimized stand-alone hybrid solar-diesel-battery systems in Buea, Cameroon. *Sci. Rep.*, 14, 1, 15543, 2024, 10.1038/s41598-024-66582-0.
14. Zhang, J., *et al.*, Experimental Study on Battery Thermal Management System for Electric Vehicles. *Energy Procedia*, 142, 2925–2930, 2017.
15. Wangsupphaphol, A., Phichaisawat, S., Nik Idris, N.R., Jusoh, A., Muhamad, N.D., Lengkayan, R.A., Systematic Review of Energy Management Systems for Battery/Supercapacitor Electric Vehicle Applications. *Sustainability*, 15, 11200, 2023.
16. Badri, Y., II, Sudabattula, S.K., Hussain, I., Malik, H., Márquez, F.P.G., Ultracapacitor as selectable energy buffer in electric vehicle application. *J. Energy Storage*, 73, 109200, 2023, http://dx.doi.org/10.1016/j.est.2023.109200.
17. Kasri, A., Ouari, K., Belkhier, Y., Bajaj, M., Zaitsev, I., Optimizing electric vehicle powertrains peak performance with robust predictive direct torque control of induction motors: A practical approach and experimental validation. *Sci. Rep.*, 14, 1, 14977, 2024, 10.1038/s41598-024-65988-0.
18. Patil, K.R., Pendse, P., Mote, A., Patil, A., Sarnobat, R., Design and Analysis of Control Strategy for an Efficient Battery Thermal Management System of an Electric Vehicle. *Int. J. Veh. Struct. Syst. (IJVSS)*, 14, 2, 2022.
19. Naveen, S.N., Gunti, S.R., Naveen Kumar, M.A., Louis, S., Sharma, U., Optimizing BTM of HV battery pack for automotive application using electro-thermal simulation, in: *2022 IEEE Sustainable Power and Energy Conference (iSPEC)*, pp. 1–5, IEEE, 2022.
20. Bhatt, A., Ongsakul, W., Pawar, J., Optimal energy management system for carbon–neutral microgrid integrating second-life batteries and crypto mining devices. *Sustain. Energy Technol. Assess.*, 64, 103686, 2024, 10.1016/j.seta.2024.103686.
21. Sahri, Y., Tamalouzt, S., Belaid, S.L., Bajaj, M., Ghoneim, S.S., Zawbaa, H.M., Kamel, S., Performance improvement of Hybrid System based DFIG-Wind/PV/Batteries connected to DC and AC grid by applying Intelligent Control. *Energy Rep.*, 9, 2027–2043, 2023, 10.1016/j.egyr.2023.01.021.
22. Rekioua, D., Mokrani, Z., Kakouche, K., Rekioua, T., Oubelaid, A., Logerais, P.O., Ghoneim, S.S., Optimization and intelligent power management control

for an autonomous hybrid wind turbine photovoltaic diesel generator with batteries. *Sci. Rep.*, 13, 1, 21830, 2023, 10.1038/s41598-023-49067-4.

23. Singh, P.P., Das, S., Wen, F., Palu, I., Singh, A.K., Thakur, P., Multi-objective planning of electric vehicles charging in distribution system considering priority-based vehicle-to-grid scheduling. *Swarm Evol. Comput.*, 77, 101234, 2023, http://dx.doi.org/10.1016/j.swevo.2023.101234.
24. Almutairi, A., Albagami, N., Almesned, S., Alrumayh, O., Malik, H., Electric vehicle load estimation at home and workplace in Saudi Arabia for grid planners and policy makers. *Sustainability*, 15, 22, 15878, 2023, http://dx.doi.org/10.3390/su152215878.
25. Afzal, M.Z., Aurangzeb, M., Iqbal, S., Pushkarna, M., Rehman, A.U., Kotb, H., Bereznychenko, V., A Novel Electric Vehicle Battery Management System Using an Artificial Neural Network-Based Adaptive Droop Control Theory. *Int. J. Energy Res.*, 2023, 1, 2581729, 2023, 10.1155/2023/2581729.
26. Ramakrishnan, V., Vishnuram, P., Yang, T., Bajaj, M., Rathore, R.S., Zaitsev, I., Design and implementation of a high misalignment-tolerance wireless charger for an electric vehicle with control of the constant current/voltage charging. *Sci. Rep.*, 14, 1, 13165, 2024, 10.1038/s41598-024-63952-6.
27. Togun, H., Basem, A., Mohammed, H.I., Sadeq, A.M., Biswas, N., Abdulrazzaq, T., Hasan, H.A., Homod, R.Z., Talebizadehsardari, P., A comprehensive review of battery thermal management systems for electric vehicles: Enhancing performance, sustainability, and future trends. *Int. J. Hydrogen Energy*, 97, 1077–1107, 2025.
28. Jena, N.K., Sahoo, S., Sahu, B.K., Naik, A.K., Bajaj, M., Misak, S., Prokop, L., Impact of a redox flow battery on the frequency stability of a five-area system integrated with renewable sources. *Energies*, 16, 14, 5540, 2023, 10.3390/en16145540.
29. Dimri, S.C., Indu, R., Bajaj, M., Rathore, R.S., Blazek, V., Dutta, A.K., Alsubai, S., Modeling of traffic at a road crossing and optimization of waiting time of the vehicles. *Alexandria Eng. J.*, 98, 114–129, 2024, 10.1016/j.aej.2024.04.050.
30. Kumar, B.A., Jyothi, B., Singh, A.R., Bajaj, M., Rathore, R.S., Berhanu, M., A novel strategy towards efficient and reliable electric vehicle charging for the realisation of a true sustainable transportation landscape. *Sci. Rep.*, 14, 1, 3261, 2024, 10.1038/s41598-024-53214-w.
31. Singh, A.R., Vishnuram, P., Alagarsamy, S., Bajaj, M., Blazek, V., Damaj, I., Othman, K.M., Electric vehicle charging technologies, infrastructure expansion, grid integration strategies, and their role in promoting sustainable e-mobility. *Alexandria Eng. J.*, 105, 300–330, 2024, 10.1016/j.aej.2024.06.093.
32. Aggarwal, S., Singh, A.K., Rathore, R.S., Bajaj, M., Gupta, D., Revolutionizing load management: A novel technique to diminish the impact of electric vehicle charging stations on the electricity grid. *Sustain. Energy Technol. Assess.*, 65, 103784, 2024, 10.1016/j.seta.2024.103784.

7

Battery Thermal Management in Hybrid Electric Vehicles

Bipasa Bimalendu Patra[1]*, Vishal Vaidya[2], Moiz Hussain[2] and Yatika Gori[3]

[1]*Electrical Engineering Department, Brainware University Barasat, Kolkata, West Bengal, India*
[2]*Electrical Engineering Department, Padmashri Dr. V. B. Kolte College of Engineering, Malkapur, Maharashtra, India*
[3]*Department of Mechanical Engineering, Graphic Era (Deemed to be University), Dehradun, Uttarakhand, India*

Abstract
Battery Thermal Management Systems (BTMS) play a pivotal role in the dynamic landscape of Electric Vehicles (EVs), where efficient energy storage and utilization are paramount. Within the context of EVs, where batteries serve as the heart of propulsion systems, the significance of BTMS cannot be overstated. These systems are intended to meticulously regulate the temperature of batteries, safeguarding against the adverse effects of thermal extremes that can compromise performance, safety, and overall battery longevity.

In the intricate ecosystem of EVs, BTMS serves as a guardian, ensuring batteries operate within their optimal temperature range. The stakes are high; overheating can trigger catastrophic thermal runaway events, posing severe safety hazards, while undercooling can impair battery efficiency and curtail lifespan. BTMS technologies offer a multifaceted approach to address these challenges, leveraging various cooling methods such as air cooling, liquid cooling, and innovative phase change materials (PCMs). Each approach brings its unique advantages and challenges, but collectively they form a comprehensive strategy to maintain battery temperature equilibrium, even under demanding conditions.

Keywords: Hybrid electric vehicles, battery temperature, battery thermal management systems, efficient energy storage

**Corresponding author*: akasvora@gmail.com; ORCID: 0000-0002-2719-4103

Ashwani Kumar, Mukesh Kumar Awasthi, Nitesh Dutt, Yogesh Kumar Singla and Sivasakthivel Thangavel (eds.) Thermal Battery Management System for Hybrid and Electric Vehicles, (133–156) © 2025 Scrivener Publishing LLC

7.1 Introduction

Beyond mere temperature regulation, BTMS in EVs faces an array of complex challenges, from managing heat generated during rapid charging to optimizing performance in diverse environmental conditions. Advanced control systems, empowered by artificial intelligence and machine learning algorithms, hold promise in tackling these challenges head-on. By continuously monitoring and analyzing real-time data, these intelligent systems can predict and preemptively adjust cooling strategies, ensuring optimal battery performance and safety throughout the EV's lifecycle.

Moreover, the evolution of BTMS is not static but rather dynamic, propelled by ongoing research and innovation. As EV adoption accelerates globally, researchers and engineers are relentlessly pursuing advancements in BTMS technology. These endeavors aim to enhance heat dissipation efficiency, simplify system complexity, and drive down costs to make EVs more accessible to a wider audience. Ultimately, the trajectory of BTMS in EVs is intertwined with the broader mission of advancing electric mobility, facilitating the transition towards sustainable transportation solutions, and laying the groundwork for a greener, more resilient future [1, 2].

7.2 Types of Systems

As the demand for high-performance HEVs grows, the development of more sophisticated and cost-effective thermal management solutions will be key to supporting the widespread adoption and sustainability of hybrid electric vehicles. Advanced thermal management systems use techniques like liquid cooling, air cooling, and phase-change materials to regulate the battery temperature, ensuring it remains within a safe and efficient operating range. These systems not only help in preserving battery life but also enhance the overall efficiency and reliability of the vehicle [3–5].

These systems integrate advanced sensors and control algorithms to monitor and adjust the battery temperature in real-time. By maintaining the battery within its optimal temperature range, these systems enhance the vehicle's performance, extend battery life, and prevent safety issues [6].

There are several types of BTMS, each with unique advantages and applications (Table 7.1):

Table 7.1 Literature review of the relevant research work.

Ref. no.	Topic	Author names	Results
[1]	Review on battery thermal management system for electric vehicles	Kim, Jinwoo Oh, Hoseong Lee	An efficient BTMS is required for EVs in order to increase both performance and safety. This study provides a comprehensive evaluation and categorization of the current BTMS investigations based on the available thermal cycle choices. Cabin air cooling system, loop liquid cooling system, with secondary and two-phase direct refrigerant cooling system are the three divisions of the BTMS with the VCC. However, BTMS without the VCC falls into the categories of cooling system -thermoelectric element, heat pipe & PCM.
[2]	Battery Thermal Management System	Venkatakrishnan S., Vikram Sudhan V.M., Sabari Kandappan S., Vishwanath S., Saravanan S., Pandiyan P.	This research suggested utilizing a motor pump to help build and optimize the battery with the cooling plate. The plate for cooling is used in this system to lower the battery pack's temperature, and as a result, the system is used to release the excess heat that the battery produces. This research observes the battery's temperature characteristics and does a comparative comparison. The cooling layer's configuration was further enhanced by the surrogate model technique. The purpose of this plate that is liquid-cooled is to reduce the excess heat produced by the battery package when the car is running or when it is being charged and discharged.

(Continued)

Table 7.1 Literature review of the relevant research work. (*Continued*)

Ref. no.	Topic	Author names	Results
			This battery temperature management technique has the potential to be highly effective at a very low cost. Consequently, a cooling plate is used to pump water that is in contact with the battery's exterior surface. The purpose of the cooling plate pumping mechanism is to lessen the transfer of heat and keep the range of battery's temperature stable throughout regular operation.
[3]	Thermal Performance of EV and HEV Battery Modules and Packs	Ahmad A. Pesaran, Ph.D. AndreasVlahinos, Ph.D. Steven D. Burch	The operating temperatures of the battery pack and the degree of temperature gradient inside it have an impact on the performance of the pack and, consequently, the performance of an EV or HEV. Because of their higher power and more aggressive charge/discharge behavior, HEV packs are more susceptible to thermal difficulties. For EVs, and especially HEVs, appropriate thermal management system design requires thermal analysis. To forecast the thermal performance of HEV modules and packs, we employed finite element analysis tools and fundamental heat transport concepts. Temperature distributions in a fictitious module and pack were obtained.

(*Continued*)

Table 7.1 Literature review of the relevant research work. (*Continued*)

Ref. no.	Topic	Author names	Results
			The pack's temperature can fluctuate greatly even at respectable air flow rates. Without any ventilation, a pack might become too hot to handle. The modules' and packs' appropriate thermal design needs to be carefully considered. According to our investigation, a HEV battery module's thermal performance was enhanced by the addition of ventilation holes. Thermal pictures of a hypothetical HEV pack and a HEV module were obtained using IR and LC thermography. These methods are helpful for evaluating thermal behavior; they might be applied to identify problem areas and offer guidance for better designs.
[4]	Review on Energy Management System of Electric Vehicles	Aruna P., Dr. Vasan Prabhu V.	Since EVs will eventually become more affordable than internal combustion engine vehicles in every way, this study essentially gives an overview of the literature on EVs. This has enormous worldwide implications since electric vehicles (EVs) will allow for lower CO2 and NOx emissions, which will lessen the environmental effect of automobiles. Additionally, the Energy Management System (EMS) is important to EVs as it extends life of the battery, makes more thermally stable, and boosts safety functionally and powertrain component reliability.

(*Continued*)

Table 7.1 Literature review of the relevant research work. (*Continued*)

Ref. no.	Topic	Author names	Results
[5]	Battery Thermal Management in EVs and HEVs: Issues and Solutions	Ahmad A. Pesaran	To maintain EV and HEV battery pack temperatures within the intended operating range, an effective thermal management system is needed. To function in both hot and cold environments, production HEVs need an active heating/cooling battery management system (BTMS). The behavior and overall thermal management of a pack are positively impacted by a module's proper thermal design. Compared to a system that uses liquid cooling or heating, an air-based thermal management system is less complex but less efficient. Air thermal management systems are usually sufficient for parallel HEVs, whereas liquid-based systems could be necessary for series HEVs and EVs to get the best possible thermal performance. Comparing NiMH batteries to Li Ion and VRLA batteries, a more complex temperature management system is needed. A robust thermal management system is also necessary for Li Ion batteries due to safety and low temperature performance issues. If the battery pack is to be liquid-cooled or air-cooled, its placement may also have a significant influence on the sort of thermal management that should be used.

(*Continued*)

Table 7.1 Literature review of the relevant research work. (*Continued*)

Ref. no.	Topic	Author names	Results
[6]	Graphene based thermal management system for battery cooling in electric vehicles	Ya Liu, Torbjörn Thiringer, Nan Wang, Yifeng Fu, Hongbin Lu, Johan Liu	As per this, we have constructed an integrated graphene system as an initial proof of concept for a future HEV/EV cooling system. The temperature near the middle of the battery under 2C and 1C drops by 11°C and 9°C, respectively, after employing such a cooling mechanism. A graphene heat spreader may efficiently lower the heat source's temperature by 4°C at a 1.3 W heating load when compared to a copper system of the same kind. Considering all of this, this work offers insight into the viability of employing graphene for HEV/EV cooling. Lastly, research is required to determine how employing such a graphene-based cooling system affects a Li-ion battery's cycle and rate performance.
[7]	Thermal Analysis of Cell Balancing for Battery Management System in Electric Vehicle	Ritwik Alok Pattnayak, Vijay T.	The method for Test case 1-c has been finished by balancing each alternate cell based on the simulation results. Test case 1-c achieves a temperature rise of less than 101 °C, far lower than the maximum rated temperature limit of 118 °C. Up to 4 hours might be needed for balancing. Enhancing the board's thermal behavior, reaching the resistor's safe working limit, attaining time balance, and boosting the Battery Management System's dependability are the results. This document provides a useful overview of the thermal study performed for several situations considered for battery management system cell balancing.

(*Continued*)

Table 7.1 Literature review of the relevant research work. (*Continued*)

Ref. no.	Topic	Author names	Results
[8]	A Review on Electric Vehicles with perspective of Battery Management System	Daisy Ranawat, M. P.R. Prasad	The electrification of vehicles has gained relevance as the world moves toward reducing carbon pollution and oil reliance, both of which are damaging to the environment and raise greenhouse gasses. This is because electrification would also lower other harmful pollutants like ozone and particulate matter. According to research institutions, widespread use of electric vehicles may reduce carbon emissions by 550 million metric tons yearly by the year 2050. EVs have a bright future since battery prices are falling quickly. There are now 300-mile completely electric vehicles that are both long range and reasonably priced.
[9]	Enhanced Battery Thermal Management Systems with Optimal Charge Control for Electric Vehicles	Rashid and T. Hofman, B. Rosca	By creating a charge control strategy that restricts the charging power from regenerative braking and rapid charging for an E-bus to minimize battery temperature excursion above the intended operating range, this study expanded on the body of research already done on battery thermal management systems. A dynamic regeneration map that adjusts to changing operating conditions is produced by the devised method, which regulates the charging power while driving.

(*Continued*)

Table 7.1 Literature review of the relevant research work. (*Continued*)

Ref. no.	Topic	Author names	Results
			The results of the simulation show that the devised technique can keep the battery operating within the intended temperature ranges. However, employing this tactic causes a loss of regenerated energy. Prediction data for an E-bus's route planning can be included to the charge control strategy as future work.
[10]	System Level Simulation Technique for Optimizing Battery Thermal Management System of an Electric Vehicle	Vipin K. Venugopal, Chandrasekaran N., Shrivatsal Sharma*	This article focuses on battery heat management system optimization using 1D modeling. In order to assure appropriate thermal load, a realistic battery model with an analogous circuit is included in the development of a system level 1D model. The final model, which derives thermal resistances from CFD data, has a high degree of accuracy in temperature prediction (max error% is 1.3%). Each simulation takes between fifteen and twenty minutes to compute. In order to maximize thermal management logic and increase vehicle range and battery life, this model is perfect.

(*Continued*)

Table 7.1 Literature review of the relevant research work. (*Continued*)

Ref. no.	Topic	Author names	Results
[11]	Review of Research about Thermal Runaway and Management of Li-ion Battery Energy Storage Systems	Xiaoqiang Zhang, Lei Li, Weiping Zhang	The causes and control of lithium-ion battery thermal runaway are interpreted in this research. The misuse of the mechanical, electrical, and thermal components of electric cars is the root cause of lithium-ion battery thermal runaway. Relevant studies on battery design and battery management systems were presented in relation to the topic of thermal runaway prevention. In conclusion, it is clear that future research will focus on studying the thermal runaway microscopically, or the internal evolution trajectory within the battery, in order to develop appropriate preventive measures based on the findings of the study, whether the research is conducted using an external car battery management system or the battery itself.

- **Air Cooling Systems:**

Passive Air Cooling: This system relies on natural convection to dissipate heat. It is simple and cost-effective but generally less effective at managing high thermal loads.

Active Air Cooling: This system uses fans to force air over the battery cells, enhancing heat dissipation. It provides better thermal control than passive systems but can be noisy and less efficient in very hot climates.

- **Liquid Cooling Systems:**

Indirect Liquid Cooling: This system circulates a coolant through channels or plates adjacent to the battery cells. The coolant absorbs heat from the battery and transfers it to a radiator where it is dissipated into the environment. Indirect liquid cooling offers efficient thermal management and is effective in maintaining uniform cell temperatures.

- **Direct Liquid Cooling:** In this system, the coolant directly contacts the battery cells, providing superior heat transfer. While more complex and potentially more expensive due to the need for leak-proof designs, it offers the highest level of thermal control, especially for high-power applications.

- **Phase-Change Material (PCM) Systems:**
PCMs absorb heat by changing phase (e.g., from solid to liquid) at a specific temperature. This process helps regulate battery temperature by absorbing excess heat during high power demands and releasing it slowly when the temperature drops. PCM systems are effective for thermal management but may require additional components to manage the weight and space constraints.

- **Thermoelectric Cooling Systems:**
These systems use thermoelectric devices, which can pump heat away from the battery cells when an electric current is applied. They offer precise temperature control and can be used for both cooling and heating. However, thermoelectric systems can be less efficient and more expensive compared to other cooling methods.

- **Refrigerant Cooling Systems:**
Similar to air conditioning systems, refrigerant cooling uses a refrigerant to absorb and dissipate heat from the battery. This method can provide very effective cooling, particularly in high-temperature environments, but it is complex and can add significant cost and weight to the vehicle.

7.3 Fundamental Heat Transfer Principle [3]

The temperature gradient inside a battery pack and its operating temperature have an impact on the performance of the pack and, in turn, the performance of an electric vehicle (EV) or hybrid electric vehicle (HEV). Because a HEV pack has larger power requirements and a more vigorous charge/discharge cycle, thermal considerations are more important. Thus, creating efficient heat management systems for EVs, especially HEVs, requires thermal analysis. We obtained temperature distributions within a hypothetical module and pack by applying finite element analysis software and basic heat transfer concepts to anticipate the thermal behavior of HEV modules and packs. Even at normal airflow rates, our findings indicated large temperature changes in the pack, and in the absence of airflow, the pack might reach dangerously high temperatures [7–11].

Module and pack thermal design must be done correctly. Our investigation showed that a HEV battery module's thermal performance was improved by the addition of ventilation holes (Figure 7.1 and Figure 7.2). Thermal photos of a HEV module and a simulated HEV pack were taken using infrared (IR) and liquid crystal (LC) thermography. These images (Figures 7.3–7.6) were helpful in assessing thermal behavior and pinpointing areas that required improvement in order to inform improved designs [12–16].

The graphic displays the schematic design for the graphene-based heat management system. A Li-ion rechargeable battery (2000 mA, 3.7 V, 40×58 mm) was utilized for the experiment (Figure 7.7). The Li-ion battery's surface was coated with a small coating of specially formulated interface material to adhere a graphene-assembled sheet. After that, this structure was put within a 1 centimeter-thick (1100 x 1300 mm) PTFE cage with a little

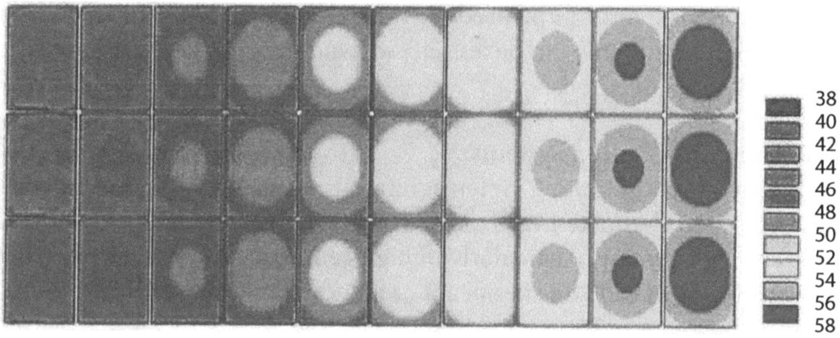

Figure 7.1 Distribution of temperature – steady-state 2-D in battery pack with air flow rate.

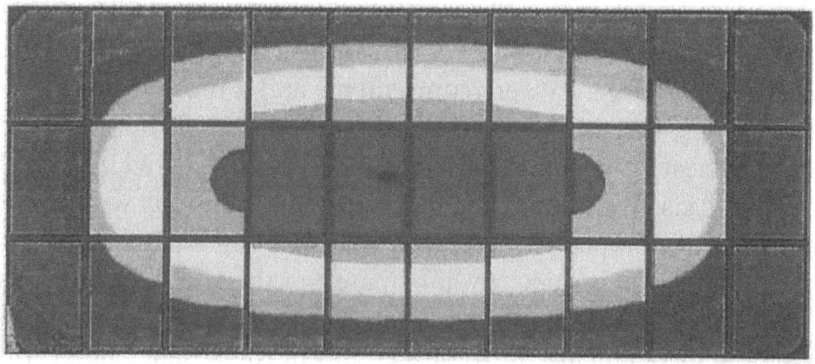

Figure 7.2 Distribution of temperature – steady-state 2-D in an enclosed battery pack.

Figure 7.3 Module of proposed HEV cross-section (purple areas – cells, red area – plastic casing, and blue area – cooling air).

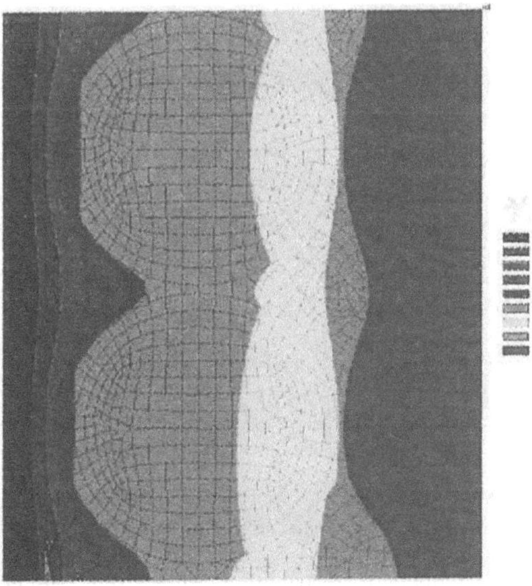

Figure 7.4 HEV module temperature distribution – steady-state without holes.

Figure 7.5 Module of proposed HEV cross-section with holes (purple areas – cells, red area – plastic casing, and blue area – cooling air).

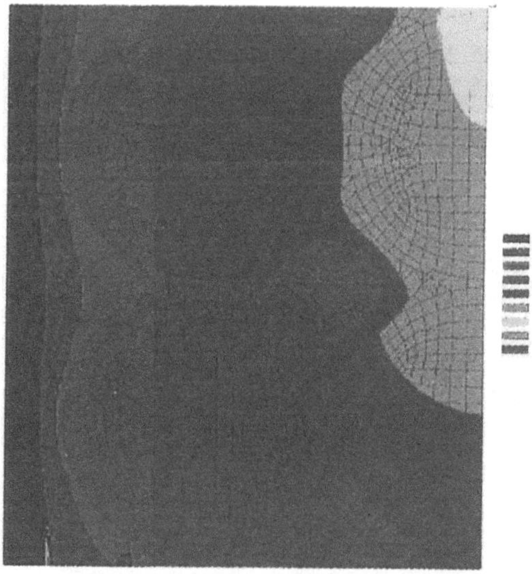

Figure 7.6 HEV module temperature distribution – steady-state with holes.

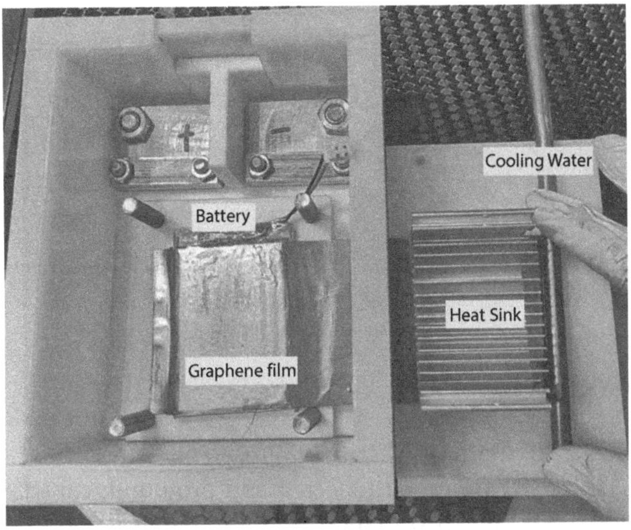

Figure 7.7 Schematic diagram of graphene based thermal management system for Li-ion battery (6).

opening to let the graphene sheet extend. After that, the PTFE box was put within a 1900 x 3000 mm polyethylene foam box that could be sealed. The graphene layer was affixed with an aluminum heat sink that was combined with a cooling water system. 17.6°C was the constant temperature of the cooling water [17–23]. An Agilent Data Acquisition Instrument (34970A) fitted with a J-type thermocouple was used to record temperature data. A NEWARE battery testing apparatus was used to charge and drain the battery. The apparatus was given 30 minutes to thermally settle before any measurements were made. SHT Smart High-Tech AB provided the graphene-assembled film, which was then physically squeezed for six hours at 45 MPa. Zeiss Supra 60 VP scanning electron microscopy (SEM) pictures (Figure 7.8) were acquired at 10 KV, and a Wyko NT 1100 optical surface profiler was used to assess the roughness of the film constructed from graphene [24, 25].

The proposed model demonstrates a graphene film integrated system as an initial proof of concept for future HEV/EV cooling solutions. Implementing this cooling system reduces the battery's core temperature by 11°C at a 2C rate and by 9°C at a 1C rate. In comparison to a copper-based system, a graphene-based heat spreader can lower the temperature of the heat source by 4°C under a 1.3 W heating load. These findings suggest that graphene could be a viable cooling solution for HEVs and EVs. However, further research is needed to explore the impact of such a graphene-based cooling system on the cycling and rate performance of Li-ion batteries, particularly at the module level [26–28].

Figure 7.8 SEM images of the graphene film with low (a) and high (b) resolution; (c) roughness distribution image of graphene assembled film.

7.4 Advantages

The advantages of Battery Thermal Management Systems (BTMS) in Electric Vehicles (EVs) are multifaceted and profound, contributing significantly to the overall performance, safety, and longevity of EV batteries. Here are some key advantages [29–33]:

1. Enhanced Battery Performance: BTMS ensures that EV batteries operate within their optimal temperature range, maximizing their performance and efficiency. By maintaining consistent temperatures, BTMS helps prevent performance degradation and ensures that the battery delivers consistent power output, thereby enhancing the driving experience.
2. Extended Battery Lifespan: Proper thermal management provided by BTMS helps to mitigate the effects of thermal stress on battery cells. By minimizing temperature fluctuations and preventing overheating or undercooling, BTMS extends the operational lifespan of EV batteries. This leads to reduced maintenance costs and enhances the long-term reliability of EVs.
3. Improved Safety: One of the critical advantages of BTMS is its role in enhancing the safety of EVs. By preventing overheating, BTMS reduces the risk of thermal runaway, which can lead to battery fires or explosions. Additionally, BTMS ensures that batteries operate within safe temperature limits, reducing the likelihood of safety incidents during charging, discharging, or driving.
4. Optimal Charging Efficiency: BTMS plays a crucial role in optimizing the charging process of EV batteries. By maintaining the battery at the optimal temperature during

charging, BTMS helps to minimize charging time and maximize charging efficiency. This ensures that EV owners can charge their vehicles quickly and conveniently, enhancing the overall usability of EVs.

5. Adaptability to Environmental Conditions: EVs operate in diverse environmental conditions, ranging from extreme heat to freezing temperatures. BTMS allows EV batteries to maintain their performance and safety across a wide range of temperatures. Whether driving in hot climates or cold winters, BTMS confirms optimizing working efficiency and reliability.
6. Energy Efficiency: BTMS contributes to the overall energy efficiency of EVs by reducing energy losses associated with temperature variations. By minimizing energy wastage due to overheating or undercooling, BTMS helps to maximize the energy stored in the battery and improve the overall energy efficiency of EVs.
7. Enhanced Driving Range: By optimizing battery performance and efficiency, BTMS indirectly contributes to extending the driving range of EVs. With a properly managed thermal environment, EV batteries can deliver consistent power output over longer distances, allowing drivers to travel further on a single charge.
8. Customizable Cooling Strategies: BTMS offers flexibility in cooling strategies, allowing manufacturers to tailor thermal management solutions to the specific requirements of different EV models. Whether employing cooling strategies, BTMS can be customized to optimize performance, safety, and efficiency for each vehicle design.

In summary, BTMS provides numerous advantages that are essential for unlocking the full potential of electric vehicles. From enhancing battery performance and safety to extending lifespan and driving range, BTMS plays a crucial role in advancing the adoption and viability of electric mobility.

7.5 Challenges

Despite the significant progressions in the technology in (EVs), several challenges persist, hindering their widespread adoption and optimal performance. These challenges encompass various aspects of BTMS technology and implementation [34–37]:

1. Heat Dissipation Efficiency: Efficient heat dissipation is essential for preventing thermal runaway and ensuring the longevity of EV batteries. However, achieving optimal heat dissipation in BTMS remains a challenge, particularly in high-power EV applications where heat generation is substantial. Improving heat dissipation efficiency without compromising system complexity or adding excessive weight remains a persistent challenge in BTMS design.
2. System Complexity and Cost: BTMS designs often involve complex configurations of cooling components, control systems, and thermal interfaces. As a result, BTMS implementation can be costly and may require additional engineering expertise and resources. Balancing the performance benefits of sophisticated BTMS designs with cost considerations remains a challenge for EV manufacturers seeking to produce cost-effective vehicles without compromising thermal management effectiveness.
3. Weight and Space Constraints: EVs are subject to stringent weight and space constraints, necessitating compact and lightweight BTMS solutions. However, integrating BTMS components into the limited space available within EVs while minimizing weight can be challenging. Striking a balance between thermal management requirements and weight/space limitations is a key challenge for BTMS designers and manufacturers.
4. Material Durability and Reliability: BTMS components are exposed to harsh operating conditions, including temperature fluctuations, mechanical stress, and exposure to corrosive substances. Ensuring the durability and reliability of BTMS materials and components under such conditions remains a challenge. Improving the lifespan and performance of materials used in BTMS, such as heat exchangers, coolant hoses, and thermal interface materials, is essential for enhancing the reliability of EVs.
5. Seamless Integration with Vehicle Systems: BTMS must seamlessly integrate with other vehicle systems, including battery management systems (BMS), powertrain components, and vehicle control systems. Achieving seamless integration requires compatibility with existing vehicle architectures, standardized communication protocols, and effective coordination between different subsystems.

Ensuring interoperability and compatibility between BTMS and other vehicle systems poses a significant challenge for EV manufacturers and system integrators.

6. Thermal Management during Fast Charging: The increasing adoption of fast-charging technologies presents unique thermal management challenges for EV batteries. Rapid charging generates significant heat, which must be efficiently dissipated to prevent thermal degradation and ensure battery safety. Developing BTMS solutions capable of effectively managing heat during fast charging without compromising battery lifespan or charging efficiency is a complex challenge that requires innovative approaches and advanced thermal management strategies.
7. Environmental Impact and Sustainability: While BTMS plays a crucial role in improving the performance and safety of EV batteries, its environmental impact must also be considered. Manufacturing BTMS components, including coolant fluids, thermal insulation materials, and heat exchangers, may involve the use of energy-intensive processes and environmentally harmful substances. Addressing the environmental impact of BTMS and promoting sustainable practices in BTMS design and manufacturing is a challenge for the EV industry as it strives to reduce its carbon footprint and environmental impact.

Addressing these challenges requires a collaborative effort involving EV manufacturers, researchers, policymakers, and industry stakeholders. By identifying and overcoming these challenges, the EV industry can further advance the development and adoption of BTMS technologies, leading to more efficient, reliable, and sustainable electric vehicles.

7.6 Recent Trends in Battery Thermal Management Systems (BTMS)

Recent trends in (BTMS) reflect the ongoing advancements aimed at enhancing battery performance, safety, and longevity. Several notable trends have emerged in BTMS technology, catering to the evolving needs of the electric mobility sector [38–41]:

1. Integration of Advanced Materials: Recent developments in BTMS focus on leveraging advanced materials to improve thermal conductivity and efficiency. Nanomaterials, such as graphene and carbon nanotubes, are being explored for their superior heat transfer properties, enabling more effective cooling of EV batteries. Additionally, the integration of high-performance composite materials in BTMS components helps to reduce weight and improve durability, thereby enhancing overall system efficiency.
2. Adoption of Liquid Cooling Systems: Liquid cooling systems are increasingly favored in EVs due to their superior heat dissipation capabilities compared to air cooling. Recent trends involve the adoption of liquid cooling systems with advanced heat exchangers and optimized coolant flow paths, enabling more efficient thermal management while minimizing system complexity and weight. Liquid cooling systems are particularly well-suited for high-power EV applications, where precise temperature control is essential to maximize battery performance and longevity.
3. Development of Intelligent Control Systems: The integration of intelligent control systems in BTMS represents a significant trend in recent years. These systems utilize Ai for optimizing thermal management in real-time. By continuously monitoring battery temperature, environmental conditions, and driving patterns, intelligent control systems can dynamically adjust cooling strategies to ensure optimal battery performance and safety. This trend aligns with the broader industry shift towards autonomous and connected vehicle technologies, enhancing the efficiency and reliability of BTMS in EVs.
4. Hybrid Cooling Solutions: Hybrid cooling solutions, including (PCMs), are gaining traction in the EV industry. These hybrid systems leverage the strengths of each cooling method to achieve more balanced and efficient thermal management. For example, integrating liquid cooling with PCM-based thermal storage can provide passive temperature regulation while minimizing energy consumption. Hybrid cooling solutions offer versatility and adaptability to diverse EV applications, providing manufacturers with flexibility in designing BTMS that meet specific performance and cost requirements.

5. Integration of Sustainable Solutions: With a growing emphasis on sustainability, recent trends in BTMS include the integration of eco-friendly and energy-efficient solutions. Biodegradable phase change materials (PCMs) and recyclable BTMS components are being developed to minimize environmental impact and support the circular economy. Additionally, energy-efficient cooling techniques, such as thermoelectric cooling and waste heat recovery, are being explored to reduce power consumption and improve overall system efficiency. These sustainable solutions align with the broader industry goal of reducing carbon emissions and promoting environmental stewardship in electric mobility.

7.7 Conclusion

In conclusion, recent trends in BTMS in EVs emphasize advancements in materials, cooling technologies, control systems, and sustainability. These trends reflect the ongoing efforts to optimize thermal management, enhance battery performance. As the power-driven mobility sector continues to evolve, continued research and innovation in BTMS will play a crucial role in shaping the future of sustainable transportation.

In conclusion, BTMS stands at the forefront of advancing electric mobility by ensuring the efficient operation and longevity of EV batteries. Through continuous research, collaboration, and innovation, BTMS will play an unmatched role in modeling the sustainable future of transit.

References

1. Kim, J., Oh, J., Lee, H., Review on battery thermal management system for electric vehicles. *Appl. Therm. Eng.*, 149, 192–212, 2019, ISSN 1359-4311, https://doi.org/10.1016/j.applthermaleng.2018.12.020.
2. Venkatakrishnan, S., Sudhan V.M., V., Kandappan S., S., Vishwanath, S., Saravanan, S., Pandiyan, P., Battery Thermal Management System. *2020 6th International Conference on Advanced Computing and Communication Systems (ICACCS)*, pp. 877–882, 2020, doi: 10.1109/ICACCS48705.2020.9074378.
3. Pesaran, A.A. and Burch, S.D., Thermal Performance of EV and HEV Battery Modules and Packs. *National Renewable Energy Laboratory (U.S.)*

International Electric Vehicle Symposium, 1997, https://worldcat.org/title/40513130.

4. Aruna, P. and Vasan, P.V., Review on Energy Management System of Electric Vehicles. *2019 2nd International Conference on Power and Embedded Drive Control (ICPEDC)*, pp. 371–374, 2019, doi: 10.1109/ICPEDC47771.2019.9036689.
5. Pesaran, A.A., Battery Thermal Management in EVs and HEVs: Issues and Solutions. *Battery Man*, 43, 5, 34–49, 2001.
6. Liu, Y., Thiringer, T., Wang, N., Fu, Y., Lu, H., Liu, J., Graphene based thermal management system for battery cooling in electric vehicles. *2020 IEEE 8th Electronics System-Integration Technology Conference (ESTC)*, pp. 1–4, 2020, doi: 10.1109/ESTC48849.2020.9229856.
7. Pattnayak, R.A. and Vijay, T., Thermal Analysis of Cell Balancing for Battery Management System in Electric Vehicle. *2020 26th International Workshop on Thermal Investigations of ICs and Systems (THERMINIC)*, pp. 1–7, 2020, doi: 10.1109/THERMINIC49743.2020.9420510.
8. Ranawat, D. and Prasad, M.P.R., A Review on Electric Vehicles with perspective of Battery Management System. *2018 International Conference on Electrical, Electronics, Communication, Computer, and Optimization Techniques (ICEECCOT)*, pp. 1539–1544, 2018, doi: 10.1109/ICEECCOT43722.2018.9001321.
9. Rashid, A., Hofman, T., Rosca, B., Enhanced Battery Thermal Management Systems with Optimal Charge Control for Electric Vehicles. *2018 Annual American Control Conference (ACC)*, pp. 1849–1854, 2018, doi: 10.23919/ACC.2018.8431216.
10. Venugopal, V.K., Chandrasekaran, N., Sharma, S., System Level Simulation Technique for Optimizing Battery Thermal Management System of an Electric Vehicle. *2019 IEEE Transportation Electrification Conference (ITEC-India)*, pp. 1–5, 2019, doi: 10.1109/ITEC-India48457.2019.ITECINDIA2019-97.
11. Zhang, X., Li, L., Zhang, W., Review of Research about Thermal Runaway and Management of Li-ion Battery Energy Storage Systems. *2020 IEEE 9th International Power Electronics and Motion Control Conference (IPEMC2020-ECCE Asia)*, pp. 3216–3220, 2020, doi: 10.1109/IPEMC-ECCEAsia48364.2020.9367968.
12. Subramanian, P. and Rengaswamy, R., Thermal Management of Lithium-Ion Batteries in Electric Vehicles: A Review. *Int. J. Sustain. Energy*, 40, 9, 873–890, 2021.
13. Sharma, A. and Sharma, S., Recent Advances in Thermal Management Systems for Electric Vehicle Batteries: An Indian Perspective. *Sustain. Energy Technol. Assess.*, 41, 100820, 2020.
14. Gupta, R. and Singh, R., Advances in Thermal Management Systems for Electric Vehicle Batteries: Insights from Indian Research. *J. Power Sources*, 437, 226923, 2019.

15. Patel, K. and Patel, D., Thermal Management Techniques for Lithium-Ion Batteries in Electric Vehicles: A Review of Indian Research. *Energy Rep.*, 4, 480–491, 2018.
16. Kumar, S. and Jain, A., Recent Developments in Thermal Management Systems for Lithium-Ion Batteries in Electric Vehicles: An Indian Perspective. *Int. J. Energy Res.*, 41, 10, 1393–1406, 2017.
17. Zhang, X. and Xu, C., Advanced Battery Thermal Management Systems for Electric Vehicles. *J. Energy Storage*, 45, 103635, 2021.
18. Lee, J. and Kim, S., Nanomaterials in Battery Thermal Management Systems: Recent Advances and Future Perspectives. *Adv. Funct. Mater.*, 30, 25, 2000362, 2020.
19. Wang, Y. and Zhao, J., Thermal Management in Battery Systems for Electric Vehicles: Current Status and Future Outlook. *Renew. Sustain. Energy Rev.*, 115, 109364, 2019.
20. Zhang, H. and Mi, C., Recent Advances in Battery Thermal Management Systems for Electric and Hybrid Vehicles. *IEEE Trans. Veh. Technol.*, 67, 11, 10786–10799, 2018.
21. Zhang, L. and Liu, J., Review on Thermal Management Systems for Lithium-Ion Batteries in Electric Vehicles. *Energy Procedia*, 105, 3751–3756, 2017.
22. Dong, G. and Liu, Y., Recent Advances in Thermal Management of Battery Systems in Electric Vehicles. *Int. J. Heat Mass Transfer*, 100, 1–12, 2016.
23. Lee, C. and Yang, L., A Review of Recent Developments in Battery Thermal Management Systems of Electric and Hybrid Electric Vehicles. *Energy Convers. Manage.*, 96, 103–115, 2015.
24. Zhang, Y. and Wang, J., Recent Advances in Thermal Management Systems for Lithium-Ion Batteries in Electric Vehicles. *J. Power Sources*, 480, 228927, 2021.
25. Kim, S. and Park, Y., Advances in Thermal Management of Lithium-Ion Batteries for Electric Vehicles. *Appl. Energy*, 279, 115853, 2020.
26. Chen, Z. and Hu, L., Recent Progress in Thermal Management Systems for Lithium-Ion Batteries in Electric Vehicles: A Review. *J. Energy Chem.*, 30, 1–14, 2019.
27. Wang, Q. and Li, H., Advances in Thermal Management of Lithium-Ion Batteries for Electric Vehicles: A Review. *Energy Storage Mater.*, 14, 60–72, 2018.
28. Li, X. and Zhang, Y., Advances in Thermal Management of Lithium-Ion Batteries for Electric Vehicles: A Review. *Energy Procedia*, 128, 545–550, 2017.
29. Zhang, L. and Liu, H., Recent Advances in Thermal Management of Lithium-Ion Batteries for Electric Vehicles. *Appl. Therm. Eng.*, 103, 1197–1208, 2016.
30. Liu, Y. and Sun, X., A Review on Thermal Management Systems for Lithium-Ion Batteries in Electric Vehicles. *Int. J. Energy Res.*, 39, 11, 1475–1499, 2015.
31. Park, J. and Lee, S., Recent Advances in Thermal Management Systems for Electric Vehicle Batteries. *Energy Rep.*, 7, 2765–2777, 2021.

32. Kim, H. and Choi, S., Thermal Management Systems for Electric Vehicle Batteries: Recent Developments and Future Perspectives. *Sustain. Energy Technol. Assess.*, 42, 100867, 2020.
33. Li, X. and Wang, Y., Advanced Thermal Management Systems for Lithium-Ion Batteries in Electric Vehicles: A Review. *Energy Rep.*, 5, 1053–1069, 2019.
34. Zhang, H. and Hu, L., Recent Advances in Thermal Management Systems for Electric Vehicle Batteries. *Renew. Energy*, 115, 895–907, 2018.
35. Wang, Q. and Liu, Y., Advances in Thermal Management Systems for Electric Vehicle Batteries: A Review. *Energy Procedia*, 142, 287–292, 2017.
36. Chen, Z. and Zhang, L., Recent Advances in Thermal Management Systems for Electric Vehicle Batteries: A Review. *Int. J. Energy Res.*, 40, 9, 1213–1229, 2016.
37. Zhao, Y. and Li, H., Thermal Management Systems for Electric Vehicle Batteries: Recent Advances and Future Directions. *J. Power Sources*, 293, 801–815, 2015.
38. Ravindran, M.A., Nallathambi, K., Vishnuram, P., Rathore, R.S., Bajaj, M., Rida, I., Alkhayyat, A., A novel technological review on fast charging infrastructure for electrical vehicles: Challenges, solutions, and future research directions. *Alexandria Eng. J.*, 82, 260–290, 2023, 10.1016/j.aej.2023.10.009.
39. Kasri, A., Ouari, K., Belkhier, Y., Bajaj, M., Zaitsev, I., Optimizing electric vehicle powertrains peak performance with robust predictive direct torque control of induction motors: A practical approach and experimental validation. *Sci. Rep.*, 14, 1, 14977, 2024, 10.1038/s41598-024-65988-0.
40. Rekioua, D., Mokrani, Z., Kakouche, K., Rekioua, T., Oubelaid, A., Logerais, P.O., Ghoneim, S.S., Optimization and intelligent power management control for an autonomous hybrid wind turbine photovoltaic diesel generator with batteries. *Sci. Rep.*, 13, 1, 21830, 2023, 10.1038/s41598-023-49067-4.
41. Singh, A.R., Vishnuram, P., Alagarsamy, S., Bajaj, M., Blazek, V., Damaj, I., Othman, K.M., Electric vehicle charging technologies, infrastructure expansion, grid integration strategies, and their role in promoting sustainable e-mobility. *Alexandria Eng. J.*, 105, 300–330, 2024, 10.1016/j.aej.2024.06.093.

8

Electric Vehicle Charging and Discharging in V2G and G2V Operation

Versha Sharma[1], Kalpana Chauhan[1]*, Rajeev Kumar Chauhan[2] and Anju Saini[3]

[1]*Department of Electrical Engineering, Central University of Haryana, Mahendragarh, India*
[2]*Department of Electrical Engineering, Dayalbagh Educational Institute (Deemed to be University), Agra, India*
[3]*Department of Mathematics, Graphic Era (Deemed to be University), Dehradun, India*

Abstract
The main objective is leveraging the electric vehicle (EV) batteries as energy storage devices within the micro- grid, enabling them to store surplus energy from the grid during low demand periods grid to vehicle (G2V) and supply energy to grid when there is peak demand for vehicle to grid (V2G). To accomplish this goal, the chapter suggests the appropriate development of infrastructure and control systems. To interact with the EVs, it suggests a micro-grid test system that includes a DC quick-charging station. Next, a simulation of the system is performed to demonstrate the off V2G and G2V's power transfer capabilities. The test findings show that the EV batteries can use both modes of operation to actively control the power inside the micro-grid. This implies that EV batteries are assist in balancing the energy within the micro-grid by absorbing excess power or providing additional power as needed. The design of the charging station ensures minimal harmonic distortion of the grid's injected current, which is crucial for maintaining the stability and quality of the grid. Moreover, the controller employed in the system exhibits satisfactory dynamic performance, particularly in terms of preserving

**Corresponding author*: kal_2312@rediffmail.com; ORCID: 000-0003-4549-8167
Versha Sharma: ORCID: 0009-0004-1774-4716
Rajeev Kumar Chauhan: ORCID: 000-0003-1357-0152
Anju Saini: ORCID: 0000-0001-7699-1187

Ashwani Kumar, Mukesh Kumar Awasthi, Nitesh Dutt, Yogesh Kumar Singla and Sivasakthivel Thangavel (eds.) Thermal Battery Management System for Hybrid and Electric Vehicles, (157–180) © 2025 Scrivener Publishing LLC

stability in the DC bus voltage. In summary, the chapter introduces and demonstrates a V2G and G2V system architecture in micro-grid employing charging of EVs. The simulation studies and test outcomes illustrate the potential advantages of utilizing EV batteries for energy management within micro-grids. They emphasize the active power regulation capabilities and stability of the system.

Keywords: Surplus energy, grid stability, dynamic performance, dc bus voltage stability, energy management, active power regulation

8.1 Introduction

The renewable energy sources like solar and wind are excellent for the environment, they are not always available. Therefore, continue to rely on fossil fuel energy to make up for the unpredictable power of solar and wind power, and must store it so that can utilize it anytime need to. Electric cars are one unanticipated source of a solution [1]. Through a tried-and-true technique known as vehicle-to-grid (V2G), EVs not only cut harmful emissions but may also assist stabilize the electrical system. The world uses this technology, which was created more than 20 years ago. With V2G, a vehicle's battery may store energy, including renewables, and safely discharge some of the power back to the grid when it is most required, as opposed to only drawing electricity from the grid [2]. When energy demand is low throughout the course of the night, a fleet, such as a line of electric bus, might be charged, allowed to operate according to schedule, and then discharged power back into the grid when parked. Technology also enables the creation of the virtual power plant by using batteries off several different vehicles [3]. This power can then be utilized to provide so-called utility services during peak hours that aid in system stabilization and avoid blackouts, or it can be traded on energy markets. By charge when electricity prices are low and having the ability to use cheap energy store within the EV batteries to even power a building, V2G also helps fleet owners save money [4]. All of this is carried out while considering battery health and making sure that a car has enough juice to carry out its regular driving tasks. By using their batteries while parked and not in use, V2G makes money and lowers the total cost of buying an electric car.

8.2 Electric Vehicle

An electric motor, as opposed to a traditional internal combustion engine that burns petrol or diesel, powers an EV (refer Figure 8.1). Motors powered

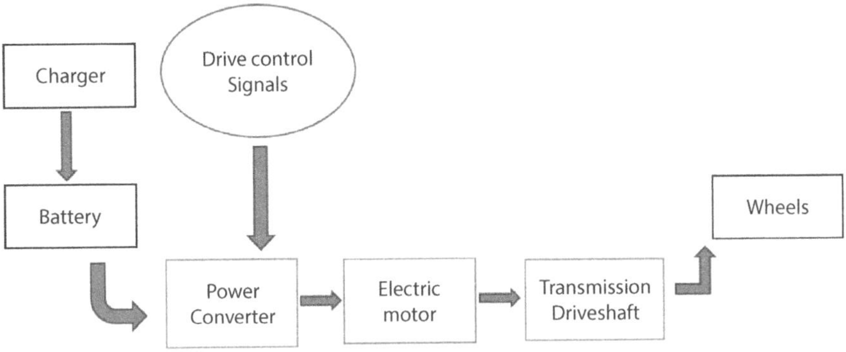

Figure 8.1 Electric vehicles.

by electricity and rechargeable batteries are used in electric cars to power the vehicle's propulsion system and other features.

Electric vehicles provide several advantages over traditional gasoline-powered cars [5]. Their tailpipe emissions are minimized or non-existent, which adds to a decrease in both pollution in the air and the production of greenhouse gases [6, 7]. In an effort to reduce dependency on fossil fuels and mitigate adverse environmental impacts, several global governments and automakers are promoting the use of electric vehicles. The Electric motors produce immediate torque, enabling swift acceleration and comfortable driving [8]. Due to the well distributed weight of the battery pack, EVs can achieve outstanding levels of performance, including quick acceleration and superior handling.

8.3 Electric Vehicle Charging System for V2G

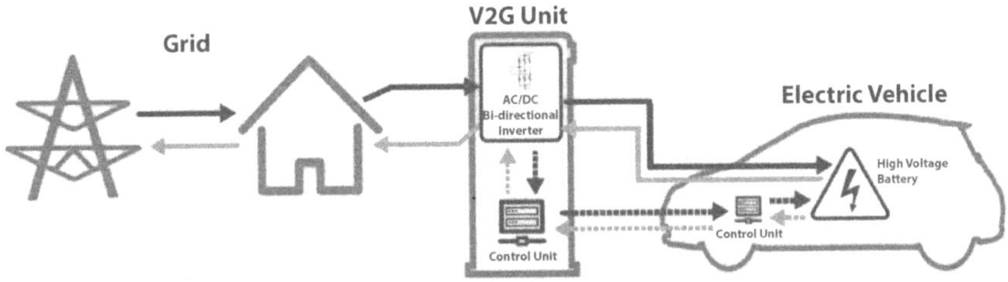

Figure 8.2 Schematic representation of V2G (source: https://www.cleantech.com/ev-charging-software-and-grid-services/, retrieved on 23.12.2019).

The vehicle to grid technology creates an option for EVs for power exchange V2G and G2V in both directions [9]. A few elements and things to think about are needed to put in place a V2G charging infrastructure (Figure 8.2). An outline of the main components at play is as follows:

8.3.1 Electric Vehicle (EV)

Start with an electric car that is capable of both charging from and discharging power into the grid (V2G). Usually, these cars are capable of power flow in both directions (i.e. charging and discharging).

8.3.2 Charging Infrastructure

In order to enable bidirectional energy flow between the EV and the grid, a charging station or charger meant for V2G functionality is required [10]. These chargers typically include extra hardware and communication features.

8.3.3 Communication Protocol

The Open Charge Point Protocol (OCPP) is one such standardized communication protocol that is required for smooth communication between the grid, the EV, and the charging infrastructure. The data and control signal interchange needed for V2G operations is made possible by OCPP.

8.3.4 Grid Interaction

The ability to connect and interact with the electrical grid is a requirement for the V2G charging technology [11]. Through this connection, the system may react to grid signals and return electricity to the grid when there is a spike in demand or when the grid needs more power.

8.3.5 Energy Management System

The task of coordinating and maximizing the electrical power flow between the EVs, the charging infrastructure, and the grid falls to an energy management system (EMS). The EMS controls EV charging and discharging by keeping an eye on grid demand [12]. It makes sure that the energy storage in EVs is utilized effectively while taking the owner's preferences and limitations into account.

8.3.6 Grid Services and Compensation

V2G-capable EVs can offer a range of grid functions, including load balancing, peak shaving, and frequency management. These services contribute to the grid's overall dependability and stabilization [13–16]. EV owners may be compensated for the energy they give to the grid in exchange, which could provide incentives for taking part in V2G schemes.

8.3.7 Regulatory Considerations

Standards compliance and regulatory considerations may arise during the implementation of V2G systems [17]. When implementing V2G infrastructure, it's critical to comprehend local laws, utility policies, and any applicable interconnection requirements.

8.4 V2G Technology

Essentially an electric vehicle is in the battery of wheels as it's a battery [18]. It can take electricity from the grid at times of low demand (i.e. low price) store that energy and then provide their energy back to the system at times of high demand (i.e. high price) [19, 20] and the consumer or the driver can get paid for doing that.

- The Vehicle-to-Grid (V2G) technology creates new possibilities for smart energy management and energy trading.
- V2G technology makes it possible for fleets and families to help the grid during times of high demand by releasing energy from electric car batteries.
- V2G enables utilities to use the renewable energy that is stored in moving cars to better coordinate loads and reduce grid congestion.

8.4.1 Working of V2G

A two-way charger is used to connect the battery packs in EV to electrical grid [21]. They keep track with tremendous regularity to determine if the supply and demand for power are in balance. Conversely, when supply exceeds demand, the frequency rises, and the REVS cells begin to charge to take on part of the additional load.

- The energy flow for charging electric vehicles is typically "Grid-to-Vehicle" in one direction. Electric vehicle batteries are used just to power vehicles.
- EV charging becomes a two-way process by bidirectional V2G charging stations. Fully charged cars have the capacity to feed stored power back into the grid at times when the demand for electricity on the system spikes.
- EV owners paying a price to the grid in exchange for the utility's services, making a profit of their EV capital while enforcing reasonable restrictions to guarantee they constantly have the battery life they require. Technology that connects vehicles to the grid is relatively new. It includes EVs that can be connected to the network, utilities or electrical systems, and information technology. It also includes EVs that are grid-capable [22, 23]. As a result, since batteries are readily available and electric vehicles can connect to the network, it is feasible to use the batteries to generate power and transfer energy from the grid to the car's battery while it is charging. This is because, as everyone is aware, electric motor vehicles use batteries as energy storage systems. To benefit both consumers of electric vehicles and the electrical grid for the power system, this technology enables a combination of energy flow and information flow (Figure 8.3).

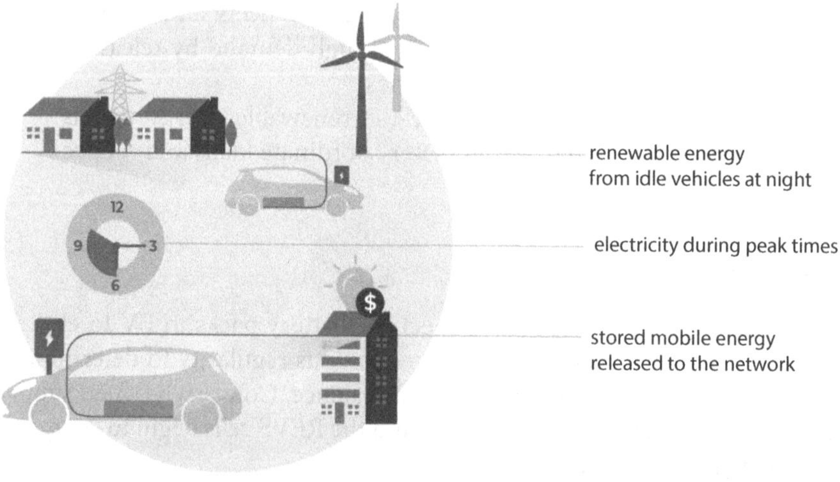

Figure 8.3 EV vehicle-to-grid.

8.4.2 Background of (V2G) Technology

"Bidirectional power flow is made possible by the innovative concept of V2G technology, which enables EVs to connect to the electrical grid. EVs with V2G technology can perform as mobile energy storage devices in addition to being a means of mobility, offering beneficial grid services [24]. This technique could lower greenhouse gas emissions, encourage the integration of renewable energy sources, and improve grid stability. This chapter gives a thorough introduction to V2G technology, going into its theoretical foundation, control schemes, and possible uses. It is an invaluable tool for comprehending the design and execution of V2G systems.

8.4.3 Advantages of Vehicle-to-Grid

Services: V2G aids in grid balancing and seamlessly integrating renewable energy sources, allowing services to get less reliant on power plants that burn fossil fuels. Utility companies may achieve their sustainability goals with the use of V2G technology.
EV Operator: V2G solutions are anticipated to be a profitable feature for utilities, therefore they have a strong motive to encourage user participation. If customers allow the utility to utilize their batteries for V2G, they will receive a reward. The overall cost of ownership will go down therefore.
Owners of EV Fleets: By returning electricity to the grid, fleet owners may generate income with their fleet and save their energy expenses by becoming more energy-active consumers. Continuous functioning is ensured, even though the accumulated energy can be utilized in the event of a power loss.
Residences and Businesses: V2G aids in the long-term preservation of clean energy so that you may use it again whenever you decide it is appropriate.

With V2G, there is no need to use more energy from the grid to balance off the short-term surges in power use in the building.

8.4.4 A New Business System in V2G

It necessitates communication between EVs and the grid operator so that we may regulate not just an energy flow from the power source to the grid but also the rate at which the batteries charge. Any car, especially an electric vehicle in this scenario, is known to be 90% idle, therefore parking it there will be excellent [25]. Therefore, if we can see all of these battery

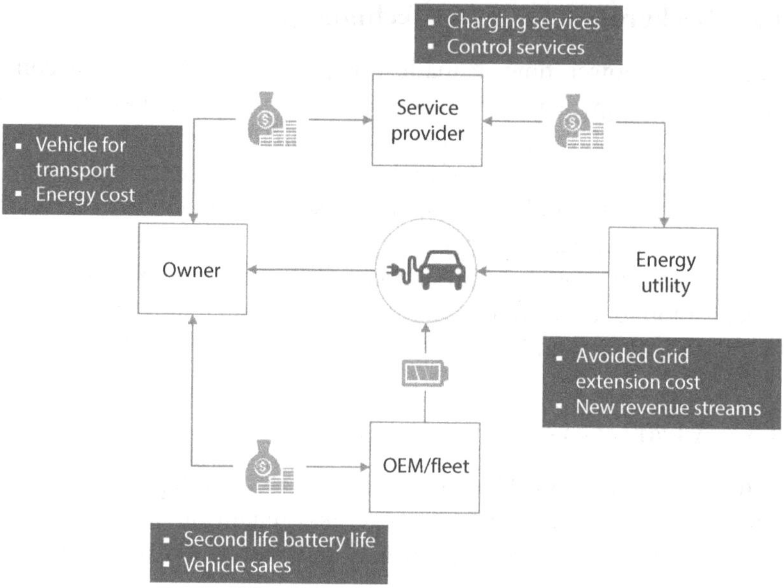

Figure 8.4 V2G is a new business model.

packs, if they are linked to the grid when they remain parked, and using the assumption that each EV can normally store anywhere between 4 and 80 kilowatt hours, we can determine that if they occupied a significant portion of the total number of cars on the road. So, this is the coordination of this can be a new business, and is a new opportunity, and a new technology, which can be under development. But now the system also must take care of the power system dynamics, and stability. So, if you bring a new generator those dynamics is different from the present grid, we must start studies such that it understands these systems to the present utility grid. So as mentioned earlier if individual EVs do the V2G action and try to connect to the grid, it will be very ineffective, because EVs can have maximum power of 85 kilowatt, but that is always on the high-end models, but typically these averages between 10 to 15 kilowatt. So, if we operate such a system, it will be very ineffective and inefficient (Figure 8.4).

8.4.5 Need for Power Aggregators

It is also advised to create a pool of EVs and make a common power aggregator, means the agency which can plug this pool of EVs and present a total

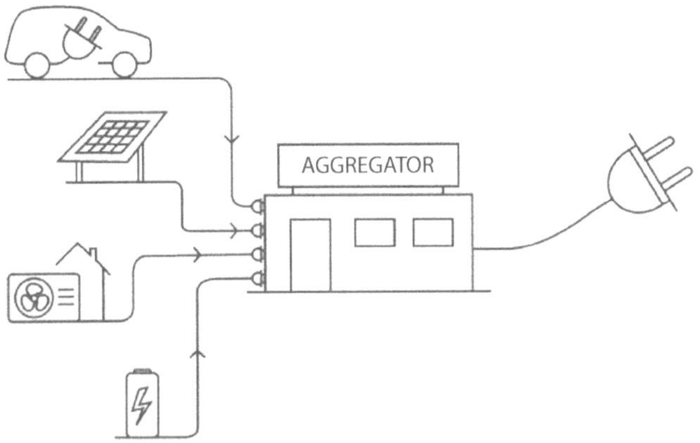

Figure 8.5 Power aggregators.

system as a one system to the grid. In this way it will be meaningful, and it will be efficient [26]. So the power aggregator must take care of internal energy flow such that it minimizes the power demand and losses. Means, some of the vehicle's mail like to charge slowly, some EVs may like to charge using fast charging, but their maybe many vehicles that are ready to generate power. So, the power aggregator should understand the requirement of the pool of EVs, such that it understands the typical power demand, which to be supported from the grid on it need can generate to the grid, such that the losses are minimum, or efficiency is maximum (Figure 8.5).

8.4.6 V2G Application in Power Systems

Now, if lot of aggregators have connected to grid it is a power generator system which can take energy capable of storing energy and capable of delivering energy. So, if such a system is connected to power systems, it can support power system in many ways. So, if the power generate generation always try to match the load demand, but there will be lot of fluctuations between the peak hours during the day and the off-peak hours during the night [27]. So the V2G operation can help the grid by charging under coordination during off peak hours or during night and it can support the grid by energy generation during daytime to support the peak load demand, or it is called peak shaving.

It can also support regulation of the grid voltage. Spinning reserve, so basically there are the systems connected to the grid, which are dormant, and they are started only during demand. So typically, these are stand-by generators or diesel generators, which are started, when there in a peak demand. So, the installation of such a system is expensive and as you know, very in efficient, and basically it takes lot of time to start up because it is mechanical system. So, if EVs under V2G application can support this by you know quickly supporting the peak demand in a very high dynamic fashion. Secondly, it will not be additional burden on the power supply system, and it will be correspondingly less expensive. The motor starting and reactive regulation are some of the areas, which is also supported by you know synchronous condenser and such kind of things in power grid. So, this demands a lot of reactive power. If the power is used as converter in these applications, it typically has an electrolytic capacitor in the DC link. This DC link capacitor is a reactive source of energy. It is possible to use the reactive power capability of all these EVs and can help the grid to support reactive power compensation. A lot of renewable power is getting added to the grid, like wind and solar, but these sources of energy intermittent in nature. It depends on solar irradiance or the wind availability. There is also this V2G and the support of EVs can be utilized such that these transients can be supported in a high dimension fashion.

8.5 Grid to Vehicle Technology

Due to the bidirectional power flow made possible by G2V technology, EVs can both draw power from and provide power to the grid. With this two-way communication and power exchange, EVs can operate as distributed energy resources and support grid stability in vehicle-to-grid (V2G) scenarios.

8.5.1 Grid-to-Vehicle (G2V) Operation

Due to the bidirectional power flow made possible by G2V technology, EVs can both consume power from and contribute to the grid. Grid-to-Vehicle (G2V) operation is made possible by this two-way communication and power exchange, allowing EVs to function as distributed energy resources and support grid stability.

8.5.2 Grid Services

EVs can offer a range of grid services, including peak shaving, load balancing, and frequency management, by utilizing G2V technology.

By responding collectively to grid needs and variations, electric vehicles (EVs) connected to the grid can function as virtual power plants, enhancing grid stability and minimizing the need for additional infrastructure.

8.5.3 Energy Storage

When renewable energy supply outpaces demand, or during off-peak hours, EVs with G2V capability can store excess electricity. By reintroducing this stored energy into the grid, conventional power plants will be less stressed, and the integration of renewable energy sources will be encouraged.

8.5.4 Demand Response

Demand response capabilities made possible by G2V technology allow grid operators to provide incentives to electric vehicle (EV) owners to modify their charging or discharging schedules in response to grid conditions. This flexibility lowers the need for fossil fuel-based power generation during peak loads and increases the overall efficiency of the grid.

8.5.5 Cost Savings and Revenue Generation

Through G2V systems, EV owners can potentially benefit from cost savings and revenue generation opportunities. For instance, they can charge their vehicles when electricity surplus power. Additionally, participation in grid services programs may provide financial incentives to EV owners.

8.5.6 Grid Resilience

G2V technology integrates a dispersed network of energy resources to improve the electrical grid's resiliency. EVs with G2V capability can be used as emergency power sources in the case of a power outage or natural disaster, providing electricity to vital infrastructure or aiding nearby communities. G2V technology implementation is still in its infancy; research, pilot programs, and standardization initiatives are still under progress. G2V technologies are improving grid operations, incorporating renewable energy sources, and fostering a more resilient and sustainable energy ecosystem as the use of electric vehicles rises.

8.6 Comparison Between V2G & G2V

8.6.1 Direction of Energy Flow

- V2G (Vehicle-to-Grid) energy flows from the EV battery back to the electrical grid. EVs act as mobile energy storage units, allowing surplus energy stored in the vehicle's battery to be fed back to the grid.
- G2V (Grid-to-Vehicle) energy flows from the electrical grid to the EV battery. EVs are charged from the grid, replenishing the energy stored in the battery.

8.6.2 Objective

- **V2G** - The primary goal of Vehicle-to-Grid is to enable bi-directional energy flow and use EV batteries as a source of grid support. The excess energy in EV batteries can be utilized during peak demand periods or to provide ancillary services like frequency regulation and load balancing to stabilize the grid.
- **G2V** - The goal of grid-to-vehicle charging is to use the electrical grid to recharge EV batteries. The primary goal is to supply EVs with power so they have enough energy to function.

8.6.3 Benefits

V2G (Vehicle-to-Grid) offers several benefits, including:

- **Grid Stabilization** - By harnessing the energy contained in EV batteries, V2G helps control peak loads, lessen the burden on the grid, and enhance system stability [28].
- **Renewable Energy Integration** - To facilitates the integration of intermittent green energy sources, such as solar and wind power, by storing excess energy and returning it to the system when needed.
- **Cost Savings** - By selling excess energy back to the system, EV owners can profit and lower the total cost of car ownership.

G2V (Grid-to-Vehicle) provides the following advantages:

- **Convenience:** By charging EVs from the grid, drivers can easily replenish the energy in their vehicles and extend their driving range without relying solely on internal combustion engines.
- **Energy Management:** G2V enables better load management by scheduling charging during off-peak hours, maximizing the utilization of available grid capacity [29].

8.6.4 Application

- **V2G** - Vehicle-to-Grid finds application in scenarios where grid stability and demand management are critical, such as in smart grids, micro-grids, and areas with high renewable energy penetration.
- **G2V**- Grid-to-Vehicle is applicable in various contexts, including residential, commercial, and public charging infrastructure, where EVs are charged from the grid to meet transportation needs.

8.7 Modeling

8.7.1 Modeling of V2G

The electrical circuit of the studied system is illustrated by Figure 8.6 and Figure 8.7. It consists on full-bridge bidirectional ac-dc power converter. There are an inductance L, switches (K_1, K_2, K_3, K_4) and a capacitor C. Table 8.1 shows the specification to simulate.

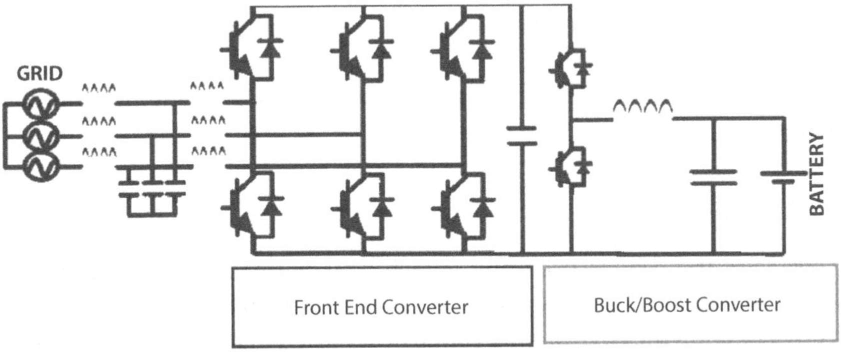

Figure 8.6 Circuit topology and SPEC.

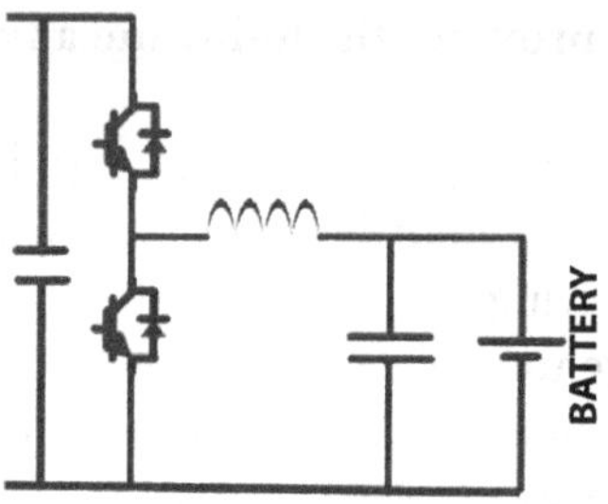

Figure 8.7 Control block.

Table 8.1 Specification of the system.

GRID Voltage = 415 V RMS, 50Hz
Inductance of the Filter = 3Mh
Capacitance = 30uF
Capacitance of Bus = 5600uF
Inductance of Buck Filter = 20mH
Output Capacitance = 0.625mF
Battery Nominal Voltage = 360V
Switching Frequency of both the converter = 10 KHz
Total Rated Power = 10KW

This is the fundamental circuit topology of a three-phase electric vehicle battery charger. At this point, an active rectifier or front converter converts AC voltage to DC. Next, a bite-directional Buck/Boost converter is used to regulate the current through the battery during the charge and discharge operations.

This controller is used to regulate the battery charging and discharging current

8.7.2 Explanation

Nowadays, the vehicle used only which runs with the petrol and diesel which increases more greenhouse gases into the environment but indirectly polluting the environment. So in order to avoid this type of problems there is introduction of the new type of vehicles which will decrease the pollution level and greenhouse gases into the atmosphere. Electrical vehicles are the best options to meet this requirement but to run the electrical vehicle need to charge the vehicle battery and to charge the battery need power. So from where this power is to be taken -to batteries or to home applications [30, 31]. Power from the grid actually sometimes it will act as a load and sometimes it will act as a source. What happen if a generation station like any power generation like thermal power plant, neutral power plant and solar generation anything. If take this in the generation station first the power must be go to grid then after from grid we are taking the power to our home applications like electrical vehicle, fan, lights etc. So here in first condition generate station to grid so here power is generating from the generating station and the power is taken by the grid the generation station generate power that's why it will act as a source and grid will take the power so that which will act as a load coming to the second condition here power is transferring from grid to loads.

So, here grid will act as a source because it gives power to the loads so the battery and remaining loads are act as a load and this great tag a source that why grid will sometime act as a load and sometime act as a source. [32] there is no need this one ever generating station just investigate the vehicle to grid operation only so hair this is power from grid to our electrical vehicle. Whenever, it is needed that mean if anyone take the grid not only one load will be there are so many loads are connected right so many vehicles also connected. An electric vehicle with sufficient battery power can supply the grid when the grid requires additional power [33, 34]. That's why because some large electrical vehicles can connect to the grid more than 10 to 15 hours mean they need to charge that much time that's why these electric vehicles are useful for the grade power management. Here when battery is charged by the grid that is grid to vehicle operation so whenever battery give power back to that operation is known as vehicle to grid vibration now [35, 36]. If you look at these two actual all batteries are in the DC form all grade are in AC nature so the buck-boost converter here while because the battery voltages are very low and the grid voltage is very high so to need to set up this DC voltage to high voltage [37, 38] a pulse converter needed when the battery power is flowing to the grid to boost up the voltage right when power is flowing from grip to vehicle mean grid to vacuum operation here grid voltages are very high our battery voltage are very low so, buck converter is used to step down the voltage his in the circuit diagram of vehicle to grid and grid to vehicle technology.

8.7.3 Simulation Model of Vehicle-to-Grid

Taking into consideration the two-arm full-bridge boost rectifier shown in Figure 8.6. Two switches make up each arm, and they must work in tandem with one another. A PWM circuit generates the switching signal u, which accepts values from the finite set {0,1}. With the use of Kirchhoff's principles and the knowledge that u can have binary values of either 1 or 0, the instantaneous model of the system under study is provided by the following equations.

$$\begin{Bmatrix} L\dfrac{di}{dt} = -ri - (2u-1)v_{dc} + v \\ C\dfrac{dv_{dc}}{dt} = (2u-1)i - i_{dc} \end{Bmatrix} \tag{8.1}$$

where i_{dc} is the dc-dc power converter's input current and v is the power grid voltage. It is more practical to take into consideration the averaged model that follows, which is produced by averaging model (1) over a single switching time, for control design purposes.

$$\begin{Bmatrix} \dot{x}_1 = -\dfrac{r}{L}x_1 - \dfrac{(2\eta-1)}{L}x_2 + \dfrac{v}{L} \\ \dot{x}_2 = \dfrac{(2\eta-1)}{C}x_1 + \dfrac{\overline{\iota}_{dc}}{L} \end{Bmatrix} \tag{8.2}$$

Where, x_1 and x_2 denote the average grid current $\overline{\iota}_{dc}$ and the average dc-bus voltage $\overline{v}_{dc}$, respectively. The control input for the model Equation (8.2) is the function η (average value of u), called duty ratio function.

In this case, x_1 and x_2 stand for the average dc-bus voltage, $\overline{v}_{dc}$, and grid current, $\overline{\iota}_{dc}$, respectively. η (average value of u), also known as the duty ratio function, serves as the control input for model Equation (8.2).

8.8 Results

The simulation is producing positive results with the voltage and current waveforms aligned, indicating that power is being injected into the grid (Figure 8.8). Aligning the voltage and current waveforms is an important aspect of power delivery and indicates the presence of real power flow.

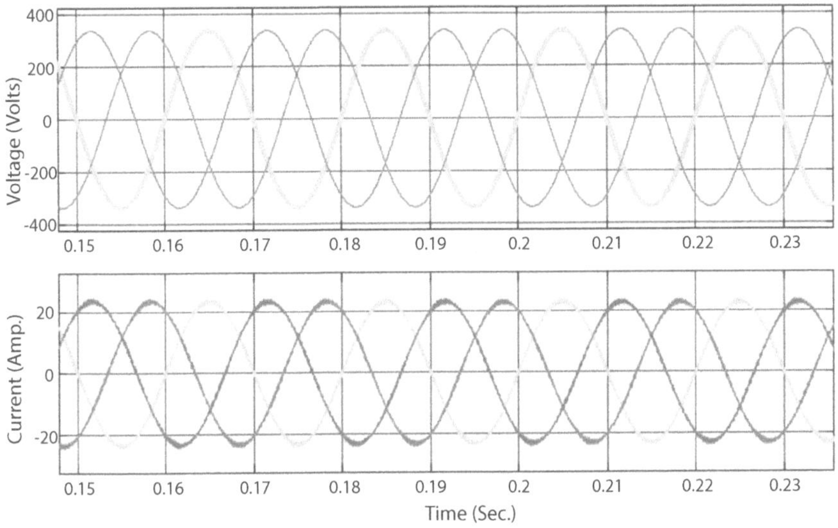

Figure 8.8 Voltage and current waveform during V2G operation.

Aligned voltage and current waveforms generally indicate that the power being generated by your simulated system is in-phase with the voltage in the grid, which means that power is being transferred effectively and contributing to the overall power flow. This alignment is crucial for efficient power transmission and distribution. By achieving aligned voltage and current waveforms, you can ensure that power is being injected into the grid as intended. It indicates that your simulation model is accurately representing the behavior of the power system and validating its functionality as in Figure 8.9.

Now set the current reference to negative and run the simulation once more in order to get the system to operate in the G2V mode, which charges the battery from the grid.

An electric vehicle (EV) is connected to the grid in order to charge its battery when it is in G2V (Grid-to-Vehicle) mode, also known as vehicle charging mode. Two main variables are involved in this process: voltage and current. The voltage and current waveforms may seem aligned but out of phase when an EV is being charged in G2V mode. Reactive parts like capacitors and inductors are present in the charging circuit, which is why there is a phase difference. There is a phase shift in the voltage and current between these reactive components. An out-of-phase voltage and current suggests a power factor that is lower than unity. The efficiency with which power is transported from the grid to the EV is shown by the power factor.

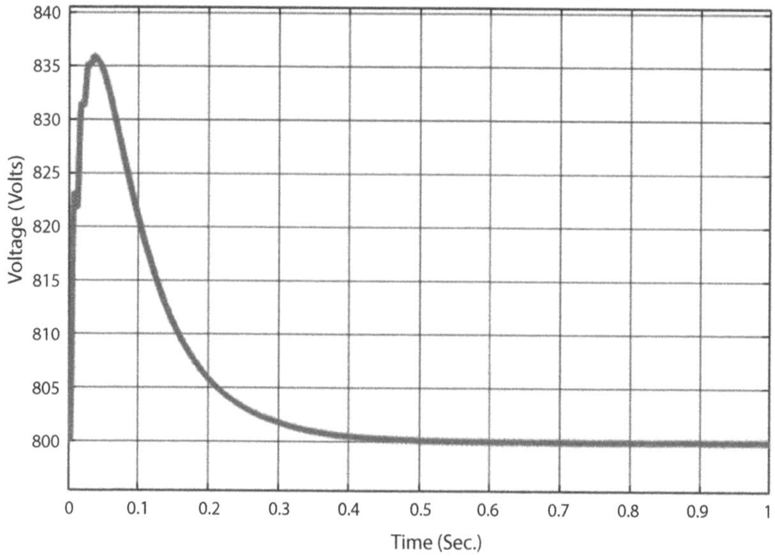

Figure 8.9 DC bus voltage during V2G operation.

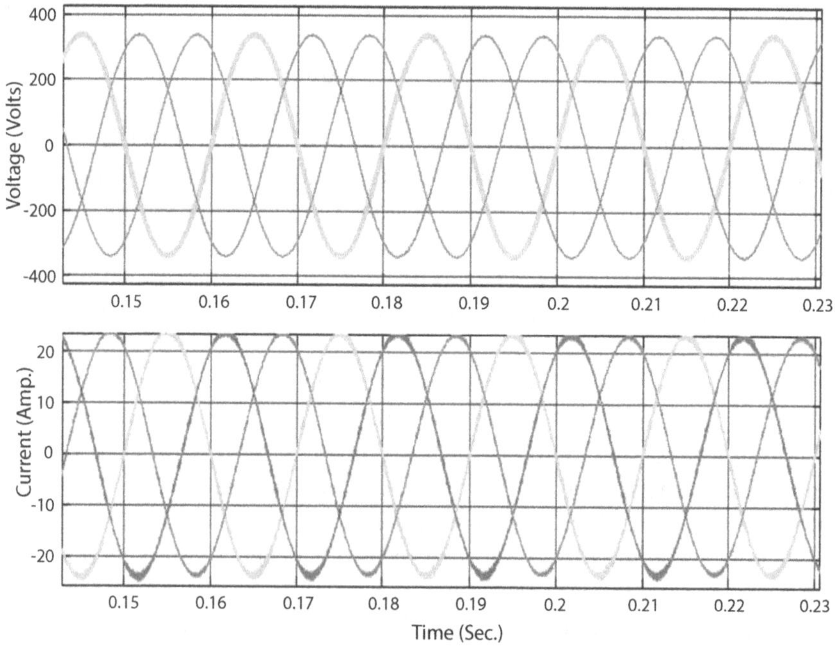

Figure 8.10 Voltage and current waveform during G2V operation.

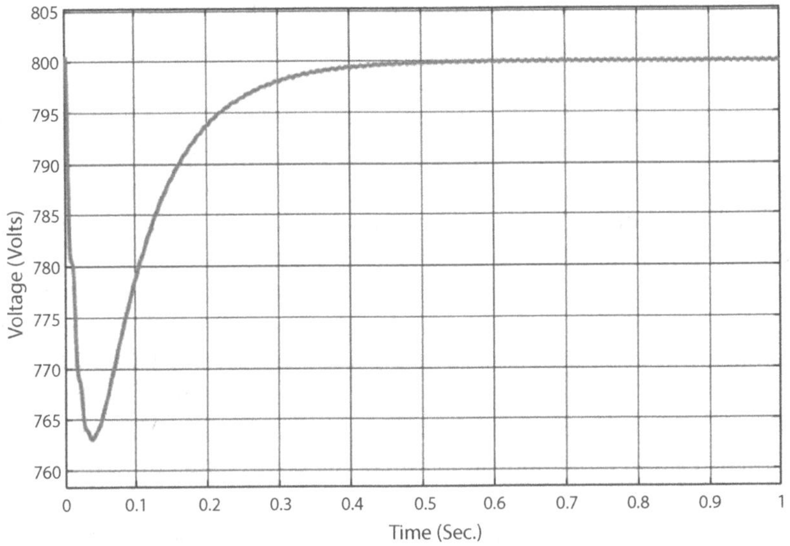

Figure 8.11 DC bus voltage during G2V operation.

If the power factor is less than unity, it means that some power is being transferred inefficiently, which means that power is being drawn from the grid instead of being used entirely to charge the car. Power electronics devices such as power factor correction (PFC) circuits can be used to increase the power factor and decrease the amount of electricity drawn from the grid. The voltage and current waveforms are aligned by these circuits, which increases their phase and makes them more.

In G2V (Grid-to-Vehicle) mode (Figure 8.10 and Figure 8.11), the power flow is from the grid to the vehicle's battery. When the power is negative, it indicates that the vehicle is drawing power from the grid to charge the battery. This is a typical scenario when the vehicle is plugged into an electrical outlet or a charging station and is actively charging its battery from the grid power supply. The bus voltage refers to the voltage level of the electrical bus that transfers power between the grid and the vehicle.

Conclusion

An explanation of model and design a DC fast charging system for a micro-grid's vehicle-to-grid (V2G) system has been provided in this chapter. The system comprises an off-board charger, a DC quick charging platform, and an inverter that is connected to the grid to facilitate the linking of electric vehicles (EVs) with the micro-grid. The control system is the central component of the charging station, allowing and regulate two-way power transmission between EVs and the grid. The efficacy of the system is demonstrated by the simulation results, which demonstrate an efficient power transfer and compliance with pertinent regulations for grid-injected current *via* the EVs. High dynamical stability is demonstrated by the developed controller's capacity to detect changes to the active power references and maintain a steady DC bus voltage. It is recommended that future research focus on creating a supervisory controller that can provide directive signals to different EV charging controllers. This supervisory controller will centrally manage and coordinate several EVs within the micro-grid, optimizing power flow and enhancing system performance.

The DC rapid charging architecture is used to create a V2G system on the micro-grid in a methodical manner. The control system of the system guarantees the successful transfer of power while also providing room for future research and development into features such as reactive power management, frequency regulation, and the creation of a monitoring controller for improved coordination.

References

1. Ng, E.C.Y., Huang, Y., Hong, G., Zhou, J.L., Surawski, N.C., Reducing vehicle fuel consumption and exhaust emissions from the application of a green-safety device under real driving. *Sci. Total Environ.*, 7931, 48602, Art, 2021.
2. Jia, Q.-S. and Long, T., A review on charging behaviour of electric vehicles: Data, model, and control. *IFAC-Papers Online*, 53, 5, 598–601, 2020.
3. Chen, C.-F., de Rubens, G.Z., Noel, L., Kester, J., Sovacool, B.K., Assessing the socio-demographic, technical, economic, and behavioural factors of Nordic electric vehicle adoption and the influence of vehicle-to-grid preferences. *Renew. Sustain. Energy Rev.*, 121, 1–39, 2020.
4. Goel, S., Sharma, R., Rathore, A.K., A review on barrier and challenges of electric vehicle in India and vehicle to grid optimisation. *Transp. Eng.*, 4, 1–14, 2021.
5. Khan, S.A. and Bohnsack, R., Influencing the disruptive potential of sustainable technologies through value proposition design: The case of vehicle-to-grid technology. *J. Clean. Prod.*, 254, 1–25, 2020.
6. Wolinetz, M., Axsen, J., Peters, J., Crawford, C., Simulating the value of electric-vehicle–grid integration using a behaviourally realistic model. *Nat. Energy*, 3, 2, 132–139, 2018.

7. Turton, H. and Moura, F., Vehicle-to-grid systems for sustainable development: An integrated energy analysis. *Technol. Forecast. Soc. Change*, 75, 8, 1091–1108, 2008.
8. Yilmaz, M. and Krein, P.T., Review of the impact of vehicle-to-grid technologies on distribution systems and utility interfaces. *IEEE Trans. Power Electron.*, 28, 12, 5673–5689, 2013.
9. Ouramdane, O., Elbouchikhi, E., Amirat, Y., Sedgh Gooya, E., Optimal sizing and energy management of microgrids with vehicle-to-grid technology: A critical review and future trends. *Energies*, 14, 14, 4166, 2021.
10. Huang, Z., Fang, B., Deng, J., Multi-objective optimization strategy for distribution network considering V2G-enabled electric vehicles in building integrated energy system. *Prot. Control Mod. Power Syst.*, 5, 1, 1–8, 2020.
11. Luo, L., Wu, Z., Gu, W., Huang, H., Gao, S., Han, J., Coordinated allocation of distributed generation resources and electric vehicle charging stations in distribution systems with vehicle-to-grid interaction. *Energy*, 192, 1–14, 2020.
12. Ahmadi, S., Arabani, H.P., Haghighi, D.A., Guerrero, J.M., Ashgevari, Y., Akbarimajd, A., Optimal use of vehicle-to-grid technology to modify the load profile of the distribution system. *J. Energy Storage*, 31, 1–8, 2020.
13. Hashim, M.S., Yong, J.Y., Ramachandaramurthy, V.K., Tan, K.M., Mansor, M., Tariq, M., Priority-based vehicle-to-grid scheduling for minimization of power grid load variance. *J. Energy Storage*, 39, 1–14, 2021.
14. Amoriya, V., Chauhan, R.K., Mittal, S., Sharma, S., Chauhan, K., Li, H., Modelling of Multi-Phase Interleaved based Universal Charging Port for Electric Vehicles. *Smart Grids Sustain. Energy*, 9, 19, 1–16, 2024.
15. Amoriya, V., Chauhan, R.K., Panit, V., Verma, S., Mittal, S., Chauhan, K., Upadhay, S., Cardena, A., Modeling of Bidirectional Electric Vehicle Charger for Grid Ancillary Services. *Int. J. Emerg. Electr. Power Syst.*, 24, 5, 61–74, 2023.
16. Shen, Y., Li, H.Y., Chauhan, R.K., Assessment of Energy Saving for Integrated CVR Control with Large Scale Electric Vehicle Connections in Microgrids, in: *Microgrids for Rural Areas: Research and Case Studies*, vol. 57–75, 2020.
17. Jian, L., Zheng, Y., Xiao, X., Chan, C.C., Optimal scheduling for vehicle-to-grid operation with stochastic connection of plug-in electric vehicles to smart grid. *Appl. Energy*, 146, 150–161, 2015.
18. Dudek, E., The flexibility of domestic electric vehicle charging: The electric nation project. *IEEE Power Energy Mag.*, 19, 4, 16–27, 2021.
19. Chauhan, R.K. and Chauhan, K., Management of renewable energy sources and battery bank for power losses optimization, in: *Smart Power Distribution Systems: Control, Communication, and Optimization*, vol. 299–320, 2018.
20. Chauhan, K. and Chauhan, R.K., Optimization of Grid Energy using Demand and Source Side Management for DC Microgrid. *J. Renew. Sustain. Energy*, 9, 1–15, 2017.

21. Venegas, F.G., Petit, M., Perez, Y., Plug-in behaviour of electric vehicles users: Insights from a large-scale trial and impacts for grid integration studies. *eTransportation*, 10, 1–44, 2021.
22. Kavousi-Fard, A., Abunasri, A., Zare, A., Hoseinzadeh, R., Impact of plug-in hybrid electric vehicles charging demand on the optimal energy management of renewable micro-grids. *Energy*, 78, 904–915, 2014.
23. Kumar, R.R. and Alok, K., Adoption of electric vehicle: A literature review and prospects for sustainability. *J. Clean. Prod.*, 253, 1–21, 2020.
24. Zhu, X., Han, H., Shi, Q., Cui, H., Zu, G., Gao, S., A multi-stage optimization approach for active distribution network scheduling considering coordinated electrical vehicle charging strategy. *IEEE Access*, 6, 50117–50130, 2018.
25. Sovacool, B.K., Axsen, J., Kempton, W., The future promise of vehicle-to-grid (V2G) integration: A sociotechnical review and research agenda. *Annu. Rev. Environ. Resour.*, 42, 1, 377–406, 2017.
26. Ponnam, V.K.B. and Swarnasri, K., Multi-objective optimal allocation of electric vehicle charging stations and distributed generators in radial distribution systems using Metaheuristic optimization algorithms. *Eng. Technol. Appl. Sci. Res.*, 10, 3, 5837–5844, 2020.
27. Tercan, S.M., Elma, O., Gokalp, E., Cali, U., An expansion planning method for extending distributed energy system lifespan with energy storage systems. *Energy Explor. Exploit.*, 40, 2, 599–618, 2021.
28. Chauhan, R.K., Chauhan, K., Guerrero, J.M., Controller design and voltage stability analysis of grid-connected DC microgrid. *J. Renew. Sustain. Energy*, 10, 1–8, 2018.
29. Chauhan, R.K. and Chauhan, K., Battery Monitoring and Control System for Photovoltaic-based DC Microgrid. *Int. J. Emerg. Electr. Power Syst.*, 20, 6, 1–12, 2019.
30. Almehizia, A.A. and Snodgrass, J.M., Investigation of V2G economical viability, in: *Proc. IEEE Texas Power Energy Conf. (TPEC)*, pp. 1–6, 2018.
31. Weldon, P., Morrissey, P., Brady, J., O'Mahony, M., An investigation into usage patterns of electric vehicles in Ireland. *Transp. Res. D Transp. Environ.*, 43, 207–225, 2016.
32. Divshali, P.H. and Evens, C., Behaviour analysis of electrical vehicle flexibility based on large-scale charging data. *Proc. IEEE Milan PowerTech*, 1, 1–6, 2019.
33. Habib, S., Khan, M.M., Abbas, F., Numan, M., Ali, Y., Tang, H., Yan, X., A framework for stochastic estimation of electric vehicle charging behavior for risk assessment of distribution networks. *Front. Energy*, 14, 2, 298–317, 2020.
34. Lee, J.H., Chakraborty, D., Hardman, S.J., Tal, G., Exploring electric vehicle charging patterns: Mixed usage of charging infrastructure. *Transp. Res. D Transp. Environ.*, 79, 1–13, 2020.
35. Tao, Y., Huang, M., Chen, Y., Yang, L., Orderly charging strategy of battery electric vehicle driven by real-world driving data. *Energy*, 193, 1–9, 2020.

36. Zhang, C., Greenblatt, J.B., MacDougall, P., Saxena, S., Jayam Prabhakar, A., Quantifying the benefits of electric vehicles on the future electricity grid in the midwestern United States. *Appl. Energy*, 270, 1–32, 2020.
37. Liu, J., Peper, J., Lin, G., Zhou, Y., Awasthi, S., Li, Y., Rehtanz, C., A planning strategy considering multiple factors for electric vehicle charging stations along German motorways. *Int. J. Electr. Power Energy Syst.*, 124, 1–12, 2021.
38. Chauhan, R.K. and Chauhan, K., Impact of demand side management system in Autonomous DC Microgrid, in: *Decision Making Applications in Modern Power Systems*, vol. Chapter 15, Aleem, S.A., Abdelaziz, A.Y., Zobaa, A., Bansal, R. (Eds.), pp. 389–410, 2019.

9

Forecasting Renewable Energy for Storage Technology Sizing Paper: A Review

Rafia Sagufta[1]*, N.P. Patidar[1] and Sidhartha Panda[2]

[1]Department of Electrical Engineering, Maulana Azad National Institute of Technology Bhopal, Bhopal, Madhya Pradesh, India
[2]Department of Electrical Engineering, Veer Surender Sai University of Technology Burla, Oddisa, India

Abstract
Due to intermittent and irregular character of the growing low-carbon power production technologies (such as wind and solar), the system operator encountered several difficulties. The most important consideration at the planning stage is the ideal size of the energy storage. Storage size and scheduling are affected by the prediction inaccuracy for renewable energy. However, from the standpoint of energy storage, precise forecasting continues to be difficult. Therefore, in order to increase the penetration of renewable energy generation, this review study investigates forecasting models with storage technique scaling. More specifically, the bulk energy services, auxiliary services, power quality management, transmission and distribution infrastructure services, and local energy management services are the categories under which the energy storage application in the power system is divided.

***Keywords*:** Battery, power system, energy storage system, forecasting, renewable energy

9.1 Introduction

The network of traditional energy systems has undergone significant alterations. Numerous factors, including the environmental issue, energy

**Corresponding author*: rafiasagufta34@gmail.com

Ashwani Kumar, Mukesh Kumar Awasthi, Nitesh Dutt, Yogesh Kumar Singla and Sivasakthivel Thangavel (eds.) Thermal Battery Management System for Hybrid and Electric Vehicles, (181–196) © 2025 Scrivener Publishing LLC

market concerns, changes in policies, and rising energy demand, have encouraged the use of non-conventional energy sources. These sources advantages might include increased dependability, decreased losses, and utilization a supplementary services. These sources are tending to as sustainable power sources in literature (RE). Yet, because they are sporadic in nature, RE sources have the opposite effect on power quality, stability, security, and the operation and control of power systems. As a result, the degree of intermittency affects how it is integrated into the grid. The RE integration issues in may be managed by energy storage systems (ESS) [1]. To store the extra RE output, energy storage devices were mostly constructed with solar systems at the household level and with wind energy at the useful level in [2, 3] residential energy storage helps to improve grid stability, peak shaving, and load shifting in addition to reducing the unpredictability that renewable energy sources present. ESS is essential in overcoming the uncertainty from which the RE has introduced. Its size ensures precise operation and control and lowers system running costs. Storage aids in the resolution of economic energy scheduling issues [4]. The main factor used to calculate the performance and size of energy storage is forecasting inaccuracy [5]. It utilized an estimate of energy demand and generation to determine the size of the ESS. To estimate the best amount of energy needed for the various storage methods, a genetic algorithm was developed [6]. The historical study has continued to look for the ideal method for calculating ESS size while taking potential advantages into account. The method places a lot of dependence on forecast errors for energy generation and consumption [7, 8]. The fundamental planning and execution of the energy storage are made possible by precise forecasting. If the forecast mistake is not taken into account, the storage system is more expensive from an economic standpoint [9, 10]. This method is used for estimating RE demand and ESS size calculated the ideal size of ESS taking forecasting and scheduling error into account, since after deployment the altering ESS infrastructure is quite challenging. Therefore, the best ESS size must be chosen in order to economically control the RE problem, including restrictions. The forecasting is crucial for various power system applications, such as renewable power dispatch, energy storage sizing, reserve size estimate, dependability evaluation, and the market for energy addition to the planning of energy storage size. It aids in estimating reserve size in [11, 12], which lowers plant capital costs and raises system dependability by taking into account anticipated energy not provided in [13], this essay focuses on the difficulty of integrating renewable energy sources, as well as on the predicting errors and energy storage capacity. The answer to the RE issue when taking RE forecasting inaccuracy into account is optimal storage size. Numerous forecasting techniques are discussed based on historical literature, including physical models, statistical models, intelligent models, and hybrid models. The fundamental procedure for calculating ESS size includes predicting inaccuracy. Storage can be used for a variety of purposes in the operation of power systems, including bulk energy supply services, auxiliary services, infrastructure services for transmission and distribution and energy management services for consumer and presumes.

9.2 Forecasting

We faced lots of challenges at the time of integration so we got an approach to managing uncertainty may be accurate forecasting of energy from renewable resources. By examining what has already happened and what is happening right now, forecasting is a technique for predicting what will happen in the future. In many works of literature, the time horizon is used in multiple different ways to illustrate the timeline prediction strategy of renewable energy. Since all of the literature is combined, which it is divided into four categories, which include predicting for the very short term, short term, medium term, and long term, as shown in below. Predicting the future for forecasting changes depends on the application. For the regulation, control, and real-time functioning of the market, action is useful. Short-term forecasting is utilized to dispatch loads for economics operations and power market operations, while long-range plantings is particularly beneficial from a maintenance planning standpoint.

Application:

1. Ultra short term-
It provides the estimations of irradiance and generated power for the next 30 min and is applied on a sliding time win doe interval.

2. Short term-
It involves creating short-term financial budgets and goals. The short period has duration of one to six months and is less than years, which include short-term elimination.

3. Medium Term-
It involves creating short-term goals and budgets. The brief term is between 1 and 6 months long and is less than a year, which involves partial elimination.

4. Long Term-
Long term forecasting is done for a period ranging from six month to five years.

Due to its intermittent nature, the challenge of precise RE forecasting is difficulty for forecasting models. An answer to the problem of predicting error may lie in the choice of forecasting model. Figure 9.1 depicts the hierarchy of forecasting models that have been employed in the literature for various RE forecasts [14–16]. Figure 9.2 shows the models of forecasting. Single model and hybrid model are the two categories into which forecasting models are often divided. While other models require comprehensive historical data, spatial correlation is more appropriate to describe the connection with other sites when complete data is not available. Curve fitting, AR (autoregressive), ARMA (auto regressive moving average), and ARIMA are examples of traditional statistical models (auto regressive integrated moving average). For short-term forecasting, ARIMA has high accuracy and can identify nonlinear patterns in the input data set. To manage RE uncertainty, the time series ARIMA model projected precise charging and discharging schedule for energy storage. Persistence models in [17] do well for short-term forecasting but do not provide findings that are suitable for long-term forecasting. Similar probabilistic models use the probability density function to solve the problem of data length and produce superior short-term forecasting outcomes. Additionally, whereas search-based optimization models are a subset of sophisticated models [15, 16, 18, 19], learning models are intelligent models or machine learning-based approaches. However, these models are having trouble with accuracy because of the large amount of data, enormous data length, over fitting, under fitting, and local minimum point [20], thus they are being combined to create a hybrid model. Hybrid models are not employed for quick decision applications because they have a large computational burden when more than two or three models are combined. For the purpose of energy scheduling and cost-benefit analysis, these models were applied to

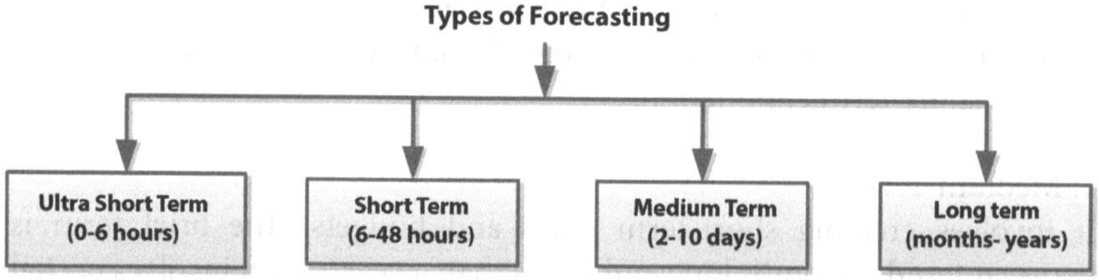

Figure 9.1 Types of forecasting.

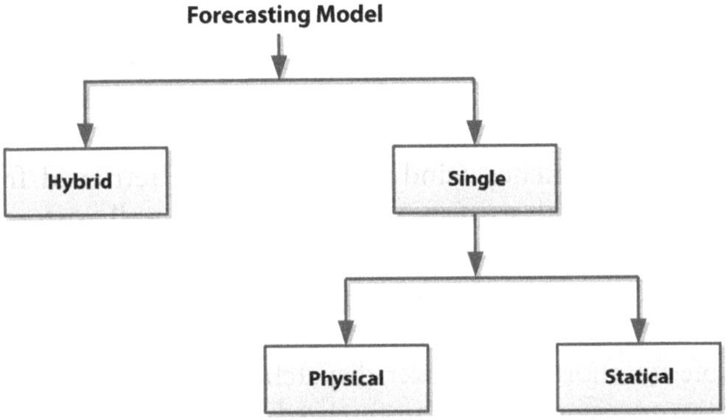

Figure 9.2 Model of forecasting.

several types of systems, including wind-storage, solar-storage, and hybrid systems (wind-solar-storage).

9.3 Energy Storage

Due to insufficient storage capacity, ESS profit is impacted by unfavorable installation, operating, and maintenance costs [21–26]. In order to overcome fore-casting difficulties, accurate forecasting provides information on the storage capacity required. The quantity and effectiveness of energy storage are dependent on the degree of predicted value uncertainty [27]. The forecasting inaccuracy also affects the values of the performance parameters, such as charging, discharging, and State of Charge(SoC) [28] system functioning is affected by the performance of the energy storage. In order to operate as efficiently as possible, stochastic dynamic programming is employed, taking battery deterioration into account. The rate of charging and discharging affects battery degradation [29]. Local and aggregate functions are both offered by energy storage. While storage aggregation has advantages when used with a prediction error model [30]. In order to establish the ideal generating output and storage capacity, a dynamic model predictive control with two stages is developed [31]. A two-stage stochastic technique for storage size that takes demand uncertainty and RE into account is presented in [5] the goal is to optimize power output while taking SoC restrictions into consideration. The variability of RE and forecast error is used to establish the storage size and dispatch variation rates system for the highest output. A Markov-Chains technique was used in to forecast the future 20 years of solar irradiance changes while accounting for imbalance tariff and battery capacity [32, 33]. According on economic analysis, best scheduling technique, and forecast uncertainties, regression using support vector is used to calculate storage size [31]. Power transmits with storage, having wind energy alliance in mind for long-term planning. If low-frequency wind components are retrieved from input data, it is valuable for short-term power dispatch as well as storage operation. Uncertainties are referred to as "unreserved energy." Power dispatch with storage, having wind energy integration in mind for long-term planning. If low-frequency wind components are retrieved from input data, it is valuable for short-term power dispatch as well as storage operation. Uncertainties are referred to as "unreserved energy." In the event of a total forecast inaccuracy, either the ESS size or the amount of unseen energy will grow. If 5% unsaved energy is permitted, the ESS's power and energy can be lowered by more than 50% [34]. The ideal size of energy storage with cost-benefit analysis is discussed in [4, 28, 31–33, 35–37].

9.4 Process of Sizing Energy System

The best storage size is established using data from RE generation, pricing, and load forecasting tools. However, each approach has advantages and disadvantages. Based on past research, four strategies are investigated in this section: probabilistic techniques, analytic techniques, direct search based techniques, and hybrid techniques [38].

9.4.1 Flow Chart of Probabilistic Techniques

Figure 9.3 shows the flow chart of probabilistic techniques. The operation of probabilistic approaches is dependent on the data resolution, which might be 15 minutes, interior-hour, or hourly. The stochastic behavior of renewable sources (such as solar and wind) is the key rationale for using this technique to predict energy storage capacity with set limitations. Model generates fabricated data and scenarios for stochastic optimization during implementation. Probabilistic models overcome the availability of insufficient data and produce good findings with a high level of confidence. One major disadvantage is that performance is dependent on the preceding data set. In many studies, probabilistic models are employed not only for data creation but also for scenario generation. Monte Carlo simulation is extensively used to produce a large number of scenarios that take into account the statistical behavior of random variables [39]. Another use of this strategy for energy storage is lowering uncertainty due to poor wind

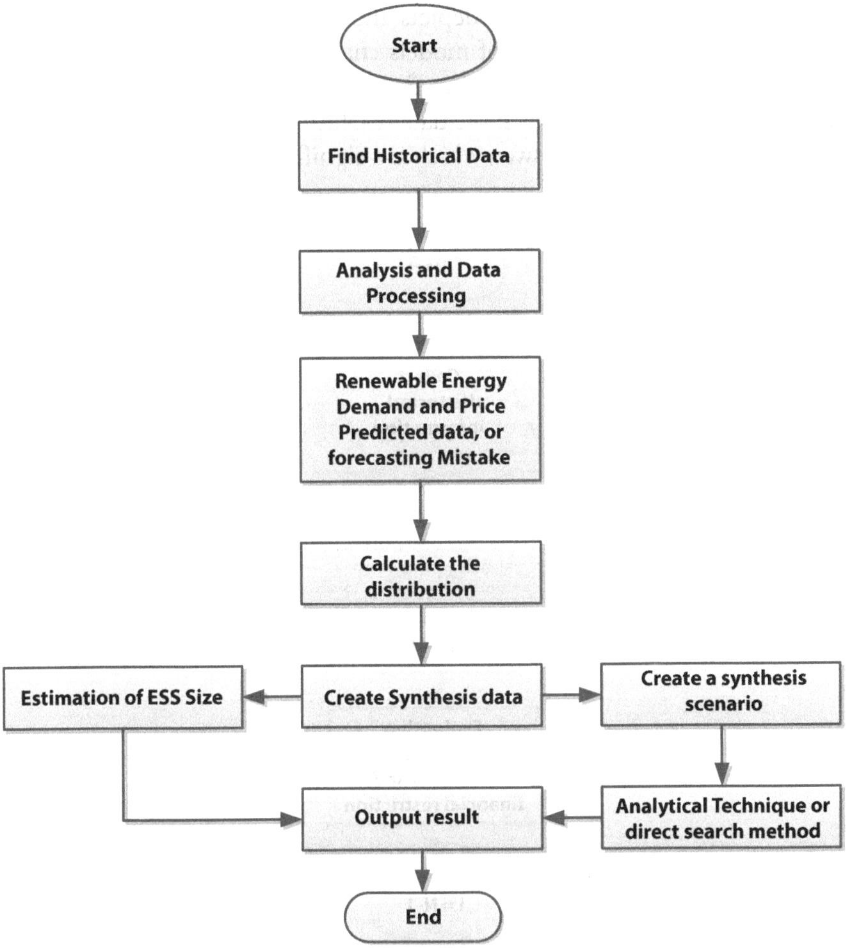

Figure 9.3 Flowchart of probabilistic techniques.

forecasts aside from these approaches. For receding operation with uncertainties, optimal size of energy storage using chance-constrained optimization [5], robust optimization [9], and stochastic control approach [14] is explored in the literature.

9.4.2 Flow Chart of Analytical Techniques

Analytical procedures are implemented by repeating the simulation after a predetermined time period, and they are based on direct computation using an intuitive criterion. In many researches, this technique is referred to as a deterministic strategy, and it is mostly employed for determining energy storage size. Figure 9.4 depicts the fundamental flow of analytic approaches. The performance of models changes under technological and economic limitations, as indicated in the image, which is used to calculate the amount of energy storage. If the data resolution is not correct, the model may produce an inferior answer, which is a significant disadvantage [20].

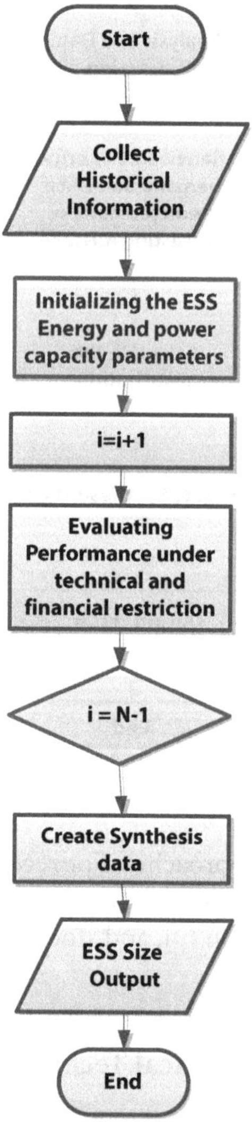

Figure 9.4 Flowchart of analytical technique.

9.4.3 Search-Based Techniques

Search-based methods get around the computational burden and rapidly deliver the best result. These techniques could be divided into two broad categories heuristic algorithms influenced by nature and mathematical algorithms. The basic flowchart of direct search-based techniques for optimizing energy storage size is shown in Figure 9.5. Storage sizing optimization appears as linear and non-linear problems in studies.

The best response is the stopping criteria in the iterative search-based optimization technique. Heuristic models can not taking complex derivatives, especially for non-linear problems. In past, research various

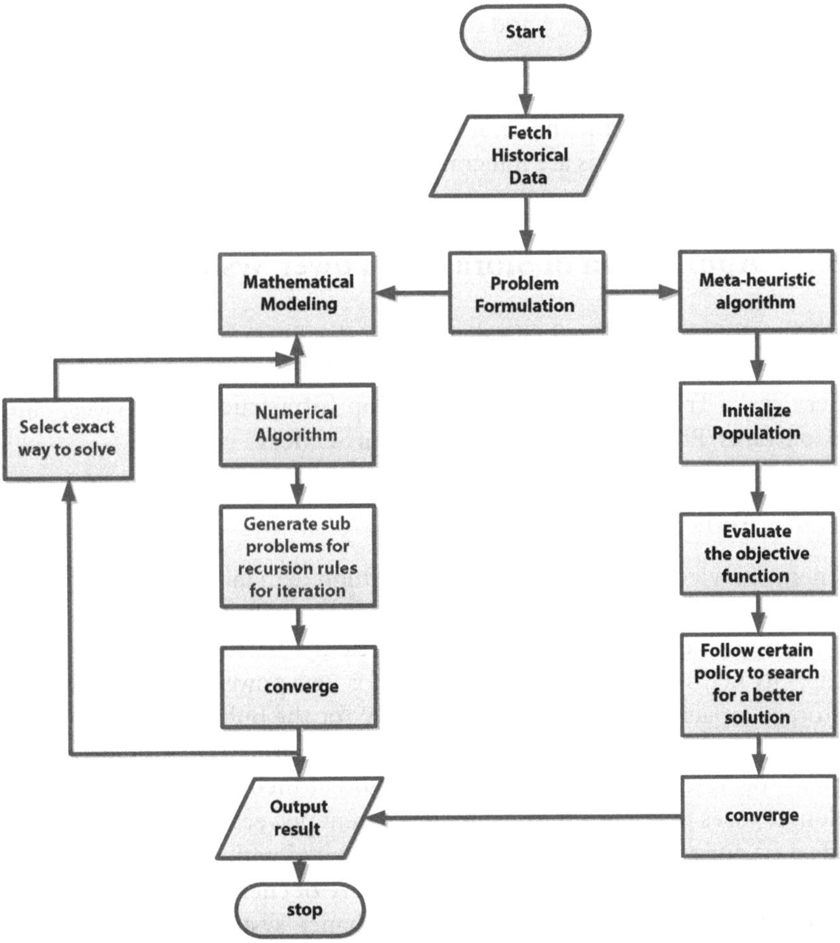

Figure 9.5 Flowchart of search-based techniques.

meta-heuristic models (GA [13], PSO [22], Tabu Search [21], BA [32]) used to determine energy storage size.

9.4.4 Hybrid Techniques

The techniques used determine the ideal size of energy storage, which were covered above, each have advantages and disadvantages. Although analytical techniques provide precise answers, they are limited by local issues or data with high resolution that can be computationally expensive. Similar towards how search-based methods struggle with high convergence for a local as opposed to a global solution The probabilistic approach slows down when several scenarios are present, though. The main motivation is to maximize performance and optimization by combining all the advantages of various methodologies. These methods are therefore known as hybrid methods and offer a comprehensive solution with a high degree of certainty. To determine the ideal storage size for solar plants, probabilistic and analytical methods are integrated.

9.5 Application of Storage in Power System

Energy storage is utilized in the energy industry for a variety of applications, which are elaborated under the headings "Bulk Energy Services," "Ancillary Services," "Transmission and Distribution Infrastructure Services," and "Consumer/Prosumer Energy Management Services."

9.5.1 Bulk Energy Services

Energy service utilizing storage helps to fulfill demand for power, including market pricing. Energy storage activates when the market price is high and fixes when the price of energy increases. Increased electric supply capacity helps to lessen the requirement for new power plants in this way. Storage capacity should be quoted in MW for the bulk energy service. In order to manage the charging and discharging process of storage, power plants operate at maximum load factor and at their highest rated condition. It poses a significant threat to the current power system network. The ideal size of energy storage should be integrated with the current system to enhance system profitability. The ideal size of energy storage should be integrated with the current system to enhance system profitability. The best storage size may be determined using day-ahead market pricing and demand predictions, which helps to optimize system performance through careful planning. Storage also extends the life of traditional power plants and lowers their cost.

9.5.2 Ancillary Services

When supply and demand are out of balance, ancillary services help preserve governance. Energy storage is employed as an additional service to sustain generation and demand fluctuations. Independent system operators employ it as an auxiliary service, and it was discovered that total benefits rose by 30%-35%. It also provides reactive power management to maintain the allowed range of system voltage fluctuation. Storage can assist the power plant's startup and also energize the transmission system.

9.5.3 Transmission and Distribution Infrastructure Services

Installing dependable energy storage can postpone or prevent the need for improved transmission and distribution infrastructure, such as increased distribution transformer power, multiplied line capacity, and new transmission line installations. The life of the equipment is increased while the burden on the current equipment is reduced. Storage can distribute the least-expensive energy to all requests and preserve system stability in the event of system congestion brought on by energy or cost terms, reducing the strain placed on the system by peak loads.

9.5.4 Consumers/Prosumers Energy Management Service

Energy storage was adopted by consumers and prosumers to preserve power quality and dependability while paying less for electricity. Storage assisted in avoiding consumption during times when power market prices were at their highest by delaying use. Consumers take part in demand response programs, which help to lower peak utility consumption and provide demand-side control. It is also beneficial to install a capacitor bank in order to reduce the peak load on a specific distribution feeder transformer, which tends to improve voltage control and energy storage.

9.5.4.1 Power Quality and Reliability Management

Controlling harmonics and maintaining stability in voltage and frequency fluctuations are two aspects of power quality management. Voltage and frequency variations are brought on by either generation-side uncertainty or RE intermittency that affects local distribution levels. Static variable compensators or virtual synchronous machines are utilized to provide reactive power compensation, and storage is included on the RE side to maintain the voltage-to-frequency ratio. Harmonics are produced by converters and inverters and can be regulated *via* current control, voltage control, or a hybrid approach. The most effective response to the aforementioned problems is superconducting magnetic energy storage (SMES).

9.5.4.2 Storage Service in Electricity Market

The functioning and decision-making processes of the power market were significantly impacted by the influence of predicting uncertainty. Game theory is utilized as a decision-making model to preserve social welfare and stabilize the functioning of the market. In terms of investment, revenue, and discount rate, the market characteristics are changing. Uncertainty in renewable energy is eliminated by these parameter control techniques. Energy storage offers a substantial number of solutions [33] and encourages prosumers to engage in the local energy market to ensure stable market functioning. The charging and discharging schedule of container which helps to reduce the difference in market clearing prices between peak and off-peak hours in the day-ahead power market. Storage charging and discharging losses have a significant influence on market costs for power. Prices on the market will rise in response to rising system losses, and vice versa. But battery health monitoring might be the answer. Due to the unstable and inflexible nature of energy output and demand, which cannot be predicted in advance, it is challenging to incorporate these factors into the market and operational framework already in place. Household loads or generation based on unpredictable renewable energy sources, like wind or solar, are examples of such unstable and low-flexibility units. As a result, there has been a lot of study done on the coupling of rigid demand and/or rigid generation to an energy storage system (ESS). In reality, ESSs can function as an internal capacity reserve and counteract the erratic nature of rigid generation and demand, promoting and facilitating dispatch ability.

9.5.5 Energy Scheduling and Cost Benefits Analysis for Renewable Energy Sources

Energy scheduling is a realistic approach to improving energy efficiency and reducing energy costs. It is a key factor that influences the amount and effectiveness of energy storage and assists in the resolution of economic energy scheduling's forecasting error. We specialize in cost-effective energy policy using renewable energy sources. Our scheduling issue is to quantify the best utilization of renewable energy, which creates a balance between the benefits to the system in general from adopting renewable energy and the cost associated with its volatility, as well as the influence of renewable energy on its volatility and its optimum use. Renewable energy generates a greater use of renewable energy in energy provision. Increasing the use of renewable energy also helps customers by dragging down the marginal cost of energy provision.

The following are the stages that a cost-benefit analysis of a distinct renewable resource goes through:

1. Effectively identifying the technical and functional requirements of the installation that the project will be used to implement.
2. The most statistical technique of the project's cost and benefit components based on all available data, including market data, experience from comparable installations, calculations, etc.
3. Expression of all costs and benefits in terms of their present value (PV), taking into account their escalation over the course of the investment's whole life cycle.
4. Evaluation of the investment's economic effectiveness metrics.

9.6 Conclusion

Planning and operating a power system with fluctuating generation is a difficult undertaking. Forecasting changing generation and demand helps to increase system adaptability. This study has examined many types of forecasting. Furthermore, certain essential strategies are employed to identify the best size of storage. The use of energy storage in power system operation and planning is also discussed. Finally, in the next phase, finish this article based on historical research:

- Synthesis of RE sources with the forecast delusion raises the yearly rate of return and lowers system operating costs.
- Accurate forecasting of RE results in ideal ESS size, which aids in effective planning to synthesis energy storage.
- Energy storage helps to increase power quality, reliability, and power system market functioning.
- Hybrid models are utilized to improve the power system performance.

References

1. Sujil, A., Kumar, R., Bansal, R., Fcm clustering based adaptive neuro-fuzzy inference system based pv and wind generation forecasting agent for energy management in a smart microgrid, in: *7th IET-RPG conf*, 2018.
2. Bahmani-Firouzi, B. and Azizipanah-Abarghooee, R., Optimal sizing of battery energy storage for micro-grid operation management using a new improved bat algorithm. *Int. J. Electr. Power Energy Syst.*, 56, 42–54, 2014.
3. Suazo-Martínez, C., Pereira-Bonvallet, E., Palma-Behnke, R., A simulation framework for optimal energy storage sizing. *Energies*, 7, 5, 3033–3055, 2014.
4. Saboori, H., Hemmati, R., Jirdehi, M.A., Reliability improvement in radial electrical distribution network by optimal planning of energy storage systems. *Energy*, 93, 2299–2312, 2015.
5. Fossati, J.P., Galarza, A., Mart´ın-Villate, A., Fontan, L., A method for optimal sizing energy storage systems for microgrids. *Renew. Energy*, 77, 539–549, 2015.
6. Wang, J., Zhou, Q., Jiang, H., Hou, R., Short-term wind speed forecasting using support vector regression optimized by cuckoo optimization algorithm. *Math. Probl. Eng.*, 1–13, 2015, 2015.
7. Exizidis, L., Kazempour, J., Pinson, P., De Greve, Z., Vallèe, F., Impact of public aggregate wind forecasts on electricity market outcomes. *IEEE Trans. Sustain. Energy*, 8, 4, 1394–1405, 2017.
8. Sheibani, M.R., Yousefi, G.R., Latify, M.A., Economics of energy storage options to support a conventional power plant: A stochastic approach for optimal energy storage sizing. *J. Energy Storage*, 33, 101892, January 2021.
9. Appino, R.R., Ordiano, J.A.G., Mikut, R., Faulwasser, T., Hagenmeyer, V., On the use of probabilistic forecasts in scheduling of renewable energy sources coupled to storages. *Appl. Energy*, 210, 1207–1218, 2018.
10. Bahramirad, S., Reder, W., Khodaei, A., Reliability-constrained optimal sizing of energy storage system in a microgrid. *IEEE Trans. Smart Grid*, 3, 4, 2056–2062, 2012.
11. Kim, W.-W., Shin, J.-S., Kim, J.-O., Operation strategy of multienergy storage system for ancillary services. *IEEE Trans. Power Syst.*, 32, 6, 4409–4417, 2017.
12. Zhang, Y., Dong, Z.Y., Luo, F., Zheng, Y., Meng, K., Wong, K.P., Optimal allocation of battery energy storage systems in distribution networks with high wind power penetration. *IET Renew. Power Gener.*, 10, 8, 1105–1113, 2016.
13. Ahmed, A. and Khalid, M., A review on the selected applications of forecasting models in renewable power systems. *Renew. Sustain. Energy Rev.*, 100, 9–21, 2019.
14. Nottrott, A., Kleissl, J., Washom, B., Energy dispatch schedule optimization and cost benefit analysis for grid-connected, photovoltaic-battery storage systems. *Renew. Energy*, 55, 230–240, 2013.

15. Awad, A.S., Fuller, J.D., El-Fouly, T.H., Salama, M.M., Impact of energy storage systems on electricity market equilibrium. *IEEE Trans. Sustain. Energy*, 5, 3, 875–885, 2014.
16. Barbosa de Alencar, D., de Mattos Affonso, C., Limaõ de Oliveira, R.C., Different Models for Forecasting Wind Power Generation: Case Study. *Energies*, 10, 12, 1-27, 2017 .
17. Mathur, F., Gupta, V., Kumar, R., *et al.*, Predictive analysis of traditional, deep learning and ensemble learning approach for short-term wind speed forecasting, in: *2020 IEEE International Conference on Computing, Power and Communication Technologies (GUCON)*, IEEE, pp. 783–788, 2020.
18. Toh, G. and Gooi, H., Incorporating forecast uncertainties into eens for wind turbine studies. *Electr. Power Syst. Res.*, 81, 2, 430–439, 2011.
19. Beltran, H., García, I.T., Alfonso-Gil, J.C., Pérez, E., Levelized cost of storage for li-ion batteries used in pv power plants for ramp- rate control. *IEEE Trans. Energy Convers.*, 34, 1, 554–561, 2019.
20. Lujano-Rojas, J.M., Dufo-Lo´pez, R., Bernal-Agustín, J.L., Catalaõ, J.P., Optimizing daily operation of battery energy storage systems under realtime pricing schemes. *IEEE Trans. Smart Grid*, 8, 1, 316–330, 2016.
21. Moya Rodriguez, J.L., Leite, J.C., Reston Filho, J.C., Different models for forecasting wind power generation: case study. *Energies*, 10, 12, 1976, 2017.
22. Lee, J., Kim, Y., Hahn, J., Seo, H., Customer energy management platform in the smart grid. *2012 14th Asia-Pacific Network Operations and Management Symposium (APNOMS)*, IEEE, pp. 1–4, 2012.
23. Baker, K., Hug, G., Li, X., Energy storage sizing taking into account forecast uncertainties and receding horizon operation. *IEEE Trans. Sustain. Energy*, 8, 1, 331–340, 2016.
24. Matos, M.A. and Bessa, R., Operating reserve adequacy evaluation using uncertainties of wind power forecast, in: *2009 IEEE Bucharest PowerTech*, IEEE, pp. 1–8, 2009.
25. Bucciarelli, M., Paoletti, S., Vicino, A., Optimal sizing of energy storage systems under uncertain demand and generation. *Appl. Energy*, 225, 611–621, 2018.
26. Delfanti, M., Falabretti, D., Merlo, M., Energy storage for pv power plant dispatching. *Renew. Energy*, 80, 61–72, 2015.
27. El-Amin, M., Grid integration challenges of wind energy: A review. *IEEE Access*, 8, 10857–10878, 2020.
28. Korpaas, M., Holen, A.T., Hildrum, R., Operation and sizing of energy storage for wind power plants in a market system. *Int. J. Electr. Power Energy Syst.*, 25, 8, 599–606, 2003.
29. Aghamohammadi, M.R. and Abdolahinia, H., A new approach for optimal sizing of battery energy storage system for primary frequency control of islanded microgrid. *Int. J. Electr. Power Energy Syst.*, 54, 325–333, 2014.

30. Pinson, P., Papaefthymiou, G., Klockl, B., Verboomen, J., Dynamic sizing of energy storage for hedging wind power forecast uncertainty, in: *2009 IEEE Power & Energy Society General Meeting*, IEEE, pp. 1–8, 2009.
31. Chakraborty, S., Senjyu, T., Toyama, H., Saber, A., Funabashi, T., Determination methodology for optimising the energy storage size for power system. *IET Generation Transm. Distrib.*, 3, 11, 987–999, 2009.
32. Ahmed, S.D., Al-Ismail, F.S., Shafiullah, M., Al-Sulaima, F.A., Grid Integration Challenges of Wind Energy: A Review. *IEEE Access*, 8, 10857–10878, 2020.
33. Xia, S., Chan, K.W., Luo, X., Bu, S., Ding, Z., Zhou, B., Optimal sizing of energy storage system and its cost-benefit analysis for power grid planning with intermittent wind generation. *Renew. Energy*, 122, 472–486, 2018.
34. Yuan, S., Kocaman, A.S., Modi, V., Benefits of forecasting and energy storage in isolated grids with large wind penetration–the case of sao vicente. *Renew. Energy*, 105, 167–174, 2017.
35. Dragicevič, T., Pandźič, H., Śkrlec, D., Kuzle, I., Guerrero, J.M., Kirschen, D.S., Capacity optimization of renewable energy sources and battery storage in an autonomous telecommunication facility. *IEEE Trans. Sustain. Energy*, 5, 4, 1367–1378, 2014.
36. Simla, T. and Stanek, W., Reducing the impact of wind farms on the electric power system by the use of energy storage. *Renew. Energy*, 145, 772–782, 2020.
37. Saini, V.K., Sharma, K.C., Prakash, V., Bhakar, R., Impact of renewable energy sources and electricvehicle penetration on generation scheduling, in: *2018 8th IEEE India International Conference on Power Electronics (IICPE)*, IEEE, pp. 1–6, 2018.
38. Saini, V.K., Kumar, R., Mathur, A., Saxena, A., Short term forecasting based on hourly wind speed data using deep learning algorithms, in: *2020 3rd International Conference on Emerging Technologies in Computer Engineering: Machine Learning and Internet of Things (ICETCE)*, IEEE, pp. 1–6, 2020.
39. Liang, X., Emerging power quality challenges due to integration of renewable energy sources. *IEEE Trans. Ind. Appl.*, 53, 2, 855–866, 2016.

10

Wireless Energy Transmission in Electrical Vehicles

Bhupender Singh[1]*, Sandeep Grover[1], Nitin Panwar[1], Sandeep Ravish[2] and Praveen Gautam[3]

[1]*Department of Mechanical Engineering, J.C. Bose University of Science and Technology, YMCA, Faridabad, Haryana, India*
[2]*Department of Civil Engineering, J.C. Bose University of Science and Technology, YMCA, Faridabad, Haryana, India*
[3]*Department of Mechanical Engineering, Raj Kumar Goel Institute of Technology, Ghaziabad, India*

Abstract
This research article examines in detail electric induction-based charging systems for electric vehicles (EVs), reviewing the latest developments, market trends, performance positive and emotional support, user experience improvement, depression, and future research. It discusses the key role of wireless charging technology in supporting the global electric vehicle transition and discusses the key issues affecting the deployment and adoption of electromagnetic induction-based charging infrastructure. This article presents the latest research, business development, and business case studies to highlight the importance of efficiency, reliability, and user-centered design in increasing customer usage and satisfaction. It highlights the need for continued innovation, collaboration and governance to realize the full potential of wireless charging technology in creating a sustainable and connected electric vehicle ecosystem. This comprehensive assessment provides a valuable resource for researchers, industry professionals, policy makers, and stakeholders to promote electronic transportation and enable the world to use wireless charging technology.

Keywords: Electromagnetic induction, wireless charging, electric vehicles, market dynamics, future prospects

**Corresponding author*: bhupee_28@jcboseust.ac.in

Ashwani Kumar, Mukesh Kumar Awasthi, Nitesh Dutt, Yogesh Kumar Singla and Sivasakthivel Thangavel (eds.) Thermal Battery Management System for Hybrid and Electric Vehicles, (197–220) © 2025 Scrivener Publishing LLC

10.1 Introduction

The global automotive industry is shifting to electric vehicles in response to environmental concerns, increased regulations and advances in battery technology. Electric vehicles (EVs) offer an alternative to traditional internal combustion engines with zero tailpipe emissions and reduced reliance on fossil fuels. However, the widespread use of electric vehicles depends on the availability of reliable and convenient charging. Although traditional plug-in charging solutions are effective, they often face issues related to user discomfort, limited space, and physical demands [1–8]. Charging systems based on electromagnetic induction are promising against these challenges by enabling wireless charging of electric vehicles by sending inductive electricity to the charging plate embedded in the ground and the receiving steering wheel in the vehicle [9–15]. The technology has many benefits, including ease of use, reduced wear and tear on components, and increased safety by eliminating the risk of electric shock. In addition, wireless charging technology (Layout of wireless energy shown in Figure 10.1) adapts to new trends in urban planning and smart city strategies, where the integration of electric vehicle charging into existing systems such as roads, parking lots and public spaces is gaining importance. Therefore, charging systems based on electromagnetic induction have attracted the attention of industry participants, policy makers and researchers who want to accelerate electric vehicles and provide efficient transportation [16–25]. This research article is designed to provide a comprehensive overview of

Figure 10.1 Layout of wireless energy transmission in EVs.

electromagnetic induction-based electric vehicle charging systems, discussing new developments, business advantages, performance considerations, improvement users, and future directions. By reviewing the current state of wireless charging technology and identifying opportunities for innovation and collaboration, this article aims to inform and encourage efforts to support the transition to electric cars and realize the full potential of wireless charging technology in a clean, green way. The future of sustainable transportation.

10.2 Technological Advancements

Technological advances in induction-based charging systems have increased the efficiency, reliability, and efficiency of electric vehicle (EV) wireless charging technology [26–33]. One area where significant progress has been made is coil design optimization. Professionals and scientists are working to create high-performance coils that provide the best magnetic properties and power conversion. These advances allow coils to reduce energy loss and increase efficiency, making charging faster and more reliable for electric car users.

Additionally, innovations in electrical energy technology have facilitated the development of more efficient and compact devices. These advancements enable higher speeds and faster charging while maintaining system efficiency and reliability. Advances in the use of electric motors are making electric induction-based charging possible to charge more to meet the needs of popular electric cars. One Smart grid integration has emerged as another important development in wireless charging technology. Integration of the smart grid can facilitate energy management, balancing and resource needs by enabling two-way communication between the power grid and the grid [34–40]. This integration optimizes charging times and grid utilization, ultimately increasing the overall performance and reliability of charging while also encouraging the integration of new renewable energy into the grid.

Additionally, advances in materials science have led to the development of new materials with improved stability, thermal conductivity, and environmental flexibility [41–50]. The development of these materials solves problems such as corrosion, temperature changes and physical damage, leading to the creation of a stronger, more reliable payment that can withstand heavy duty and is durable. Together, these technologies make charging systems based on electromagnetic induction an important part of the transition to electric vehicles and pave the way for a clean, green future in transportation. Continued research and innovation in these areas will be critical to further developing the technology, solving remaining challenges and leading to the widespread use of these wireless charging solutions for electric vehicles.

10.3 Methodology and Modeling

In terms of complexity, the concept car is simple. In terms of construction, it's a piece of cake compared to a gas-powered car. Two motors and their controller, an inverter circuit, a battery pack, relays, an Arduino UNO and a speed controller are simple devices; both controllers share an accelerator for operation. When using the brakes the engine will be stopped by the brake switch. When you reverse the car to the engine, the engine will turn off. To achieve this, two levels and two control lines need to be switched while the vehicle is in motion. The reverse button is conveniently located above the steering wheel. It uses magnetic resonance imaging technology wireless power transmission (WPT) to eliminate negative wires. In fact, WPT uses the same principle as inductive power transfer and has been researched and developed for at least three decades [51–59]. WPT as a field has made significant progress in recent years. Power (mill watts) kilowatts, transmission power increased from a few millimeters to hundreds of millimeters with higher efficiency. With these improvements, WPT has become attractive for sustainable and powerful EV charging applications [60–69]. The technology discussed at this conference that could be used for wireless charging of electric vehicles can be found in the field of universal power transfer (WPT). Problems such as low transportation, high cost, and weak electrical charge can be easily solved by using WPT. Electric vehicles (EVs) have reached critical mass and battery technology is the limit. We hope that researchers will be inspired by the state's results and use this motivation alongside WPT and EV.

10.3.1 Block Diagram and Working Principle

Figure 10.2 shows the block diagram of the system and Figure 10.3 shows the circuit diagram of the system.

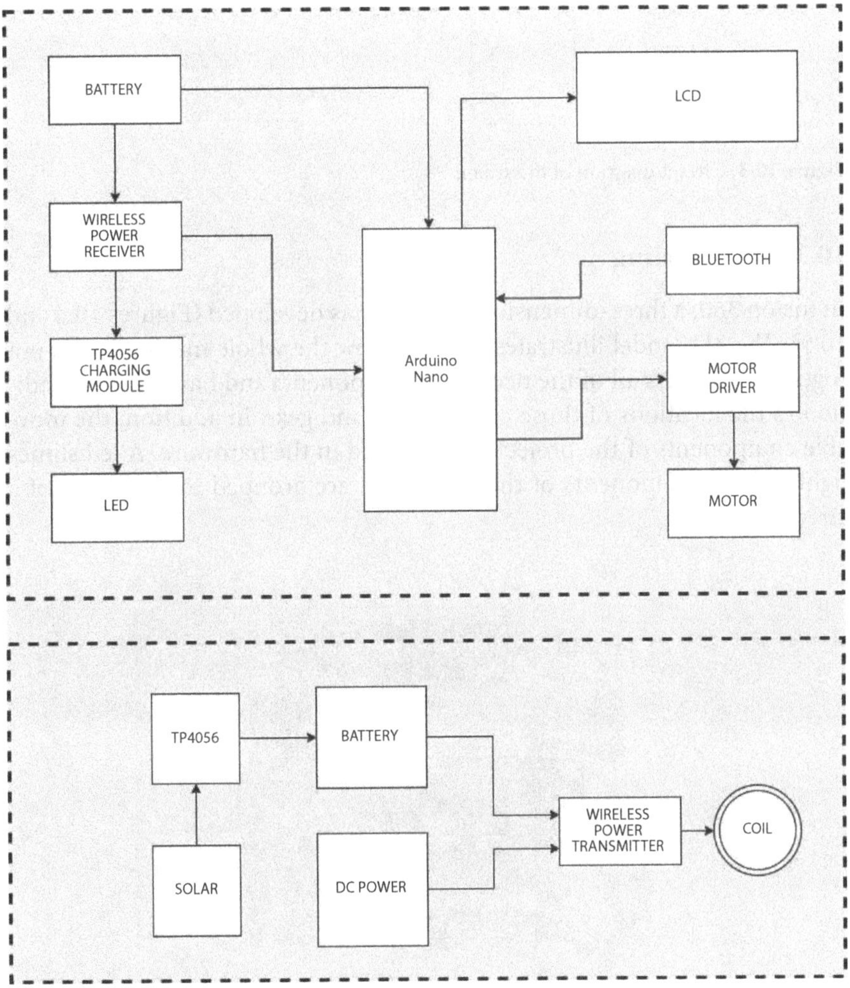

Figure 10.2 Block diagram of the system.

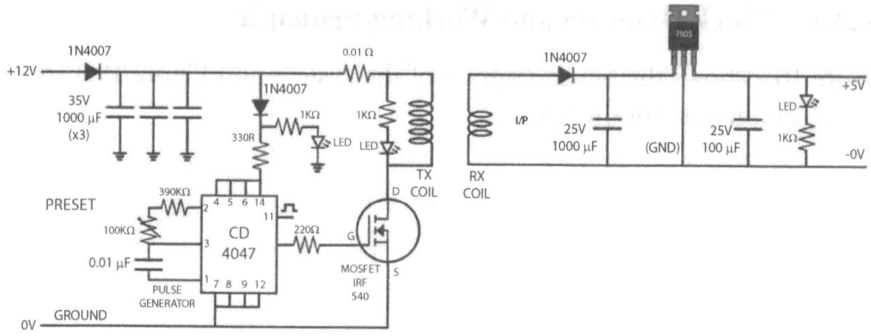

Figure 10.3 Circuit diagram of the system.

10.3.2 Modeling

In fusion 360, a three-dimensional model was developed (Figures 10.4 and 10.5). The 3D model illustrates in detail how the whole mechanism is put together. It covers all of the necessary components and hardware, in addition to the locations of those components and gear. In addition, the movable components of the project are included in the hardware. A led shines light on the components of the car, which are grouped in the shape of a circular.

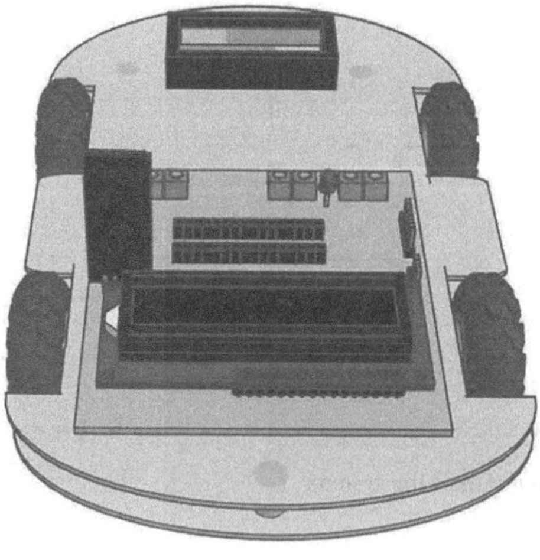

Figure 10.4 3D prototype for EV.

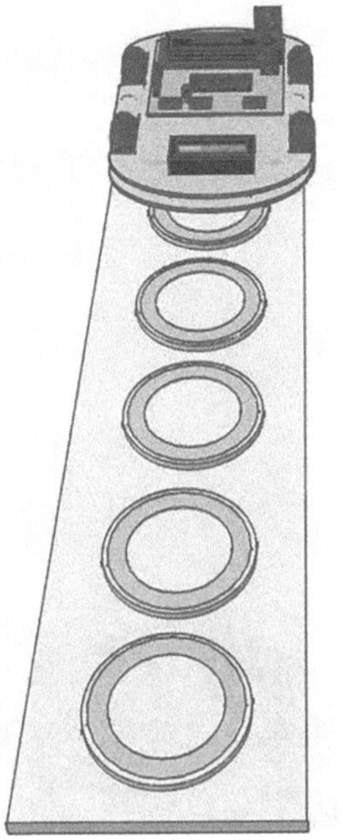

Figure 10.5 3D prototype for road.

10.4 Components

Figures 10.6–10.9 presents the different components used in an electrical vehicles. Arduino Uno is the best choice when it comes to learning how to code for embedded systems. If you're just starting out, the Arduino Uno board is a good board to compete with. The full PCB layout and schematics of this open Arduino platform are available online. An interesting Arduino Uno home automation and IoT project course may also appear in a web search. The Arduino Software Development Environment (IDE) is used to create all Arduino boards and is available for free. The Arduino Software Development Kit (IDE) can be downloaded to Windows, Linux and Mac computers. There are a number of third-party mobile applications that can be used to power Arduino boards.

5V 2A High Current Wireless Charging Module Transmitter Receiver Charging Coil Module is small, easy to use, efficient and low cost; this makes it ideal for the design and production of various small electronic

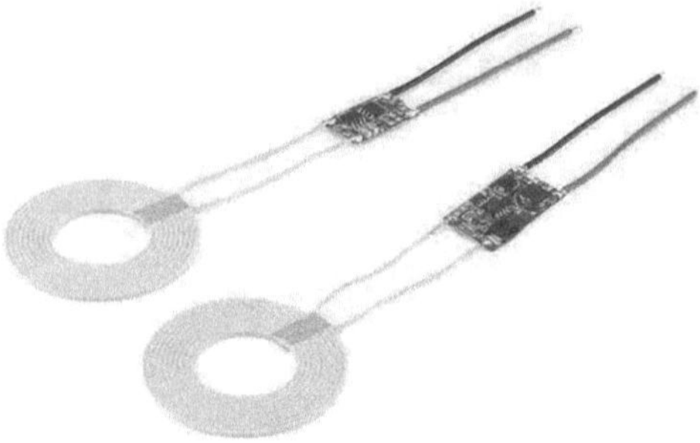

Figure 10.6 Arduino Uno.

Figure 10.7 Wireless coil.

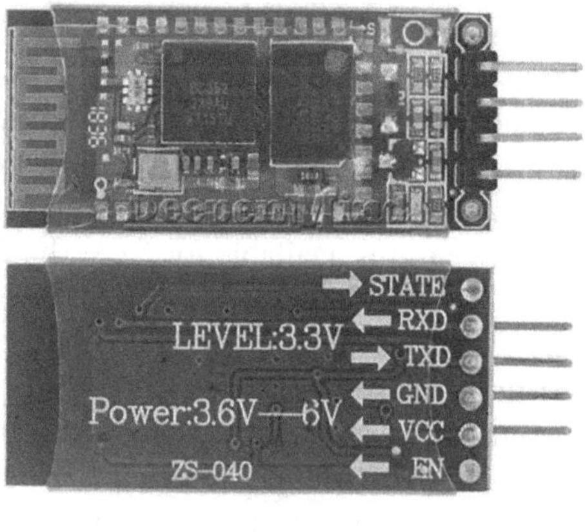

Figure 10.8 Arduino bluetooth module (HC-06).

Figure 10.9 LCD display.

devices, wireless charging and electronics. Wireless charging of mobile phones, video game consoles, tanks, digital audio players, digital still and video recorders, electric shavers, and many other applications benefit from this technology. The packaging is fully insulated, waterproof and dust-free.

HC-06 is a Class 2 slave Bluetooth module designed for transparent wireless communication. When paired with a key Bluetooth device such as a PC, smartphone or tablet, its functions become transparent to users. All data received from the serial port is instantly transmitted wirelessly. LCD Display with a yellow and blue lighting. It works wonderfully with Arduino-based creations. This 16×2 standard LCD with a yellow/blue backlight is simple to connect to an Arduino or other microcontroller. Displayed information can be either plain text or numeric values read from the sensors (such as temperature or pressure).

10.5 Market Dynamics and Adoption Trends

The market share of electromagnetic induction charging systems for electric vehicles (EVs) is influenced by many factors, including regulatory requirements, consumer preferences, technological advances and business participation. As the global automotive industry continues its transition to electric energy, the need for reliable and convenient payment options continues to increase. Government efforts to reduce greenhouse gas emissions and promote clean transportation have created opportunities for business growth and innovation by encouraging investment in electric vehicles.

Consumer preferences also play an important role in influencing the use of wireless charging technology. Electric vehicle drivers value simplicity, reliability and ease of use when choosing a solution. As a result, there is a growing demand for wireless charging systems that can be integrated into daily life, such as home appliances and public charging stations in parking lots. Additionally, improvements in user experience such as smartphone app integration, payment options, and payment notifications have increased the appeal of induction-based charging systems for customers.

Technological advancements and industry partnerships are driving innovation and expansion in the wireless charging industry [70–75]. Automakers, technology companies, and electronics and infrastructure developers have come together to create interoperable solutions, simplify deployment, and create connectivity for EV drivers. This partnership drives adoption and business growth by supporting the design, collaboration and optimization of the payments ecosystem.

Global adoption demonstrates interest in wireless payments across the region, with governments, municipalities and private companies investing in infrastructure and encouraging EV adoption. Additionally, emerging trends such as urbanization, shared services, and electric fleets are increasing the demand for wireless charging in urban environments, business environments, and public transportation [76–78].

Looking to the future, the market for payments based on electricity usage will continue and develop. Factors such as technology development, management support, customer education and infrastructure investment will play a key role in shaping the future of wireless charging technology. By responding to market trends and changes, stakeholders can seize opportunities, overcome challenges, and transition to cleaner, safer transportation that is better than using wireless charging technology.

10.6 Efficiency and Sustainability Considerations

Efficiency and stability are important factors in the development and application of electric induction charging systems for electric vehicles (EVs). As the automotive industry transitions to electricity to reduce environmental impact, it is important to ensure that electric charging meets sustainability targets and operates efficiently.

An important function of wireless charging technology is energy transfer. Efforts are currently being made to optimize the design of the charger coil and control the power supply to minimize power loss during induction. High-efficiency chokes and power electronics provide more efficient power conversion, shortening charging time and reducing energy consumption. Additionally, advances in resonance-based charging systems can further improve power transfer by providing greater flexibility and efficiency in the integration of coils.

Additionally, sustainability issues are not limited to energy use but also include the environmental impacts of housing payments. Integrating renewable energy sources such as solar and wind into wireless charging systems can reduce dependence on fossil fuels and reduce carbon emissions associated with charging. Electromagnetic induction-based charging systems can reduce carbon footprint and contribute to overall environmental sustainability by using clean energy.

Life assessment plays an important role in assessing the sustainability of housing payments. The assessment considers the environmental impact of construction materials throughout their lifecycle, from production and installation to operation and end-of-life disposal. By conducting a lifecycle assessment, stakeholders can identify opportunities to reduce environmental impact, improve resource use, and reduce waste throughout the payment lifecycle. one In addition, durable materials and structure are important in reducing environmental impact and extending the payment life. The use of recycled materials, environmentally friendly construction methods and production processes can reduce the carbon footprint of construction projects and support sustainability goals.

In general, performance and safety considerations are important for the design, installation and operation of electromagnetic induction-based charging systems for electric vehicles. By prioritizing energy efficiency, integrating renewable energy, making life assessments, and using products and practices, stakeholders can create a safer and more environmentally friendly ecosystem that supports the transition to electric vehicles.

10.7 User Experience and Convenience

Improving user experience and comfort is key to operating electric induction-based charging systems for electric vehicles (EVs). User-friendly design, seamless integration with the vehicle engine and personalized services play an important role in making EV drivers comfortable and free.

An important aspect of improving user experience is user-centered design. Payment design should be user-friendly to include products such as ease of use, accessibility and beauty. Payment information needs to be placed in convenient and accessible locations in parking lots, car parks and within the city for easy access by EV drivers. Additionally, the user interface should be intuitive and user-friendly, providing clear instructions and feedback to guide users through the checkout process.

The seamless connection with the car engine is another key factor in improving user experience. Wireless charging systems must communicate seamlessly with onboard systems to provide instant charge status updates, alerts, and vehicle-to-grid integration capabilities. Integration with smartphones and in-car infotainment systems allows EV drivers to monitor charging status, schedule payments, and easily manage payment options from their mobile phone or vehicle dashboard. One personalized services that meet the individual needs and preferences of using electric vehicles and strengthen the user experience. Providing personalized payment information, preferences and support from a customer account or loyalty program allows EV drivers to customize and streamline their payments. Additionally, value-added services such as concierge payment, electronic payment and priority payment support enhance the overall experience and differentiate payment centers in the market.

Overall, the importance of user experience and convenience is key to the adoption and satisfaction of electromagnetic induction charging systems for electric vehicles in driving cars. By providing a user-friendly design process, integration with the car engine, and personalized services, participants can encourage the use of electric vehicles by creating an interactive and entertaining experience.

10.8 Reliability and Durability Challenges

Ensuring the reliability and durability of electromagnetic induction-based charging systems is important in maintaining user confidence, reducing downtime and extending the life of charging systems. However, to achieve the high reliability and long-term performance of wireless charging technology, many challenges need to be solved.

A significant challenge is material selection and environmental changes. Charging infrastructure is affected by many environmental factors such as temperature changes, humidity, and physical wear and tear. Therefore, choosing durable materials that can withstand harsh environments is important for your products to last a long time. Additionally, corrosion-resistant processes and protection measures can reduce the impact of the environment on the payment of the structure and extend its service life and reliability.

Another challenge relates to thermal management and distribution. Wireless charging systems generate heat during operation, which can affect performance and reliability if not properly controlled. Adhering to good thermal management practices such as heat sinks, fans, and cooling systems is crucial to dissipating excess heat and maintaining efficient operation. By providing proper heating, housing payments can operate efficiently over long periods of use and when needed.

In addition, regular maintenance and proper maintenance are important to detect and solve potential problems before they turn into serious problems. Adopting monitoring technologies such as remote monitoring and diagnostics allows operators to detect early signs of equipment failure and plan maintenance accordingly. In-house payments that cover maintenance needs can reduce downtime, maximize uptime, and improve overall reliability.

Additionally, compliance and compliance with regulatory standards are essential to ensure the security and reliability of payments. Compliance with industry standards such as ISO 15118 and SAE J2954 ensures compatibility, compatibility and security across different devices and different tools. Additionally, regular security audits, audits and certifications help verify compliance with regulatory requirements and industry best practices, thus ensuring customers and stakeholders have confidence in the reliability and security of payments.

In summary, reliability and durability challenges are important to ensure the long-term and success of electromagnetic induction charging systems for electric vehicles. Through the use of quality materials, thermal management, efficiency and compatibility management, stakeholders can increase the reliability, durability and safety of electric charging to accelerate EV adoption and prepare for the future of transportation.

10.9 Future Scopes

As electromagnetic induction charging systems for electric vehicles (EVs) continue to evolve, future research should focus on significant advances to solve key challenges and open new opportunities for updated and improved design. Here are some great tips for research:

Advanced Coil Design: Research new coil geometries, materials and configurations to optimize the magnetic field, increase energy transfer and improve charging.

Dynamic Charging System: Develop a wireless charging system that can charge electric vehicles while driving, allowing continuous charging on highways and highways with heavy traffic and supporting long journeys.

Bidirectional Power Transfer: Researching bidirectional power transfer can complete the integration of vehicle-to-grid (V2G) and vehicle-to-home (V2H), making the electric vehicle the power of the mobile phone and contributing to a stable and strong grid.
Wireless Charging Infrastructure: Research into scalable, cost-effective, and interoperable wireless charging infrastructure solutions to support the expansion and adoption of induction-based charging systems in a variety of applications and environments.
Energy Management Systems: Develop advanced energy management systems and processes to optimize charging times, coordinate charging, and instantly balance energy needs and products to be efficient and resilient.
Smart Grid Integration: The wireless charging system is integrated with the smart grid to ensure uninterrupted communication, data exchange and charging management in buildings, vehicles and networks, and to support improvement plans and demand responses.
Standardization and interoperability: Continue efforts towards standardization and interoperability of wireless charging protocols, interfaces and communication protocols to ensure compatibility, easy integration and seamless user interface between payment systems and vehicle platforms.
Safety and Security Improvement: Best security research, bug detection and reliability engineering techniques to increase security, resilience and usability, increase the strength of wireless charging infrastructure, reduce risks and increase user trust.
Cost Reduction Strategies: Research manufacturing processes, materials, and equipment to reduce the cost of wireless charging infrastructure and increase cost, scalability, and accessibility for customers and suppliers' stakeholders.
User-centered design: Integrating user input, preferences, and behavioral analytics into the design and development of wireless charging systems to create an intuitive, user-friendly, and personalized experience that increases adoption and satisfaction.

By focusing on these research directions, stakeholders can overcome remaining challenges, stimulate innovation, and help support clean, green, and longer transportation by accelerating energy efficiency for electric vehicles.

10.10 Market Expansion and Global Collaboration

Expanding market access and promoting global cooperation for the use of electric vehicles (EVs) is important for driving electric vehicle adoption worldwide and accelerating the transition to EVs. There are several key strategies that can be used to facilitate business expansion and collaboration:

Global Market Analysis: Conduct a global market analysis to identify emerging markets, growth opportunities, and market trends for wireless payments. Analyzing factors such as regulatory frameworks, customer preferences, infrastructure needs and the competitive landscape can help stakeholders prioritize business strategies and investment decisions.
Collaboration and Collaboration: Create collaboration, coordination, and collaboration among industry stakeholders including automakers, suppliers, technology, utilities, infrastructure, and government agencies. Collaboration facilitates sharing, resource utilization, and collaboration to solve different problems, spur innovation, and accelerate business growth.
International Standards Development: Participate with international standards bodies and regulatory bodies to develop international standards, cooperative agreements and regulatory frameworks for wireless charging technology. Standardization promotes compatibility, reliability and scalability across different industries and supports global adoption and integration of electromagnetic induction-based charging systems.
Cross-border distribution: Promote cross-border distribution and cooperation to improve communications for electric vehicle operators traveling between different countries and regions. Harmonizing specifications, regulations and standards can facilitate cross-border payments and promote international travel and trade. Business support and policy support: Advocate for government, municipal and regulatory bodies to provide business support, financial support and introduce policies to encourage investment in wireless charging infrastructure and promote electric vehicles. Incentives such as tax credits, grants, subsidies and rebates help reduce housing costs, stimulate consumer spending and stimulate business demand.
Public-Private Partnerships: Public-private partnerships (PPPs) are encouraged to leverage public and private sector efforts and resources in the design, financing, and operation of wireless charging infrastructure. PPP enables greater efficiency and effectiveness by encouraging collaboration, risk sharing and innovation in infrastructure projects.
Technology transfer and capacity building: Promote technology transfer and capacity building initiatives to promote knowledge exchange, skills development, and technology transfer between area code design and construction. The capacity building program develops local skills, builds capacity and engages regional stakeholders in the development and deployment of wireless charging technology. By using these strategies and promoting

international cooperation, stakeholders can overcome economic challenges, stimulate demand, build better relationships and connect the ecosystem to induction-based charging systems, ultimately accelerating the transition to electric vehicles and achieving a clean, greener transportation future.

Conclusions

In summary, charging systems based on electromagnetic induction hold great promise in the development of electric vehicle (EV) charging infrastructure. From this extensive research, it is clear that these systems have many benefits such as optimization, simplicity and stability. Technological advances such as optimized coil design, smart grid integration and user-oriented features reveal the potential of wireless charging technology to increase vehicle power.

Moreover, the strong market and international cooperation, as well as the importance of energy management, show many ways to complement and expand wireless charging infrastructure. By solving trust issues, improving technology management, and building global partnerships, stakeholders can overcome challenges and enable the world to transition to electric cars. One essentially, charging systems based on electrical induction represent a significant step towards a clean and portable future. Through continued innovation, collaboration and investment, the vision of efficient mobility through wireless charging technology can be realized; this could lead to an era of cleaner air, reduce greenhouse gas emissions and improve energy security for future generations.

References

1. Singh, S., Hasarmani, T., Holmukhe, R., Wireless transmission of electrical power overview of recent research & development. *Int. J. Comput. Electr. Eng.*, 4, 207–211, 2012, 10.7763/ijcee.2012.v4.480.
2. Kumar, S., Paliwal, H., Vyas, S., Sekhor, S., Vikramaditya, R., Dynamic wireless power transfer in electric vehicles. *J. Phys. Conf. Ser.*, 1854, 012014, 2021, 10.1088/1742-6596/1854/1/012014.
3. Ahmed, A. and Yadav, A.K., Parametric Analysis of Wastewater Electrolysis for Green Hydrogen Production: A Combined RSM, Genetic Algorithm, and Particle Swarm Optimization Approach. *Int. J. Hydrogen Energy*, 59, 51–62, 2024.

4. Ahmad, A., Yadav, A.K., Hasan, S., Enhanced production of methane enriched biogas through intensified co-digestion process and its effective utilization in a biodiesel/biohydrogen fueled engine with duel injection strategies: ML-RSM based an efficient optimization approach. *Int. J. Hydrogen Energy*, 65, 671–686, 2024.
5. Ahmed, A., Yadav, A.K., Singh, A., Singh, D.K., A machine learning–genetic algorithm based models for optimization of intensification process through microwave reactor: A new approach for rapid and sustainable synthesis of biodiesel from novel Hiptage benghalensis seed oil. *Fuel*, 374, 132470, 2024.
6. Khan, O., Khan, M.P., Yadav, A.K., A Comparative Analysis of Various Nanocomposites for the Remediation of Heavy Metals from Biomass Liquid Digestate using Multi-Criteria Decision Methods. *Biomass Bioenergy*, 186, 107281, 2024.
7. Khan, O., Alsaduni, I., Equbal, A., Parvez, M., Yadav, A.K., Performance and Emission Analysis of Biodiesel Blends Enriched with Biohydrogen and Biogas in Internal Combustion Engines. *Process Saf. Environ. Prot.*, 183, 1013–1037, 2024.
8. Ahmed, A., Yadav, A.K., Singh, A., Singh, D.K., A machine learning-response surface optimization approach to enhance the performance of diesel engine using novel blends of aloe vera biodiesel with MWCNT nanoparticles and hydrogen. *Process Saf. Environ. Prot.*, 186, 738–755, 2024, doi: 10.1016/j.psep.2024.04.013.
9. Ahmed, A., Yadav, A.K., Singh, D.K., Singh, A., A Comprehensive Machine Learning-Coupled Response Surface Methodology Approach for Predictive Modeling and Optimization of Biogas Potential in Anaerobic Co-Digestion of Organic Waste. *Biomass Bioenergy*, 180, 106995, 2024.
10. Ahmad, A., Yadav, A.K., Singh, A., Singh, D.K., Ağbulut, U., A hybrid RSM-GA-PSO approach on optimization of process intensification of linseed biodiesel synthesis using an ultrasonic reactor: Enhancing biodiesel properties and engine characteristics with ternary fuel blends. *Energy*, 288, 129077, 2024.
11. Ahmed, A., Yadav, A.K., Hasan, S., Enhancing Efficiency and Environmental Impacts of a Nano-Enhanced Oleander Biodiesel-Biohydrogen Dual Fuel Engine Equipped with EGR, through Operational Parameter Optimization using RSM-MOGA Technique. *Int. J. Hydrogen Energy*, 78, 1157–1172, 2024.
12. Khan, O., Parvez, M., Yahya, Z., Alhodaib, A., Yadav, A.K., Hoang, A.T., Ağbulut, U., Waste-to-energy power plants: Multi-objective analysis and optimization of landfill heat and methane gas by recirculation of leachate. *Process Saf. Environ. Prot.*, 186, 957–968, 2024.
13. Khanam, S., Khan, O., Ahmad, S., Sherwani, A.F., Khan, Z.A., Yadav, A.K., Ağbulut, U., A Taguchi-based hybrid multi-criteria decision-making approach for optimization of performance characteristics of diesel engine fuelled with blends of biodiesel-diesel and cerium oxide nano-additive. *J. Therm. Anal. Calorim.*, 149, 8, 3657–3676, 2024.

14. Ahmed, A., Yadav, A.K., Singh, A., Enhancing waste cooking oil biodiesel yield and characteristics through machine learning , response surface methodology, and genetic algorithm for optimal utilization in CI Engines. *Int. J. Green Energy*, 21, 1345–1365, 2024.
15. Singh, A.A., Singh, R.K., Yadav, A.K., Saha, A., Dewangan, A., Singh, B., Experimental investigations on the electrochemical machining of D3 die steel material and multi-objective parameter optimization. *Surf. Rev. Lett.*, 31, 2450074, 2024.
16. Ahmed, A., Yadav, A.K., Hasan, S., Biogas as a Sustainable and Viable Alternative Fuel for Diesel Engines: A Comprehensive Review of Production, Purification, Economic Analysis, and Performance Evaluation. *J. Process Mech. Engg. Part E*, 2024. https://doi.org/10.1177/09544089241255930.
17. Khan, F., Karimi, M.N., Khan, O., Yadav, A.K., Alhodaib, A., Gürel, A.E., Ağbulut, U., A hybrid MCDM optimization for utilization of novel set of biosynthesized nanofluids on thermal performance for solar thermal collectors. *Int. J. Thermofluids*, 22, 100686, 2024.
18. Alhodaib, A., Yahya, Z., Khan, O., Equbal, A., Equbal, M.S., Parvez, M., Idrisi, M.J., Sustainable coatings for green solar photovoltaic cells: performance and environmental impact of recyclable biomass digestate polymers. *Sci. Rep.*, 14, 1, 11221, 2024.
19. Singh, B., Yadav, A.K., Saha, A., Fabrication and Characterization of AA6082-T6 Produced by Friction Stir Additive Manufacturing. *Surf. Rev. Lett*, 2024. doi:10.1142/S0218625X25500222.
20. Equbal, A., Equbal, A., Ahmad, R., Kumar, M., Khan, O., Yadav, A.K., RSM based statistical approach to enhance the compressive performance of ABS P400 parts fabricated *via* FDM Technology. *J. Surf. Rev. Lett*, 2024. doi:10.1142/S0218625X25500210.
21. Dewangan, A., Ahmad, A., Yadav, A.K., Innovative research on waste tire recycling for sustainable biofuel production: Assessment of its usability on multi-cylinder diesel engine employing constant injection of oxyhydrogen and biogas through a premixing device. *Int. J. Hydrogen Energy*, 84, 155–163, 2024.
22. Dewangan, A., Mallick, A., Yadav, A.K., Islam, S., Saleel, C.A., Shaik, S., Ağbulut, U., Production of oxy-hydrogen gas and the impact of its usability on CI engine combustion, performance, and emission behaviors. *Energy*, 278, 127937, 2023.
23. Khan, O., Parvez, M., Kumari, P., Yadav, A.K., Akram, W., Ahmad, S., Idrisi, M.J., Modelling of compression ignition engine by soft computing techniques (ANFIS-NSGA-II and RSM) to enhance the performance characteristics for leachate blends with nano-additives. *Sci. Rep.*, 13, 1, 15429, 2023.
24. Ahmad, A., Yadav, A.K., Singh, A., Singh, D.K., Multi-response optimization of a microalgae-spirulina-fueled VCR diesel engine: a comprehensive RSM-GA approach. *Environ. Dev. Sustain.*, 27, 1–29, 2023.

25. Ahmed, A., Yadav, A.K., Singh, A., Process optimization of spirulina microalgae biodiesel synthesis using RSM coupled GA technique: a performance study of a biogas powered dual fuel engine. *Int. J. Environ Sci. Tech.*, 21, 169–188, 2023.
26. Khan, O., Ali, V., Parvez, M., Alhodaib, A., Yahya, Z., Yadav, A.K., Ağbulut, U., Exploring the performance of biodiesel-hydrogen blends with diverse nanoparticles in diesel engine: A hybrid machine learning K-means clustering approach with weighted performance metrics. *Int. J. Hydrogen Energy*, 78, 547–563, 2024.
27. Ahmed, A., Yadav, A.K., Singh, A., Application of machine learning and genetic algorithm for prediction and optimization of Biodiesel Yield from Waste Cooking Oil. *Korean J. Chem. Engg.*, 40, 2941–2956, 2023.
28. Ahmed, A., Yadav, A.K., Singh, A., Enhancement of Biogas Yield from Dual Organic Waste Using Hybrid Statistical Approach and Its Effects on Ternary Fuel Blend (Biodiesel/n-butanol /Diesel) Powered Diesel Engine Environmental Progress & Sust. Energy, 42, 5, e14163, 2023, doi.org/10.1002/ep.14163.
29. Ahmed, A., Yadav, A.K., Singh, A., Optimization of Cavitation-Assisted Biodiesel Production and Fuel Properties from Neochloris Oleoabundans Microalgae oil using Genetic Algorithm and Response Surface Methodology. *J. Process Mech. Engineering Part E*, 238, 4, 1821–1835, 2023, doi.org/10.1177/09544089231159832.
30. Yadav, M., Karimi, M.N., Yadav, A.K., Effect of Graphene Oxide Nanoparticles Dispersed Biodiesel on Combustion, Performance, and Emission Characteristics of a Diesel Engine. *J. Process Mech. Engineering Part E*, 2023, https://doi.org/10.1177/09544089231209139.
31. Dwivedi, S.P., Yadav, A.K., Saxena, A., Dwivedi, V.K., Tribo-mechanical, physical and thermal behaviour of Al/Si3N4 composite with and without the addition of Cu, Ni and Cr entropy elements. Proceedings of the Institution of Mechanical Engineers. *Part E J. Process Mech. Eng.*, 2023, 09544089231189663.
32. Das, D., Yadav, A.K., Ahmed, A., Numerical Simulation of the Combustion Chamber and Performance Analysis of Microalgae-Based Biodiesel/Diesel-Powered CI Engine: An Axisymmetric Flow Approach. *J. Process Mech. Engineering Part E*, 2023, https://doi.org/10.1177/09544089231218121.
33. Dewangan, A., Mallick, A., Yadav, A.K., Ahmad, A., Alqahtani, D., Islam., S., Combined Effect of Operating Parameters and Nano Particles on Performance of a Diesel Engine: Response Surface Methodology coupled Genetic algorithm Approach. *ACS Omega*, 8, 27, 24586–24600, 2023.
34. Singh, D., Singh, S., Yadav, A.K., Khan, O., Dewangan, A., Alkahtani, M.Q., Islam, S., From Theory to Practice: A Sustainable Solution to Water Scarcity by Using a Hybrid Solar Distiller with a Heat Exchanger and Aluminum Oxide Nanoparticles. *ACS Omega*, 37, 8, 33543–33553, 2023.

35. Ahmed, A., Yadav, A.K., Singh, A., Optimization of Biogas Yield from Anaerobic Co-digestion of Dual waste for environmental sustainability: ANN, RSM, and GA, Approach. *Int. J. Oil Gas Coal Tech.*, 33, 1, 75, 2023.
36. Ahmed, A., Yadav, A.K., Singh, A., An environmental impact assessment and optimization study of biodiesel production from microalgae. *Int. J. Global Warming*, 31, 3, 294–313, 2023.
37. Ahmed, A., Yadav, A.K., Singh, A., Biodiesel yield optimization from a third-generation feedstock (microalgae spirulina) using a hybrid statistical approach. *Int. J. Ambient Energy*, 44, 1, 1202–1213, 2023.
38. Ahmed, A., Yadav, A.K., Singh, A., Modelling and optimisation of VCR diesel engine parameters fuelled with microalgae methyl-ester: a RSM-coupled Taguchi approach. *Int. J. Ambient Energy*, 44, 1, 2498–2506, 2023.
39. Ahmed, A., Yadav, A.K., Singh, A., Biodiesel Production from Mahua Oil: Characterization, Optimization, and Modeling with a Hybrid Statistical Approach. *Int. J. Ambient Energy*, 44, 1, 2600–2609, 2023.
40. Khan, T.A., Khan, T.A., Yadav, A.K., Khan, M.E., Experimental Studies on Utilization of Neochloris Oleoabundans Microalgae Biodiesel as an Alternative Fuel for Diesel Engine. *Int. J. Ambient Energy*, 44, 1, 115–123, 2023.
41. Singh, D., Yadav, A.K., Kumar, A., Samsher, Energy matrices and life cycle conversion analysis of N-identical hybrid double slope solar distiller unit using Al2O3 nanoparticle. *J. Water Environ. Nanotechnol.*, 8, 3, 267–284, 2023.
42. Khan, T.A., Yadav, A.K., Khan, T.A., A hydrodynamic cavitation-assisted system for optimization of biodiesel production from green microalgae oil using a genetic algorithm and response surface methodology approach. *Environ. Sci. Pollut. Res.*, 29, 49465–49477, 2022.
43. Dewangan, A., Yadav, A.K., Mallick, A., Singh, S., Comparative study of Manilkara Zapota and Karanja based Biodiesel Properties and Its Effect on Multi-Cylinder Diesel Engine Characteristics. *Energy Sources Part A*, 44, 2, 5143–5153, 2022.
44. Yadav, A.K., Khan, T.A., Khan, T.A., Kumar, S., Methyl ester of Gmelina arborea oil as a substitute for petroleum diesel: an experimental study on its performance and emissions in a diesel engine. *Energy Sources Part A*, 43, 11, 1307–1314, 2021.
45. Khan, O., Yadav, A.K., Khan, M.E., Parvez, M., Characterization of Bio-Ethanol Obtained from Eichhornia Crassipes Plant; Its Emission and Performance Analysis on CI Engine. *Energy Sources Part A*, 43, 14, 1793–1803, 2021.
46. Khan, I.A., Singh, S.K., Pal, A., Yadav, A.K., Efficient Production of Biodiesel from Cannabis Sativa oil using Intensified Transestericatión (Hydrodynamic Cavitation) Method. *Energy Sources Part A*, 42, 20, 2461–2470, 2020.

47. Dewangan, A., Yadav, A.K., Mallick, A., Tyagi, A., Optimisation of biodiesel production and engine performance from Simarouba oil in compression ignition engine. *Int. J. Oil Gas Coal Technol.*, 25, 3, 357–376, 2020.
48. Khan, M.E., Khan, T.A., Yadav, A.K., Pal, A., A comprehensive review on performance, emission and combustion characteristics of diesel engines fuelled with algae oil and methyl ester. *WEENTECH Proc. Energy*, 6, 2, 160–168, 2020.
49. Yadav, A.K., Khan, M.E., Pal, A., Custard Apple Seed (Annona Squamosa) Oil as a Promising Feedstock for Biodiesel Production using Advanced Techniques and Its Experimental Investigations on an Agricultural Diesel Engine. *Int. J. Oil Gas Coal Tech.*, 20, 4, 473–492, 2019.
50. Yadav, A.K., Pal, A., Ghosh, U., Gupta, S.K., Comparative Study of Biodiesel Production Methods from Yellow Oleander Oil and its Performance Analysis on an Agricultural Diesel Engine. *Int. J. Ambient Energy*, 40, 2, 152–157, 2019.
51. Khan, I.A., Singh, S.K., Yadav, A.K., Sharma, D., Enhancement in the Performance of a Diesel Engine Fueled with Pongamia Methyl Ester and n-Butanol as Oxygenated Additive. *Int. J. Ambient Energy*, 40, 8, 842–846, 2019.
52. Yadav, A.K., Khan, O., Khan, M.E., Utilization of High FFA Landfill Waste (Leachates) as a Feedstock for Sustainable Biodiesel Production: Its Characterization and Engine Performance Evaluation. *Environ. Sci. Pollut. Res.*, 25, 32, 32312–32320, 2018.
53. Yadav, A.K., Dewangan, A., Mallick, A., Effect of n-Butanol and Diethyl Ether Additives on Performance and Emission Characteristics of a Diesel Engine Fuelled with Diesel-Pongamia Biodiesel Blends. *J. Energy Eng.*, 144, 6, 04018062, 2018.
54. Vinay, S.B. and Yadav, A.K., Optimization of Performance and Emission Characteristics of CI Engine Fuelled with Mahua Oil Methyl Ester-Diesel Blend using Response Surface Methodology. *Int. J. Ambient Energy*, 44, 6, 674–85, 2020.
55. Yadav, A.K., Dewangan, A., Mallick, A., Synthesis and Stability Study of Biodiesel from Kachnar Seed Oil. *J. Energy Eng.*, 144, 5, 04018053, 2018.
56. Yadav, A.K., Pal, A., Dubey, A.M., Experimental Studies on Utilization of Prunus Armeniaca L. (Wild Apricot) Biodiesel as an Alternative Fuel for CI Engine. *Waste Biomass Valorization*, 9, 10, 1961–1969, 2018.
57. Yadav, A.K., Khan, M.E., Pal, A., Ultrasonic Assisted Production of Biodiesel from Karabi Oil using Heterogeneous Catalyst. *Biofuels*, 9, 1, 101–112, 2018.
58. Yadav, A.K., Vinay, Singh, B., Optimization of Biodiesel Production from Annona Squamosa Seeds Oil Using Response Surface Methodology and Its Characterization. *Energy Sources Part A*, 40, 9, 1051–1059, 2018.
59. Dewangan, A., Yadav, A.K., Mallick, A., Current Scenario of Biodiesel Development in India: Prospects and Challenges. *Energy Sources Part A*, 40, 20, 2494–2501, 2018.

60. Yadav, A.K., Khan, M.E., Pal, A., Performance and Emission Characteristics of a Stationary Diesel Engine fuelled by Schleichera Oleosa oil Methyl Ester (SOME) Produced through Hydrodynamic Cavitation Process, Egypt. *J. Pet.*, 27, 1, 89–93, 2018.
61. Yadav, A.K., Khan, M.E., Pal, A., Performance, Emission and Combustion Characteristics of an Indica Diesel Engine Operated with Yellow Oleander (Thevetia Peruviana) Oil Biodiesel Produced through Hydrodynamic Cavitation Method. *Int. J. Ambient Energy*, 39, 4, 365–371, 2018.
62. Dewangan, A., Mallick, A., Yadav, A.K., Effect of Metal Oxide Nanoparticles and Engine Parameters on the Performance of a Diesel Engine A Review. *Mater. Today Proc.*, 21, 1722–27, 2020.
63. Yadav, A.K., Khan, M.E., Pal, A., Dubey, A.M., Experimental Investigations of Performance and Emissions Characteristics of Kusum (Schleichera Oleosa) Biodiesel in a Multi-Cylinder Transportation Diesel Engine. *Waste Biomass Valor*, 8, 4, 1331, 2017.
64. Yadav, A.K., Khan, M.E., Pal, A., Biodiesel Production from Oleander (Thevetia Peruviana) Oil and Its Performance Testing on a Diesel Engine. *Korean J. Chem. Eng.*, 34, 2, 340–345, 2017.
65. Dubey, A.M., Kumar, A., Yadav, A.K., Wear behaviour of friction stir weld joint of cast Al (4–10%) Cu alloy welded at different operating parameters. *J. Mater. Process. Tech.*, 240, 87–97, 2017.
66. Khan, O., Khan, M.E., Yadav, A.K., Sharma, D., The Ultrasonic Assisted Optimization of Biodiesel Production from Eucalyptus Oil. *Energy Sources Part A*, 39, 13, 1323–1331, 2017.
67. Dewangan, A. and Yadav, A.K., Wax Deposition during Production of Waxy Crude Oil and Its Remediation. *Pet. Sci. Tech.*, 35, 18, 1831–1838, 2017.
68. Yadav, A.K., Khan, M.E., Pal, A., Kaner biodiesel production through hybrid reactor and its performance testing on a CI engine at different compression ratios. *Egypt. J. Pet.*, 26, 2, 525–532, 2017.
69. Yadav, A.K., Khan, M.E., Pal, A., Dubey, A.M., Performance and emission characteristics of a transportation diesel engine operated with non-edible vegetable oils biodiesel. *Case Stud. Therm. Eng.*, 8, 236–244, 2016.
70. Yadav, A.K., Khan, M.E., Pal, A., Biodiesel Production from Nerium Oleander (Thevetia Peruviana) Oil through Conventional and Ultrasonic Irradiation Methods. *Energy Sources Part A*, 38, 23, 3447–3452, 2016.
71. Yadav, A.K., Khan, M.E., Pal, A., A Comparative Study on Ultrasonic Cavitation and Mechanical Stirring Method towards Efficient Production of Biodiesel from Non-Edible Oils and Performance Testing on a C.I. Engine. *Int. J. Environ. Waste Manag.*, 18, 4, 349–367, 2016.
72. Dewangan, A., Mallick, A., Yadav, A.K., Kumar, R., Combustion-generated pollutions and strategy for its control in CI engines: A review. *Mater. Today Proc.*, 21, 1728–1733, 2020.
73. Yadav, A.K., Khan, M.E., Pal, A., Sharma, D., Optimization of biodiesel production from bitter groundnut oil using Taguchi method and its performance

and emissions characteristics on a 4-cylinder Tata Indica engine. *Int. J. Sust. Agri. Mgt. Inf.*, 1, 285–300, 2015.
74. Khan, F., Khan, O., Parvez, M., Ahmad, S., Yahya, Z., Alhodaib, A., Yadav, A.K., Ağbulut, U., K-means clustering optimization of various quantum dots and nanoparticles-added biofuels for engine performance, emission, vibration, and noise characteristics. *Therm. Sci. Eng. Prog.*, 54, 102815, 2024.
75. Khan, O., Parvez, M., Yadav, A.K., *et al.*, Investigation of the Thermal Properties of Lauric Acid Melting as a Phase Change Material Using Priority based K-Means Clustering Method. *Energy Storage*, 6, 6, e70026, 2024, doi:10.1002/est2.70026.
76. Sharma, G., Dewangan, A.K., Yadav, A.K., Ahmed, A., Current status of research on hydrogen generation, storage and transportation technologies: A state-of-the-art reviewtowards sustainable energy Process. *Saf. Environ. Prot.*, 191, Part A, 1445–1460, 2024.
77. Singh, S., Singh, B., Kumar, S., Yadav, A.K., Temperature-dependent dynamic hysteresis scaling of ferroelectric hysteresis parameters of lead free [(Ba 0.825 + xCa0.175-x) (Ti1- x Snx)O3]ceramics. *Ferroelectrics*, 551, 1, 133–142, 2019.
78. Dubey, A.M., Ram, R., Yadav, A.K., Ion cell performance using single particle representation of battery electrode. *J. Electro Anal. Chem.*, 768, 11–17, 2016.

11

Efficient Battery Thermal Management System in High-Performance Applications

Ashwani Kumar[1]*, Mukesh Kumar Awasthi[2], Nitesh Dutt[3] and Neha Singh Raghuvanshi[4]

[1]*Department of Mechanical Engineering, Technical Education Department, Uttar Pradesh, Kanpur, India*
[2]*Department of Mathematics, Babasaheb Bhimarao Ambedkar University, Lucknow, India*
[3]*Department of Mechanical Engineering, COER University, Roorkee, Uttarakhand, India*
[4]*Department of Applied Science & Humanities, ABES Engineering College, Ghaziabad, Uttar Pradesh, India*

Abstract

Maintaining optimal battery temperature is critical for maximizing performance, lifespan, and safety in high-demand applications. This chapter explores advancements in battery thermal management systems (BTMS) designed high performance application scenarios. Chapter examine three key cooling approaches: air-based systems, utilizing forced convection for heat dissipation; liquid-based systems, employing coolants for more efficient heat transfer; and phase change materials (PCMs), leveraging their latent heat storage capacity for passive thermal regulation. Additionally, the integration of intelligent control strategies with machine learning is discussed, enabling proactive system adjustments based on real-time battery data. By analyzing these advancements, the chapter highlights the path towards efficient BTMS design, ensuring optimal battery performance and reliability in high-performance applications.

**Corresponding author*: drashwanikumardte@gmail.com; ORCID: 0000-0003-4099-935X
Mukesh Kumar Awasthi: ORCID: 0000-0002-6706-5226
Nitesh Dutt: ORCID: 0000-0002-0405-3850
Neha Singh Raghuvanshi: ORCID: 0000-0002-8437-6261

Ashwani Kumar, Mukesh Kumar Awasthi, Nitesh Dutt, Yogesh Kumar Singla and Sivasakthivel Thangavel (eds.) Thermal Battery Management System for Hybrid and Electric Vehicles, (221–240) © 2025 Scrivener Publishing LLC

Keywords: Battery thermal management system (BTMS), high-performance applications, air-based cooling, liquid-based cooling, phase change materials (PCMs), machine learning

11.1 Introduction

In the process of high-performance applications, batteries have very critical roles. From electric vehicles (EVs) to drones, and high-powered electronics to the silent guardians of grid stability, batteries, enables these technologies to push the boundaries of performance [1, 2]. These remarkable devices act like tiny powerhouses, storing and releasing energy to fuel these high-performance applications. Boyes [1] and Liu *et al.* [2] have investigated the batteries performance and advance material application for thermal managements.

Li *et al.* [3] investigated the electrochemical properties of drone batteries. The mechanism of battery's function lies in its electrochemical potential. During discharge, internal chemical reactions occur within the battery, generating an electrical current. This current, characterized by its voltage and amperage, delivers the usable power to the application. The capacity of a battery, typically measured in Ampere-hours (Ah), signifies the total amount of electrical charge it can deliver before needing recharge. High-performance applications often necessitate batteries with exceptional energy density (Wh/kg), which refers to the amount of energy stored per unit mass. This translates to a battery that can pack a greater punch in terms of power delivery without adding excessive weight, a crucial factor for applications like EVs and drones where weight directly impacts range and flight time (Estrada-Martinez *et al.* [4]).

The four different examples of high performance applications for understanding the battery energy density are as follows: Case 1, for EVs, the battery capacity directly translates to driving range. A larger capacity battery allows the vehicle to travel farther on a single charge. Case 2, in drones, higher energy density batteries enable extended flight times, critical for tasks like aerial photography, search and rescue operations, and commercial deliveries. Case 3, in high-powered electronics, uninterrupted operation hinges on the battery's ability to deliver sufficient current to sustain demanding workloads (powerful laptops running complex programs for hours). Case 4, grid storage applications, on the other hand, rely on batteries with exceptional capacity to store excess renewable energy during periods of low demand and release it back to the grid when required, ensuring grid stability and efficient energy utilization (Sun *et al.* [5]).

While batteries offer the potential for exceptional performance, their operation in high-performance applications presents a unique set of challenges. One of the most significant concerns is ***battery degradation***. During discharge and recharge cycles, internal chemical and physical changes occur within the battery, leading to a gradual decrease in its capacity and overall performance. This degradation is often accelerated by factors like high operating temperatures, deep discharge cycles (completely draining the battery), and high charge/discharge rates. Schmuch *et al.* [6] have Analyzes battery performance metrics for EVs. In EVs, battery degradation translates to a reduction in driving range over time. Similarly, for drones, it translates to shorter flight times before requiring recharge. Another critical challenge in high-performance applications is ***thermal management***. Battery performance and lifespan are highly dependent on temperature. Operation at elevated temperatures can accelerate degradation processes, reduce capacity, and even pose safety risks. Conversely, operating at very low temperatures can also hinder performance. Khalilzadeh *et al.* [7] reviewed the battery modeling for EVs, including battery capacity. In EVs, battery performance can suffer in extreme cold weather, leading to reduced driving range. Similarly, drones operating in cold environments may experience shorter flight times due to sluggish battery performance. Therefore, maintaining batteries within an optimal temperature range is crucial for maximizing performance and ensuring safety in high-performance applications.

Wang *et al.* [8] discusses advancements in electrode materials for higher energy density. A high-performance application has driven the development of advanced battery technologies with real time thermal management systems. ***Lithium-ion (Li-ion) batteries*** have emerged as the dominant technology due to their high energy density, relatively long lifespan, and ability to deliver high currents. However, ongoing research and development efforts continue to explore alternative battery chemistries with the potential to surpass the limitations of Li-ion.

11.2 Battery Thermal Management System (BTMS)

Batteries are vital components in various applications, including electric vehicles (EVs), drones, portable electronics, and energy storage systems in grids. Dunn *et al.* [9] have studied the battery types used for grid storage and their capacity. They have concluded that battery performance and safety are highly dependent on temperature. BTMS plays a critical role in maintaining optimal battery temperature, ensuring efficient operation, extended lifespan, and enhanced safety. BTMS is used to regulate battery temperature within a specific range, typically between 20°C and 45°C. This is achieved through cooling and heating mechanisms. During charging and discharging, batteries generate heat due to internal resistance. The BTMS employs cooling techniques to remove excess heat and prevent overheating. This can involve air cooling, liquid cooling, or phase-change materials (PCMs) [2]. Liu *et al.* [2] studied battery performance enhancement and temperature management using Phase change Material (PCM). In cold environments, battery performance can be hampered. The BTMS can incorporate heating elements to ensure the battery reaches optimal operating temperature before use.

11.2.1 Functions of BTMS

Enhanced Battery Performance: Lithium-ion batteries, the workhorses of modern electronics and electric vehicles (EVs), are susceptible to temperature fluctuations. They can function within a broad range, but their optimal performance lies within a narrow window using BTMS. By maintaining the battery pack at ideal temperature, the BTMS significantly enhances battery performance in several ways. The BTMS ensures optimal operation of the electrochemical reactions occurring within the battery cells. At moderate temperatures, these reactions proceed efficiently, maximizing the battery's ability to deliver power (Modine Electric Vehicle, 2023 [12]). Deviations from this range, however, can hinder these reactions. Excessive heat accelerates battery degradation, reducing capacity and increasing the risk of thermal runaway – a potentially dangerous situation where battery failure triggers a self-propagating heat event (CIC energiGUNE, 2023 [13]). Conversely, cold temperatures limit the battery's ability to generate power and accept a charge (MathWorks, 2023 [14]). The BTMS actively counters these detrimental effects by employing cooling or heating mechanisms, ensuring the battery operates within its ideal temperature window. This temperature regulation translates to real-world benefits. A well-managed battery delivers consistent power output, maximizing the performance of the device it powers, be it a laptop or an electric car (Laserax, 2023 [15]). In addition, by preventing excessive heat buildup, the BTMS extends battery life by mitigating degradation (National Renewable Energy Laboratory, 2007 [16]). In essence, the BTMS acts as a silent guardian, ensuring optimal battery function and maximizing its overall performance.

Extended Battery Lifespan: Lithium-ion batteries are susceptible to performance and lifespan degradation under extreme temperatures. Excessive heat, in particular, accelerates the chemical reactions within the battery cell, leading to faster capacity loss and increased safety risks (Laserax, 2023 [15]). BTMS plays a critical role in mitigating this thermal stress, thereby extending the overall lifespan of the battery. During operation, batteries naturally generate heat due to internal resistance. This heat generation becomes more significant during charging and discharging cycles, especially at high currents. Additionally, ambient temperature can significantly impact battery performance. A BTMS actively regulates the battery's operating temperature by employing various cooling or heating techniques. Common BTMS methods include air or liquid cooling systems, which transfer heat away from the battery pack. In colder environments, heating elements within the BTMS may be used to maintain optimal battery temperature (Modine Electric Vehicle, 2023 [12], MathWorks, 2023 [14]). By maintaining the battery within a desired temperature range (typically between 15°C and 35°C), a BTMS offers several benefits. It reduces the rate of chemical degradation within the battery cells, leading to a significant extension of the battery's overall lifespan [16]. Additionally, a BTMS ensures optimal battery performance by preventing excessive heat-induced capacity limitations and power fade. This translates to improved efficiency and consistent operation of the device the battery powers [13].

Improved Safety: Lithium-ion batteries, while having high energy density, are susceptible to performance and safety issues at extreme temperatures. BTMS plays a critical role in mitigating these concerns, particularly by preventing thermal runaway, a dangerous chain reaction. During operation, batteries generate heat due to internal resistance. Excessive temperatures accelerate degradation processes within the battery, reducing lifespan and capacity. More worrisome is the potential for thermal runaway. When a battery cell reaches a critical temperature, a self-heating cycle can initiate. This runaway generates even more heat, causing the cell to vent flammable electrolytes and potentially ignite surrounding components [12]. A properly designed BTMS actively regulates battery temperature, preventing this hazardous scenario. By maintaining the battery within its optimal operating range (typically between 15–35°C), a BTMS ensures safe and efficient function. This is achieved through various cooling techniques, such as air or liquid-based heat exchangers, or passive methods using heat sinks. The BTMS also monitors battery temperature and communicates with the vehicle control unit to adjust cooling or heating strategies as needed [13, 14]. Effective thermal management promotes battery longevity, optimizes performance, and significantly reduces the risk of catastrophic thermal runaway incidents [16, 17].

Faster Charging: During charging, lithium-ion batteries experience an increase in temperature due to internal resistance. Excessive heat can degrade battery health and limit charging speeds. The BTMS mitigates this issue by employing various cooling techniques to maintain optimal battery temperatures. By preventing excessive heating, BTMS allows for safer and faster charging of lithium-ion batteries.

11.3 Types of BTMS

Maintaining optimal temperature is crucial for battery performance and safety. Battery Thermal Management Systems (BTMS) play a vital role in achieving this by dissipating heat generated during charging and discharging cycles. Figure 11.1 shows five Types of BTMS. Each offers distinct advantages, and disadvantages depending on the application are stated in Table 11.1.

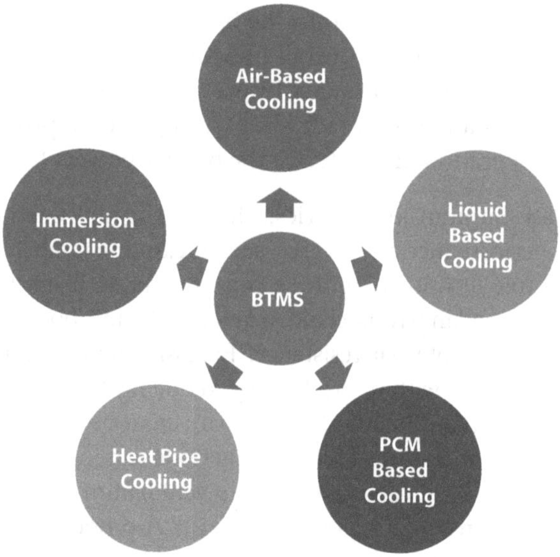

Figure 11.1 Different types of battery thermal management systems (BTMS) [41–44].

Table 11.1 Different types of battery thermal management systems [45–52].

Type of BTMS	Key highlights	Applications	Advantages	Disadvantages
Air-Based Cooling [7]	Uses fans to circulate air over battery pack for heat dissipation.	Laptops, Portable electronics	Simple design, Low cost	Limited cooling capacity, Not suitable for high-performance applications

(*Continued*)

Table 11.1 Different types of battery thermal management systems [45–52]. (*Continued*)

Type of BTMS	Key highlights	Applications	Advantages	Disadvantages
Liquid-Based Cooling [8]	Pumps coolant (water or glycol mixture) through channels or plates in contact with batteries.	Electric Vehicles (EVs), High-power electronics	Efficient heat transfer, Scalable for large battery packs	More complex design, Higher weight and cost compared to air cooling
Phase Change Materials (PCM) [2]	Utilizes materials that absorb and release heat during phase changes (solid to liquid or vice versa). Integrated within or around battery pack.	Wearable electronics, Drones	Passive cooling, Lightweight	Lower cooling capacity compared to active systems, Limited control over temperature range
Heat Pipe Cooling	Utilizes sealed tubes containing a working fluid that evaporates in a hot zone, transfers heat, and condenses in a cool zone.	EVs (limited applications), High-performance electronics	Efficient heat transfer over long distances, Relatively lightweight	More complex design compared to basic air cooling, May require additional heat sink for optimal performance
Immersion Cooling [4]	Submerges batteries in a dielectric coolant with high heat capacity.	Large scale energy storage systems	Excellent heat transfer, Uniform temperature distribution	High complexity, Increased maintenance due to potential leaks, Safety concerns with some coolants

11.3.1 Air-Based Cooling

This simpler and more cost-effective approach utilizes fans to convect ambient air over the battery pack. Airflow removes heat from the battery surface, keeping temperatures within a safe range [17]. The main advantage of Air-based cooling is Lower cost and simpler design, Easier maintenance and repairs and Lightweight and less bulky compared to liquid-based systems. But this cooling method is less efficient heat transfer compared to liquids, limiting its effectiveness in high-power applications like electric vehicles (EVs) with large battery packs. This cooling method has difficulty achieving uniform temperature distribution within the battery pack, which can lead to performance and safety concerns and it requires significant airflow, which can increase noise levels and energy consumption from the fans [19].

11.3.2 Liquid-Based Cooling

This method employs a liquid coolant, such as water or a glycol mixture, to absorb heat directly from the battery cells. The coolant is circulated through a network of channels or plates within the battery pack, transferring heat to a radiator where it is dissipated to the surrounding air [17]. Liu *et al.* [20] investigated that the main advantage of Liquid-based cooling is Superior heat transfer efficiency compared to air, enabling precise temperature control for high-power applications like EVs and more uniform temperature distribution across the battery pack, improving battery health and performance. It operation is Quieter due to reduced reliance on fans. It has disadvantages such as higher complexity and cost due to additional components like pumps, radiators, and coolant, increased weight and bulk compared to air-based systems. Potential leakage risks that can damage the battery or pose safety hazards.

Taeyoung *et al.* [21] have investigated the Liquid Versus Air Cooling in BTMS. They have concluded that the choice between air-based and liquid-based cooling hinges on three key factors: battery power and size, cost and complexity, and space limitations. For low-power applications like laptops and portable electronics, where compact size is crucial, air cooling offers a simpler and more affordable solution. However, in high-power applications like electric vehicles (EVs) with large battery packs, superior temperature control is essential. Here, liquid cooling takes precedence despite its increased complexity and cost. The long-term benefits of liquid cooling, such as improved battery performance and safety, can outweigh the initial investment for these demanding applications.

11.3.3 Phase Change Materials (PCM)

Air-based and liquid-based cooling are the dominant approaches. PCMs are a class of materials with a unique property. They absorb and release large amounts of latent heat during a solid-to-liquid phase transition. Wazeer *et al.* [22] studied that PCM based battery cooling methods offer unique advantages and are particularly well-suited for specific applications. In a BTMS, PCMs are incorporated near the battery pack. During operation, heat generated by the battery raises the PCM's temperature, triggering a solid phase change to liquid. This absorbs the excess heat, maintaining optimal battery temperature. As the battery cools down, the PCM solidifies again, releasing the stored heat. Sevugan *et al.* [23] have studied the advantage & limitations of PCM based BTMS. They study that a major advantage of Phase Change Material (PCM) based cooling lies in its passive operation. This eliminates the need for noisy fans or bulky pumps, simplifying the system's design. Additionally, PCMs offer precise thermal control by maintaining a relatively constant temperature during their phase change. This characteristic is particularly beneficial for battery health. Furthermore, PCM based systems are lightweight and compact compared to liquid cooling solutions, making them ideal for applications with space constraints. However, some limitations exist with PCMs. Their heat transfer capacity is lower compared to liquid coolants, and uneven heat distribution within the battery pack can occur if PCM placement isn't optimized. Choosing the right PCM with a phase change temperature suited for the specific battery application is also crucial for optimal performance.

11.3.4 Heat Pipe Cooling

Heat pipes offer a unique solution for battery thermal management. These sealed tubes house a wick structure, a working fluid, and a vapor space. Heat from the battery vaporizes the working fluid within the wick. This vapor travels to the cooler end of the heat pipe, condenses, releasing its stored thermal energy. The condensate then utilizes the wick structure to flow back to the hot end, completing the cycle. Chavan *et al.* [24] studied the heat pipe and liquid cold plate based BTMS. They concludes that heat pipe cooling design have several advantages like highly efficient heat transfer over long distances, passive operation requiring minimal maintenance, and a flexible design that adapts to various battery pack configurations. However, there are limitations to consider. Heat pipes are more complex and expensive compared to simpler air-based cooling systems. They also require careful design and working fluid selection to ensure optimal performance at specific temperatures. Additionally, performance can degrade over time due to factors like the accumulation of non-condensable gases within the sealed tube.

11.3.5 Immersion Cooling

Immersion cooling offers a unique approach to battery thermal management. Liu *et al.* [20] study the immersion cooling and states that the entire battery pack is submerged in a dielectric liquid possessing high thermal conductivity. This liquid directly absorbs heat from the battery cells,

Table 11.2 Comparative studies between different types of BTMS [51–56].

Feature	Air-based cooling	Liquid-based cooling	Phase change materials (PCM)	Heat pipe cooling	Immersion cooling
Cooling Mechanism [57, 58]	Convection (fans blow air over batteries)	Conduction (liquid absorbs heat and transfers it elsewhere)	Heat absorption/release through phase change	Conduction (heat pipe transfers heat away from batteries)	Conduction (liquid directly contacts batteries)
Complexity	Low	Moderate	Low	Moderate	High
Efficiency	Low	High	Moderate	Moderate	High
Cost	Low	Moderate to High	Moderate	Moderate to High	High
Temperature Control	Limited	Precise	Passive (within PCM operating range)	Moderate	Precise
Weight and Size	Lightweight and compact	Bulkier than air-based	Moderate	Moderate	Bulky
Maintenance	Low	Moderate	Low	Moderate	Moderate
Suitability [59, 60]	Low to moderate power applications (laptops)	High power applications (data centers, EVs)	Low to moderate power with limited space	Moderate power applications requiring good temperature control	High power applications requiring precise temperature control [61–63]

resulting in exceptional thermal control and uniformity throughout the pack. This method proves highly beneficial for high-power battery applications due to its superior heat transfer efficiency, enabling precise temperature management. Additionally, the consistent thermal distribution minimizes thermal stress on the battery cells, improving their overall health and lifespan. Immersion cooling can even enhance fire safety compared to traditional battery systems that rely on flammable organic electrolytes. However, this technique comes with its own set of challenges. The need for specialized cooling loops and containment vessels for the dielectric liquid increases the complexity and cost of the system. Furthermore, immersion cooling tends to be bulkier and heavier compared to other BTMS options. Selecting a compatible dielectric liquid is crucial, as it should not degrade the battery components over time.

Table 11.1 summarizes the above BTMS and Table 11.2 concludes that different BTMS technologies offer promising alternatives based on applications for efficient and reliable battery thermal management. Selecting the most suitable approach depends on a careful evaluation of application-specific requirements.

11.4 BTMS Challenges and Future Trends

BTMS are an essential element for ensuring the safe, efficient, and reliable operation of batteries. Their role becomes even more crucial as battery technology progresses towards faster charging, higher energy densities, and new chemistries. While significant advancements have been made, challenges remain that need to be addressed. However, by overcoming these hurdles and exploring promising future trends, BTMS development will play a key role in unlocking the full potential of battery-powered technologies, paving the way for a cleaner and more sustainable future.

- **Balancing Efficiency and Weight:** BTMS components, like pumps, radiators, and piping in liquid-based systems, add weight to the overall battery system. This is particularly critical for applications like electric vehicles (EVs) where weight directly impacts range. Future BTMS designs will need to focus on lightweight materials and innovative cooling techniques to achieve optimal efficiency without compromising weight.
- **Uniform Temperature Distribution:** Maintaining consistent temperature across all battery cells within the pack is crucial for optimal performance and lifespan. Uneven

temperatures can lead to cell degradation and potential safety hazards. Optimizing heat transfer pathways through the battery pack using innovative designs and materials is a key area of development in BTMS.

- **Adapting to Advanced Battery Technologies:** As battery chemistries evolve, with the potential for faster charging and higher energy densities, their thermal management requirements will also change. BTMS need to be adaptable to accommodate these evolving needs and ensure efficient and safe operation with next-generation batteries.
- **Integration with Battery Packs:** Seamless integration of BTMS components within the battery pack design itself holds significant potential. This can improve overall efficiency by reducing heat transfer losses and optimizing space utilization within the battery pack.
- **Advanced Materials:** The development of new materials with superior thermal conductivity is another exciting frontier. These materials can be incorporated into heat pipes, fins, or even the battery pack structure itself to enhance heat transfer capabilities and achieve more uniform temperature distribution.
- **Smart Control Systems:** The future of BTMS lies in intelligent control systems. These systems will utilize an array of sensors to monitor battery health and temperature in real-time. By employing machine learning algorithms, these systems can optimize BTMS operation based on the specific needs of the battery, further improving efficiency, safety, and overall battery lifespan.

By addressing current challenges and embracing these future trends, BTMS will play a critical role in enabling the widespread adoption of next-generation batteries and accelerating the transition towards a cleaner and more sustainable future [64, 65].

11.5 Future Generation of High Performance Batteries

Solid-state batteries [10, 11, 18]: Hassan *et al.* [10], Winter *et al.* [11] and Janek *et al.* [18] have studied the progress in solid state lithium batteries and its revolutionizing application in electric vehicles. They have Discusses solid-state batteries, a potential future technology for high energy density and also Explores the potential of solid-state batteries for EVs, highlighting energy density benefits. These batteries employ a solid electrolyte instead of the traditional liquid electrolyte found in Li-ion batteries. This offers several advantages, including improved safety, faster charging times, and potentially higher energy density. However, solid-state battery technology is still under development and faces challenges in terms of cost and scalability.

Lithium-sulfur (Li-S) batteries: Sun *et al.* [5] study focuses on Li-S batteries and energy density Li-S batteries boast a theoretical energy density significantly exceeding that of Li-ion batteries. However, technical challenges related to cycle life and capacity fade need to be addressed before they become viable for widespread use in high-performance applications.

Redox flow batteries: These batteries store energy in liquid electrolytes, separate from the electrodes. This allows for independent scaling of energy capacity (determined by the electrolyte volume) and power delivery (determined by the electrode size). Redox flow batteries are particularly suited for grid storage applications due to their inherent scalability and long lifespan. Dunn *et al.* [9] have studied the electrical energy storage for the grid.

11.6 Intelligent Control Strategies for Proactive BTMS

Traditional BTMS with pre-programmed responses struggle to keep up with the dynamic demands of high-performance batteries in applications like EVs and large-scale energy storage. The answer lies in intelligent control strategies that integrate machine learning (ML) for proactive adjustments based on real-time battery data. Modern batteries are equipped with a wealth of sensors that continuously monitor factors like cell voltage (indicating health and charge), temperature (crucial for safety and efficiency), current flow (charging/discharging rate), and even ambient temperature. By analyzing this real-time data, intelligent control systems can gain a deeper understanding of the battery's current state and predict its future thermal behavior.

Machine learning algorithms become even more powerful when trained on historical battery data alongside these real-time readings. Yang *et al.* [25] studied ML application in BTMS and concludes that, ML enabled BTMS allows them to:

- Identify complex relationships between various battery parameters to predict future temperature trends.
- Forecast thermal behavior, essentially predicting how the battery temperature will evolve in the near future based on historical data and current conditions.

- Optimize cooling strategies by anticipating future thermal needs. This proactive approach allows the system to adjust cooling mechanisms, like fan speed or coolant flow, to maintain optimal battery temperature.

The benefits of this proactive approach are numerous. By anticipating cooling needs, the system avoids unnecessary operation of cooling components, leading to improved efficiency and reduced energy consumption. Proactive control also enhances safety by identifying potential overheating events before they occur, preventing thermal runaway scenarios. Ultimately, maintaining optimal temperatures through proactive cooling extends battery life and improves overall battery health [26–34].

Beyond efficiency and safety, machine learning offers even more possibilities for BTMS. Predictive maintenance, for example, analyzes historical data on battery degradation and temperature patterns to predict potential maintenance needs before a critical failure. Furthermore, ML models can be used for adaptive cooling control, dynamically adjusting cooling strategies based on specific situations like driving patterns in EVs or charging/discharging cycles in energy storage systems. Fault detection and diagnosis are also possible, with ML algorithms analyzing sensor data to identify anomalies that might indicate potential battery faults, allowing for early intervention and preventing further damage [35–39].

While the integration of ML offers significant advantages, challenges remain. The effectiveness of ML models heavily relies on the quality and quantity of data used for training. Insufficient or inaccurate data can lead to unreliable predictions. Additionally, running complex ML algorithms on embedded systems within BTMS might require additional processing power and resources. Perhaps the most crucial consideration is safety and trust. As BTMS rely more on AI-powered control, ensuring the safety and reliability of these algorithms becomes paramount. Robust validation and testing are essential before deploying ML models in real-world applications.

The integration of intelligent control strategies with machine learning represents a significant leap forward in BTMS development. By leveraging real-time data and powerful machine learning algorithms, BTMS can become more proactive, efficient, and ultimately contribute to a more sustainable future for battery-powered technologies. As research in this field continues, we can expect even more sophisticated ML models and control strategies to emerge, pushing the boundaries of battery performance, safety, and lifespan in high-demand applications [40].

11.7 Conclusion

The chapter emphasizes the critical role of Battery Thermal Management Systems (BTMS) in maintaining optimal battery health and safety, particularly in demanding applications. It highlights into five prominent cooling strategies: air-based, liquid-based, PCM-based systems, Heat Pipe Cooling and Immersion Cooling and each offering distinct advantages for specific scenarios. Furthermore, the chapter explores the potential of integrating intelligent control systems with machine learning, paving the way for proactive BTMS adjustments based on real-time battery data. By analyzing these advancements, the chapter equips readers with the knowledge needed to design efficient BTMS solutions, ensuring optimal battery performance and reliability in high-performance applications.

References

1. Boyes, E.D., Battery technology for electric vehicles. *Sustain. Mater. Technol.*, 18, 11–18, 2018.
2. Liu, J., Song, W., Zhang, Z., Li, H., Liu, C., Phase change materials for lithium-ion battery thermal management. *Energy Storage Mater.*, 18, 283–297, 2019. [Liu *et al.*, 2019] (Although not directly mentioned, this reference provides context on battery performance and temperature).
3. Li, Z., Xu, J., Chen, X., Zhou, L., Fang, J., Electrochemical energy storage devices for drone applications. *Nano Res.*, 11, 7, 3329–3360, 2018.
4. Estrada-Martinez, A., Macken, D., Bhatt, B., Lithium ion battery modeling and simulation for battery management systems - A review. *Renew. Sustain. Energy Rev.*, 94, 800–817, 2018.
5. Sun, Y., Liu, Z., Li, J., High-energy-density lithium-sulfur batteries: Material design, modeling, and engineering challenges. *Chem. Rev.*, 121, 12, 7431–7497, 2021.
6. Schmuch, R., Wagner, R., Hörmann, G., Pouškouřová, M., Richter, P., Straessler, P., Performance and cost targets for automotive Li-ion batteries: A techno-economic analysis. *Energies*, 11, 3, 613, 2018.
7. Khalilzadeh, A., Forsberg, C., Jónsson, E., A review of electric vehicle battery modeling and simulation: Focus on lithium-ion batteries. *J. Power Sources*, 471, 228002, 2020.
8. Wang, Q., Li, B., Zhang, X., Liu, Z., Dou, H., Recent progress in electrode materials for high-energy-density lithium-ion batteries. *Adv. Energy Mater.*, 9, 18, 1803843, 2019.

9. Dunn, B., Kamath, H., Tarascon, J.M., Electrical energy storage for the grid: A battery technology update. *Science*, 334, 6059, 928–935, 2011, [Dunn *et al.*, 2011] (Explains battery types used for grid storage and their capacity).
10. Hassan, M.H., Song, Y., Yang, Z., Progress in solid-state lithium batteries. *Adv. Mater.*, 31, 18, 1807334, 2019.
11. Winter, M., Janek, J., Besenhard, J.O., Will solid-state batteries revolutionize electric vehicles? *Adv. Mater.*, 28, 31, 6944–6951, 2016.
12. Modine Electric Vehicle, How It Works: Battery Thermal Management System, 2023, March 2, https://www.modineev.com/how-it-works-battery-thermal-management-system/ (Accessed, May 4, 2024).
13. CIC energiGUNE, Battery Thermal Management Systems (BTMS) for mobility applications, 2023, February 22, https://cicenergigune.com/en/mobility-storage (Accessed, May 4, 2024).
14. MathWorks, Battery Thermal Management System - MATLAB & Simulink, 2023, April 12, in.mathworks.com.
15. Laserax, All You Need to Know About Battery Thermal Management, 2023, January 12, https://www.laserax.com/sites/default/files/public/assets/product-datasheet/files/battery-laser-welding-machine_0.pdf (Accessed, May 4, 2024).
16. National Renewable Energy Laboratory, Battery Thermal Management System Design Modeling, 2007, https://www.nrel.gov/docs/fy07osti/40446.pdf (Accessed, May 4, 2024).
17. Li, W., Zhou, Y., Zhang, H., Tang, X., A Review on Battery Thermal Management for New Energy Vehicles. *Energies*, 16, 13, 4845, 2023, https://doi.org/10.3390/en16134845.
18. Janek, J. and Zeier, W.G., Challenges in speeding up solid-state battery development. *Nat. Energy*, 8, 230–240, 2023, https://doi.org/10.1038/s41560-023-01208-9.
19. Garud, K.S., Tai, L.D., Hwang, S.-G., Nguyen, N.-H., Lee, M.-Y.A., Review of Advanced Cooling Strategies for Battery Thermal Management Systems in Electric Vehicles. *Symmetry*, 15, 1322, 2023, https://doi.org/10.3390/sym15071322.
20. Liu, J., Chen, H., Huang, S., Jiao, Y., Chen, M., Recent Progress and Prospects in Liquid Cooling Thermal Management System for Lithium-Ion Batteries. *Batteries*, 9, 400, 2023, https://doi.org/10.3390/batteries9080400.
21. Taeyoung, *et al.*, Li-Ion Battery Pack Thermal Management: Liquid Versus Air Cooling. *J. Therm. Sci. Eng. Appl.*, 11, 2021009, 1–9, 2019, https://doi.org/10.1115/1.4041595.
22. Wazeer, A., Das, A., Abeykoon, C., Sinha, A., Karmakar, A., Phase change materials for battery thermal management of electric and hybrid vehicles: A review. *Energy Nexus*, 7, 100131, 2022, https://doi.org/10.1016/j.nexus.2022.100131.
23. Sevugan, P.A., Pradeep, M., Krishnaswamy, A., Karunamurthy, K., Battery Thermal Management System for electric vehicles using Phase Change

Materials. *Mater. Today Proc.*, 51, Part 3, 1604–1611, 2022, https://doi.org/10.1016/j.matpr.2021.10.468.

24. Chavan, U., Prajapati, O., Hujare, P., Lithium ion battery thermal management by using coupled heat pipe and liquid cold plate. *Mater. Today Proc.*, 80, Part 2, 382–388, 2023, https://doi.org/10.1016/j.matpr.2022.10.185.
25. Yang, R., *et al.*, Machine Learning for Electric Vehicle Battery Management System: A Critical Review. *Mach. Learn. Appl. Battery Manage. Systems J. Power Sources*, 479, 228700, 2020.
26. Verma, V., Thangavel, S., Dutt, N., Kumar, A., Weerasinghe, R. (Eds.), *Highly Efficient Thermal Renewable Energy Systems: Design, Optimization and Applications*, CRC Press, Boca Raton, FL USA, 2024, https://doi.org/10.1201/9781003472629 ISBN 9781032595641.
27. Kundu, A., Kumar, A., Dutt, N., Meena, C.S., Singh, V.P., Introduction to Thermal Energy Resources and Its Smart Application, in: *Thermal Energy Systems: Design, Computational Techniques and Applications*, A. Kumar, V.P. Singh, C.S. Meena, N. Dutt, (Eds.), pp. 1–15, CRC Press, Boca Raton, FL, USA, 2023, http://dx.doi.org/10.1201/9781003395768-1.
28. Dewangan, A.K., Moinuddin, S.Q., Cheepu, M., Sajjan, S.K., Kumar, A., Thermal Energy Storage: Opportunities, Challenges and Future Scope, in: *Thermal Energy Systems: Design, Computational Techniques and Applications*, A. Kumar, V.P. Singh, C.S. Meena, N. Dutt, (Eds.), pp. 17–28, CRC Press, Boca Raton, FL, USA, 2023, http://dx.doi.org/10.1201/9781003395768-2.
29. Kundu, A., Kumar, A., Dutt, N., Singh, V.P., Meena, C.S., Modelling and Simulation of Thermal Energy System for Design Optimization, in: *Thermal Energy Systems: Design, Computational Techniques and Applications*, A. Kumar, V.P. Singh, C.S. Meena, N. Dutt, (Eds.), pp. 103–140, CRC Press, Boca Raton, FL, USA, 2023, http://dx.doi.org/10.1201/9781003395768-7.
30. Kumar, A., Gori, Y., Meena, C.S., Singh, V.P., Sharma, V.K., Design and analysis of heavy vehicle medium duty transmission gearbox system, in: *Additive Manufacturing in Industry 4.0: Methods, Techniques, Modeling and Nano Aspects*, vol. Chapter 4, V.K. Sharma, A. Kumar, M. Gupta, V. Kumar, D.K. Sharma, S.K. Sharma, (Eds.), pp. 67–86, CRC Press, Boca Raton, FL, USA, 2022, doi: DOI: 10.1201/9781003360001-4.
31. Singh, V.P., Kumar, A., Meena, C.S., Dwivedi, G., Introduction to Energy-Efficient Vehicles for Sustainable Transportation: Transitions and Challenges, in: *Energy Efficient Vehicles Technologies and Challenges*, vol. Chapter 01, V.P. Singh, A. Kumar, C.S. Meena, G. Dwivedi (Eds.), pp. 01–18, CRC Press, Boca Raton, FL, USA, 2024, DOI: https://doi.org/ 10.1201/9781003464556-1.
32. Kumar, A., Jaiswal, H., Patil, P.P., Parametric Optimization of Vibrating Heavy Vehicle Medium Duty Transmission Gearbox Housing Using Response Surface Method. *Int. J. Veh. Struct. Syst.*, 9, 3, 149–153, 2017, doi: 10.4273/ijvss.9.3.04.
33. Singh, V.P. and Kumar., A., Techno-Economic and Future Aspects of the HEV-EV-FCV: Decarbonization, Digitalization and Sustainability, in: *Energy*

Efficient Vehicles Technologies and Challenges, vol. Chapter 11, V.P. Singh, A. Kumar, C.S. Meena, G. Dwivedi (Eds.), pp. 212–231, CRC Press, Boca Raton, FL, USA, 2024, DOI: https://doi.org/ 10.1201/9781003464556-11.

34. Kumar, A. and Patil, P.P., Modal Analysis of Heavy Vehicle Truck Transmission Gearbox Housing Made From Different Materials. *J. Eng. Sci. Technol.*, 11, 2, 252–266, 2016.
35. Singh, V.P., Kumar, A., Meena, C.S., Dwivedi, G. (Eds.), *Energy Efficient Vehicles: Technologies and Challenges*, CRC Press, Boca Raton, FL USA, 2024, https://doi.org/10.1201/9781003464556 ISBN: 9781032548111.
36. Kumar, A., Jain, R., Patil, P.P., Dynamic Analysis of Heavy Vehicle Medium Duty Drive Shaft Using Conventional and Composite Material. *IOP Mater. Sci. Eng.*, 149, 1–11, 2016, DOI: 10.1088/1757-899X/149/1/012156.
37. Kumar, A., Singh, V.P., Meena, C.S., Dutt, N., *Thermal Energy Systems: Design, Computational Techniques and Applications*, CRC Press, Boca Raton, FL, USA, 2023, https://doi.org/10.1201/9781003395768.
38. Kumar, A. and Patil, P.P., Dynamic Vibration Analysis of Heavy Vehicle Truck Transmission Gearbox Housing Using FEA. *J. Eng. Sci. Technol. Rev.*, 7, 4, 66–72, 2014, ISSN: 1791-2377.
39. Upadhyay, R.K., Singh, V.P., Kumar, A., vol. Chapter 05, V.P. Singh, A. Kumar, C.S. Meena, G. Dwivedi (Eds.), pp. 74–95 CRC Press, Boca Raton, FL, USA, 2024, DOI: https://doi.org/ 10.1201/9781003464556-5.
40. Kumar, A., Low to High Speed Transient Structural and Thermal Temperature Measurement of Oil-Lubricated Multi-Speed Heavy Vehicle Transmission Gearbox System Based on FEA, in: *Advanced Numerical Simulations in Mechanical Engineering*, pp. 1–21, IGI Global, Hershey, PA 17033, USA, 2018, DOI: 10.4018/978-1-5225-3722-9.ch001. ISBN13: 9781522537229.
41. Amoussou, I., Paddy, E.Y., Agajie, T.F., Ibrahim, F.S., Agajie, E.F., Nsanyuy, W.B., Mohammadi, S.A.D., Enhancing residential energy access with optimized stand-alone hybrid solar-diesel-battery systems in Buea, Cameroon. *Sci. Rep.*, 14, 1, 15543, 2024, 10.1038/s41598-024-66582-0.
42. Razaviyn, Z., Heidari, M., Thangavel, S., Verma, V., Kumar, A., Yadav, A.K., Marine Energy Harvesting from Fluid Flow *via* Vortex Induced Vibrations. *Int. J. Thermofluids*, 25, 101015, January 2025, https://doi.org/10.1016/j.ijft.2024.101015.
43. Jena, N.K., Sahoo, S., Sahu, B.K., Naik, A.K., Bajaj, M., Misak, S., Prokop, L., Impact of a redox flow battery on the frequency stability of a five-area system integrated with renewable sources. *Energies*, 16, 14, 5540, 2023, 10.3390/en16145540.
44. Hamed, S.B., Hamed, M.B., Sbita, L., Bajaj, M., Blazek, V., Prokop, L., Ghoneim, S.S., Robust optimization and power management of a triple junction photovoltaic electric vehicle with battery storage. *Sensors*, 22, 16, 6123, 2022, 10.3390/s22166123.
45. Rawat, T., Niazi, K.R., Gupta, N., Sharma, S., Multi-objective techno-economic operation of smart distribution network integrated with reactive

power support of battery storage systems. *Sustain. Cities Soc.*, 75, 103359, 2021. http://dx.doi.org/10.1016/j.scs.2021.103359.

46. Gupta, A., Singh, R.K., Paul, J., Sharma, M., Joshi, S., Battery-Operated Electric Vehicles (BOEVs) Adoption in India: Analysis of Barriers and Strategies. *IEEE Trans. Eng. Manage.*, 71, 7076–7087, 2023, http://dx.doi.org/10.1109/TEM.2023.3253621.
47. Aggarwal, S., Singh, A.K., Rathore, R.S., Bajaj, M., Gupta, D., Revolutionizing load management: A novel technique to diminish the impact of electric vehicle charging stations on the electricity grid. *Sustain. Energy Technol. Assess.*, 65, 103784, 2024, 10.1016/j.seta.2024.103784.
48. Oubelaid, A., Kakouche, K., Belkhier, Y., Khosravi, N., Taib, N., Rekioua, T., Tuka, M.B., New coordinated drive mode switching strategy for distributed drive electric vehicles with energy storage system. *Sci. Rep.*, 14, 1, 6448, 2024, 10.1038/s41598-024-56209-9.
49. Ravindran, M.A., Nallathambi, K., Vishnuram, P., Rathore, R.S., Bajaj, M., Rida, I., Alkhayyat, A., A novel technological review on fast charging infrastructure for electrical vehicles: Challenges, solutions, and future research directions. *Alexandria Eng. J.*, 82, 260–290, 2023, 10.1016/j.aej.2023.10.009.
50. Dimri, S.C., Indu, R., Bajaj, M., Rathore, R.S., Blazek, V., Dutta, A.K., Alsubai, S., Modeling of traffic at a road crossing and optimization of waiting time of the vehicles. *Alexandria Eng. J.*, 98, 114–129, 2024, 10.1016/j.aej.2024.04.050.
51. Kumar, B.A., Jyothi, B., Singh, A.R., Bajaj, M., Rathore, R.S., Berhanu, M., A novel strategy towards efficient and reliable electric vehicle charging for the realisation of a true sustainable transportation landscape. *Sci. Rep.*, 14, 1, 3261, 2024, 10.1038/s41598-024-53214-w.
52. Singh, A.R., Vishnuram, P., Alagarsamy, S., Bajaj, M., Blazek, V., Damaj, I., Othman, K.M., Electric vehicle charging technologies, infrastructure expansion, grid integration strategies, and their role in promoting sustainable e-mobility. *Alexandria Eng. J.*, 105, 300–330, 2024, 10.1016/j.aej.2024.06.093.
53. Kanouni, B., Badoud, A.E., Mekhilef, S., Bajaj, M., Zaitsev, I., Advanced efficient energy management strategy based on state machine control for multi-sources PV-PEMFC-batteries system. *Sci. Rep.*, 14, 1, 7996, 2024, 10.1038/s41598-024-58785-2.
54. Bagga, P., Das, A.K., Wazid, M., Rodrigues, J.J., Choo, K.K.R., Park, Y., On the design of mutual authentication and key agreement protocol in internet of vehicles-enabled intelligent transportation system. *IEEE Trans. Veh. Technol.*, 70, 2, 1736–1751, 2021, 10.1109/TVT.2021.3050614.
55. Sahri, Y., Tamalouzt, S., Belaid, S.L., Bajaj, M., Ghoneim, S.S., Zawbaa, H.M., Kamel, S., Performance improvement of Hybrid System based DFIG-Wind/PV/Batteries connected to DC and AC grid by applying Intelligent Control. *Energy Rep.*, 9, 2027–2043, 2023, 10.1016/j.egyr.2023.01.021.
56. Bhatt, A., Ongsakul, W., Pawar, J., Optimal energy management system for carbon–neutral microgrid integrating second-life batteries and crypto

mining devices. *Sustain. Energy Technol. Assess.*, 64, 103686, 2024, 10.1016/j.seta.2024.103686.

57. Rekioua, D., Mokrani, Z., Kakouche, K., Rekioua, T., Oubelaid, A., Logerais, P.O., Ghoneim, S.S., Optimization and intelligent power management control for an autonomous hybrid wind turbine photovoltaic diesel generator with batteries. *Sci. Rep.*, 13, 1, 21830, 2023, 10.1038/s41598-023-49067-4.
58. Das, S., Thakur, P., Singh, A.K., Singh, S.N., Optimal management of vehicle-to-grid and grid-to-vehicle strategies for load profile improvement in distribution system. *J. Energy Storage*, 49, 104068, 2022, http://dx.doi.org/10.1016/j.est.2022.104068.
59. Singh, P.P., Das, S., Wen, F., Palu, I., Singh, A.K., Thakur, P., Multi-objective planning of electric vehicles charging in distribution system considering priority-based vehicle-to-grid scheduling. *Swarm Evol. Comput.*, 77, 101234, 2023, http://dx.doi.org/10.1016/j.swevo.2023.101234.
60. Thapliyal, S., Wazid, M., Singh, D.P., Das, A.K., Islam, S.H., Robust authenticated key agreement protocol for internet of vehicles-envisioned intelligent transportation system. *J. Syst. Archit.*, 142, 102937, 2023, 10.1016/j.sysarc.2023.102937.
61. Almutairi, A., Albagami, N., Almesned, S., Alrumayh, O., Malik, H., Electric vehicle load estimation at home and workplace in Saudi Arabia for grid planners and policy makers. *Sustainability*, 15, 22, 15878, 2023, http://dx.doi.org/10.3390/su152215878.
62. Badri, Y., II, Sudabattula, S.K., Hussain, I., Malik, H., Márquez, F.P.G., Ultracapacitor as selectable energy buffer in electric vehicle application. *J. Energy Storage*, 73, 109200, 2023, http://dx.doi.org/10.1016/j.est.2023.109200.
63. Kasri, A., Ouari, K., Belkhier, Y., Bajaj, M., Zaitsev, I., Optimizing electric vehicle powertrains peak performance with robust predictive direct torque control of induction motors: A practical approach and experimental validation. *Sci. Rep.*, 14, 1, 14977, 2024, 10.1038/s41598-024-65988-0.
64. Afzal, M.Z., Aurangzeb, M., Iqbal, S., Pushkarna, M., Rehman, A.U., Kotb, H., Bereznychenko, V., A Novel Electric Vehicle Battery Management System Using an Artificial Neural Network-Based Adaptive Droop Control Theory. *Int. J. Energy Res.*, 2023, 1, 2581729, 2023, 10.1155/2023/2581729.
65. Ramakrishnan, V., Vishnuram, P., Yang, T., Bajaj, M., Rathore, R.S., Zaitsev, I., Design and implementation of a high misalignment-tolerance wireless charger for an electric vehicle with control of the constant current/voltage charging. *Sci. Rep.*, 14, 1, 13165, 2024, 10.1038/s41598-024-63952-6.

12

Energy Management of PV Battery Supercapacitor-Based System

Nishant Thakkar*, Nidhi Mishra, Priyanka Paliwal, Tripta Thakur and Anoop Arya

Department of Electrical Engineering, Maulana Azad National Institute of Technology, Bhopal, Madhya Pradesh, India

Abstract

The issue of global energy scarcity has become increasingly pressing, with traditional energy sources proving to be both finite and environmentally detrimental. As a result, there has been a marked shift towards utilizing renewable energy sources (RES) such as solar, wind, and hydroelectric power. RES offer a more sustainable and clean alternative to traditional power systems and are rapidly gaining momentum worldwide. Therefore, in this paper a mathematical model of 100 kW PV system integrated with MPPT (Maximum power point tracking) controller has been presented. Additionally, a hybrid energy storage system consisting of batteries and supercapacitors is used to ensure the efficient use of the renewable energy while maintaining system reliability. The energy management system is included to harness the renewable energy efficiently, while also maintaining the robustness and reliability of power system. The system analysis is carried out in the Matlab/Simulink environment, which is a widely-used tool for simulating and analyzing complex systems.

***Keywords*:** MPPT, supercapacitor, P&O, boost converter, PV system

**Corresponding author*: nishantthakkar1109@gmail.com; ORCID: 0000-0003-3686-1630
Nidhi Mishra: ORCID: 0009-0002-0689-0318
Priyanka Paliwal: ORCID: 0000-0002-1093-3766
Tripta Thakur: ORCID: 0000-0002-3735-2398
Anoop Arya: ORCID: 0000-0002-9735-511X

Ashwani Kumar, Mukesh Kumar Awasthi, Nitesh Dutt, Yogesh Kumar Singla and Sivasakthivel Thangavel (eds.) Thermal Battery Management System for Hybrid and Electric Vehicles, (241–250) © 2025 Scrivener Publishing LLC

12.1 Introduction

The increasing demand for energy has caused the shortage of fossil fuel like oil and petroleum. Energy system around the world has adopted the large and small scale sources of energy like wind, tidal energy, solar and biomass to reduce the worldwide energy issues. The challenges in global energy and the effect they have on environment have resulted in efficient advancement of micro-grid technologies and renewable energy sources [1]. In case of PV system, everything depends on solar power. So it becomes very important to harvest the power continuously independent of variations in irradiance and temperature and controller should be design in such a way [2]. The converter that is efficient if it has very less switching loss. To extract the maximum power MPPT controller is designed for PV system. In [3] PID controller is adopted to enhance output of PV system, it is followed by GA method to synchronize the PID controller to get the exact voltage with respect to the reference. In [4] various types of comparison is presented for perturb and observation techniques for different environmental conditions under the conditioning unit of boost converter. In [5] MPPT is controlled based on the fuzzy logic and maximum power is tracked by GA method which significantly improve the transient and steady state response. Various MPPT methods are proposed which are suitable for changing weather conditions [6]. There are various topologies for DC-DC converter and MPPT to harvest maximum power from PV modules [7]. The most important techniques used for MPPT are perturb and observation method [8], predictive control technique [9], incremental conductance method [10], neural network technique [11], intelligent fuzzy logic control method [12]. Amongst all the methods, P&O is most commonly used. Most of the research is based on extracting maximum power from PV cells but not on obtaining power continuously under 0 °C temperature. If PV system is not connected to grid in case of standalone PV system, then it is really imperative to have a robust energy storage system for backup. Battery is most frequently employed storage element and they also have very high energy density but they lack in power density. Supercapacitor is used to overcome this limitation since they have higher power density than battery and high energy density than ordinary capacitor. Tani *et al.*, presents an analysis of batteries and ultracapacitors for energy management in decentralized systems in different loading condition [13]. Jian *et al.*, presents ground based supercapacitor for energy reduction and reduction in peak demand [14]. In this paper, MPPT technique is used along with DC-DC converter to allow PV modules to generate maximum power continuously irrespective of environmental conditions like temperature and solar irradiation. To control the maximum power,

P&O technique is proposed. This technique is easy to implement, has low cost equipment, and has good performance.

12.2 Modeling of PV System

12.2.1 PV System

PV system converts sunlight to electrical energy. The voltage of each PV cell is around 0.5 to 0.7V. Since each cell produces very small amount of power so they have to be integrated into series and parallel to form a PV array. Further, PV panel are connected in series and in parallel to form a PV array to produce desired output voltage. PV system is modeled as current source across a diode that exhibits non-linear I-V characteristics. The parameter for mathematical modeling and equation governing the PV system is given below:

$$I = I_{PV} - I_O[\exp\left(\frac{V + R_S I}{V_t a}\right) - 1] - \frac{V + R_s I}{R_P} \tag{12.1}$$

where, I_{PV} is current from incident light, I_O is reverse saturation current, V_t is called thermal voltage of array, a is ideality factor, R_S and R_p are series and parallel resistance respectively.

12.2.2 Hybrid Energy Storage System (HESS)

In this paper a hybrid energy storage system of battery storage and supercapacity is used to support PV system (Figure 12.1). The charge rate of the battery can be expressed using Equation (12.2) and behavior of a supercapacitor during charging and discharging can be modeled using Equation (12.3):

$$\mathrm{dQ/dt = I} \tag{12.2}$$

where dQ/dt is the rate of change of charge stored in the battery and I is the charging current.

$$\mathrm{i = C * dV/dt} \tag{12.3}$$

where i is the current flowing into or out of the supercapacitor, and dV/dt is the rate of change of voltage across its terminals.

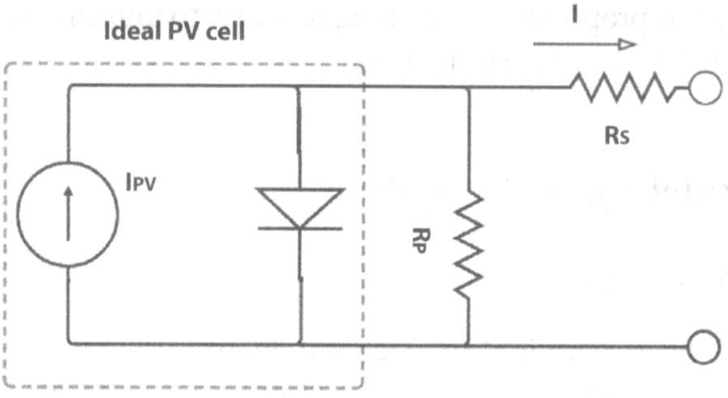

Figure 12.1 Photovoltaic model.

12.2.3 MPPT

Since the temperature and solar irradiance varies with time, the PV system is adjusted to exploit the maximum power. MPPT algorithm of perturb and disturb method is employed to regulate the duty cycle of boost convertor connected to PV panel [8]. The method of perturb and observe measures the current and voltage of PV system to measure the maximum power at a given instance. In this technique, to cause changes in power of PV array, a small perturbation is introduced in the system. The output power of PV is calculated continuously and compared with power of previous state [8]. The flow chart of MPPT algorithm of P&O is shown in Figure 12.2.

12.3 Simulation Model

The parameters of PV array are presented in Table 12.1. The Simulink model of 100 kW grid connected PV system is exhibited in Figure 12.3. Figure 12.4 shows the output power from PV system for different irradiance levels. Figure 12.5 shows the Simulink model for energy storage system comprising of battery and supercapacitor.

12.4 Results and Discussion

The standard condition in this simulation for the test is taken 25 °C for temperature and 1000 W/m^2 for irradiance. The simulation result presented in Figure 12.4 shows that from the beginning t = 0 sec to time t = 0.3 sec, the duty cycle of the DC/DC boost converter is kept fixed at D = 0.5. This results in PV voltage equal to 250 V whereas the output power of the PV array

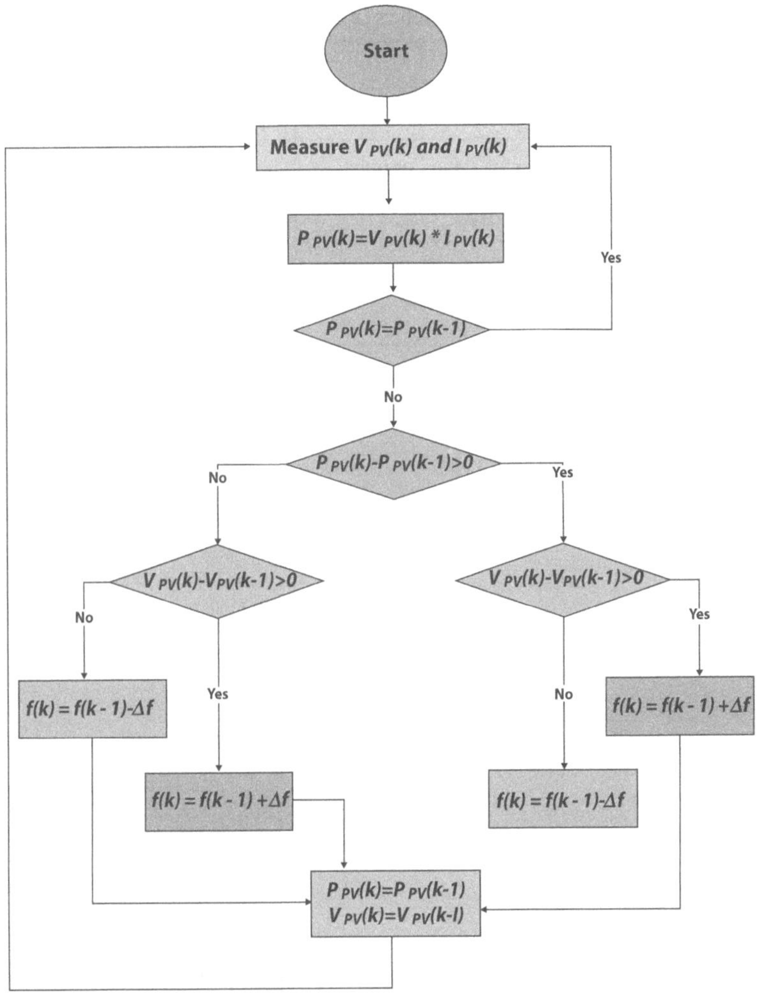

Figure 12.2 Flow chart of P&O algorithm.

is 96 kW, at the same time, power is transferred from supercapacitor to battery and grid at a rate of 1000 W/sec for 2 sec as seen in Figure 12.6. The maximum power at the given irradiance is 100.7 kW. MPPT is enabled at time t = 0.3 sec. It controls the voltage of the PV array by changing the duty cycle to get the maximum output power which is 100.7 kW and results in 0.453 duty cycle. From the time t = 0.3 sec to the time t = 0.5 sec, duty cycle varies between 0.450 and 0.459, the voltage of PV array is 273.5V, but the power still remains the same. From the time t = 0.5 sec to t = 1.0 sec, irradiance is taken down to 250 W/m^2 from 1000 W/m^2. This results in decreasing

Table 12.1 Parameters of PV array.

Parameter	Value
Irradiance	1000W/m2
Number of series panels	5 panels
Temperature	25 °C
Number of parallel panels	66 panels
MPP current	5.58 A

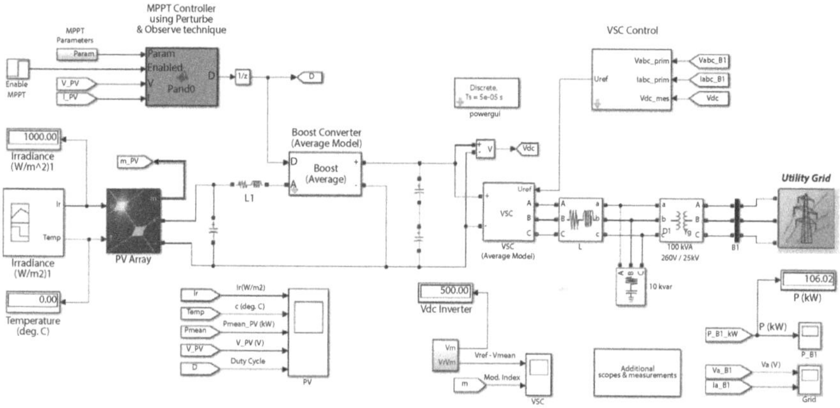

Figure 12.3 100 kW grid connected PV array.

the power output from the PV generators from a lavel of 100 kW to 25 kW. It is clear from the scope result that the maximum power is only tracked by the MPPT controller when the irradiance values remain constant. Irradiance remains 250 W/m^2 from time t = 1.0 sec to time t = 1.5 sec and the duty cycle changes from 0.466 to 0.474. During this time PV voltage remains 265 V and the mean PV voltage is 24.4 kW. Irradiance is again set back to 1000 W/m^2 from time t = 1.5 sec to time t = 6.0 sec while the temperature is varied between 50 °C and 0 °C to analyze the effect of temperature on the PV system. From the scope result, it is clear that at the minimum temperature, that is, at 0 °C maximum output power of PV which is 107.5 kW is achieved.

In the long run, as seen in Figure 12.6, the power captured is of 10 sec cycle. Correspondingly, the power and current delivered from battery and supercapacitor is found to be a particular behavior of their own. However,

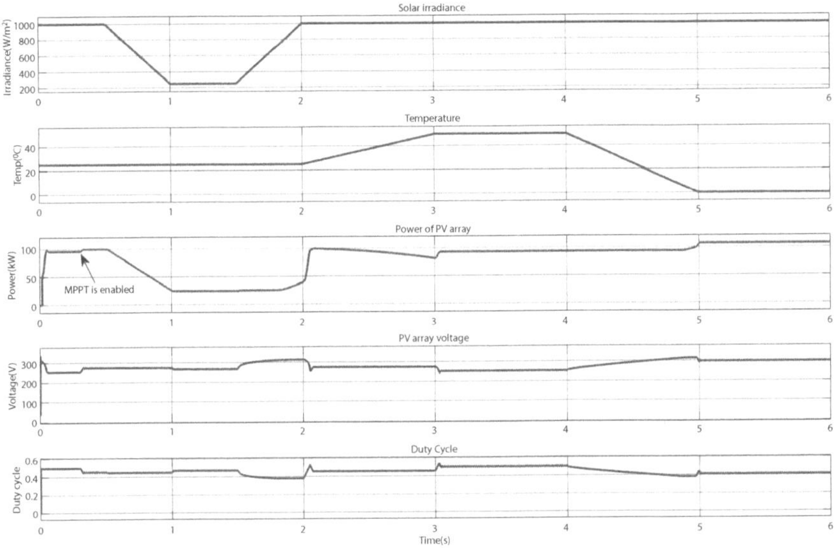

Figure 12.4 Output power of PV system.

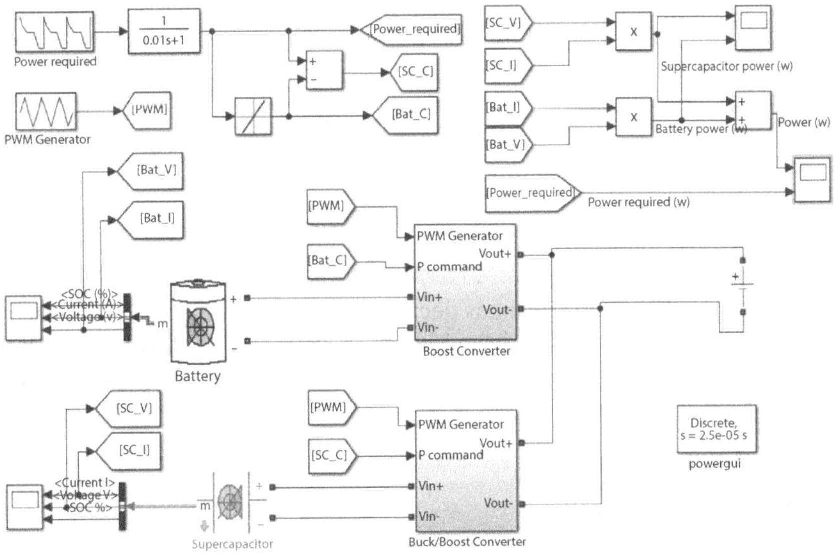

Figure 12.5 Simulink model of battery and supercapacitor.

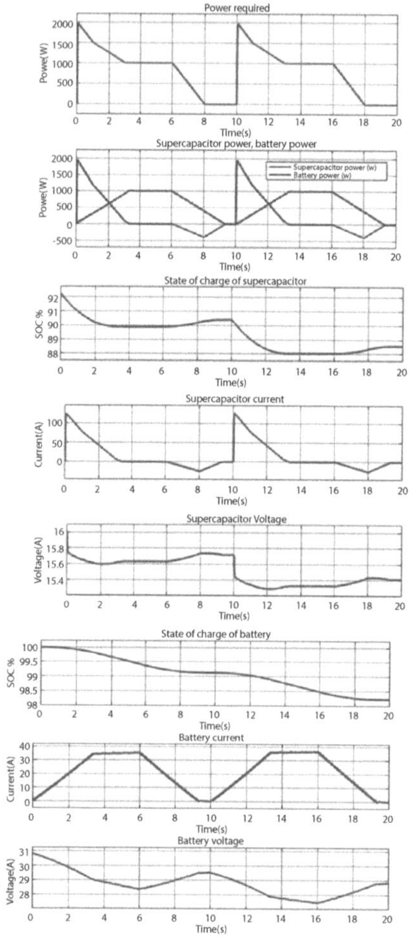

Figure 12.6 Simulation results.

the *SOC* of the supercapacitor has decreased by approximately 2% for each load cycle (10 sec period) and the SoC of the battery has also reduced by approximately 1% as compared to the SoC obtained in the previous cycle, this is seen in Figure 12.6.

12.5 Conclusion

This paper represents the modeling and simulation of 100 kW PV system connected to the MPPT which uses the technique of P&O algorithm for extracting maximum power irrespective of changing weather conditions. This is done in order to increase the power that is injected into the grid, in this way, the power of PV cells is exploited fully, and there is reduction in losses during the energy transfer from PV system to the electric utility grid. The properties of supercapacitor are verified. The complementary nature of battery and supercapacitor is shown *via* simulation result.

To improve the power quality of energy generated by RES, energy storage systems play a vital role. In future, new technologies for battery and SC are anticipated to evolve in following directions:

- Considerable improvement in their efficiency, space requirement, cost, and lifespan.
- The improvement in HESS technology is expected to develop toward intelligent and autonomous microgrid which will be connected to large scale electricity grid.
- Small scale HESS is also expected to develop reliable, easy maintenance, efficient and robust for standalone microgrid for isolated and remote area.

References

1. Jagadeesan, G.M., *et al.*, A two-stage single-phase grid-connected solar-PV system with Simplified Power Regulation. *Chin. J. Electr. Eng.*, 8, 1, 81–92, 2022.
2. Moreira, A., *et al.*, Allocation and sizing of photovoltaic systems to reduce power losses and economic aspects using a new PSO approach. *IEEE Lat. Am. Trans.*, 20, 6, 977–985, 2022.
3. Arandian, B., *et al.*, An effective optimization algorithm for parameters identification of photovoltaic models. *IEEE Access*, 10, 34069–34084, 2022.
4. Alahmadi, A.A., *et al.*, Hybrid wind/PV/battery energy management-based intelligent Non-Integer Control for smart DC-Microgrid of smart university. *IEEE Access*, 9, 98948–98961, 2021.
5. Kumar, M. and Tyagi, B., A robust adaptive decentralized inverter voltage control approach for solar PV and storage-based Islanded Microgrid. *IEEE Trans. Ind. Appl.*, 57, 5, 5356–5371, 2021.
6. Xu, S., *et al.*, Single-phase grid-connected PV system with Golden Section Search-based MPPT algorithm. *Chin. J. Electr. Eng.*, 7, 4, 25–36, 2021.
7. Sangwongwanich, A. and Blaabjerg, F., Mitigation of interharmonics in PV systems with Maximum Power Point Tracking Modification. *IEEE Trans. Power Electron.*, 34, 9, 8279–8282, 2019.

8. Moon, S., *et al.*, A new low-cost centralized MPPT controller system for Multiply Distributed Photovoltaic Power Conditioning Modules. *IEEE Trans. Smart Grid*, 6, 6, 2649–2658, 2015.
9. Bollipo, R. B., Mikkili, S., Bonthagorla, P. K., Hybrid, optimization, intelligent and classical PV MPPT techniques: review. *CSEE J. Power Energy Syst.*, 7, 1, 9–33, 2020.
10. Ibrahim, A.-W., *et al.*, PV maximum power-point tracking using modified particle swarm optimization under partial shading conditions. *Chin. J. Electr. Eng.*, 6, 4, 106–121, 2020.
11. Roman, E., *et al.*, Intelligent PV module for grid-connected PV systems. *30th Annual Conference of IEEE Industrial Electronics Society, 2004*, IECON, 2004.
12. Sahu, H.S., *et al.*, A novel approach for direct MPP estimation of a PV module under different irradiation conditions. *IEEE Trans. Energy Convers.*, 36, 4, 3127–3136, 2021.
13. Mahmoud, Y. and El-Saadany, E., Accuracy improvement of the ideal PV model. *IEEE Trans. Sustain. Energy*, 6, 3, 909–911, 2015.
14. Bialasiewicz, J.T., Renewable energy systems with photovoltaic power generators: Operation and modeling. *IEEE Trans. Ind. Electron.*, 55, 7, 2752–2758, 2008.

13

Advancement in Differential Locking Technology for Electric Vehicles

Bhupender Singh[1*], Nitin Panwar[1], Arvind Gupta[1], Sanjeev Sharma[2], Sandeep Ravish[3] and Praveen Gautam[4]

[1]Department of Mechanical Engineering, J.C. Bose University of Science and Technology, YMCA Faridabad, Haryana, India
[2]Mechanical Engineering Department, Amity University, Gurugram, Haryana, India
[3]Department of Civil Engineering, J.C. Bose University of Science and Technology, YMCA Faridabad, Haryana, India
[4]Department of Mechanical Engineering, Raj Kumar Goel Institute of Technology, Ghaziabad, India

Abstract
Traditional differentials transmit approximately the same torque to the wheels in a straight line, but problems occur when the wheels rotate. The current equipment has some drawbacks, which the project will try to overcome. When the vehicle gets stuck in a pit or muddy area, it is extremely difficult to move the vehicle because in open differentials, power is transmitted *via* least resistance route, thus the wheel that is stuck keeps spinning and enough torque is not there on the wheel at surface with good traction, thus the vehicle goes nowhere. A locking mechanism is introduced here in which a servo motor rotates a cam and actuates a dog plate which slides on the dog plate which is fixed with the sun gear, hence locking the sun gear with the differential. This locking system can be turned on/off manually or automatically depending on the driving situation. The incorporation of e-Diff technology into Electric Vehicles (EVs) represents a significant advancement in automotive engineering. This chapter intends to replace traditional mechanical differentials with a more efficient electronic version for electric vehicles. By smoothly incorporating electronic control systems; the suggested e-Diff intends to address the obstacles that mechanical counterparts provide. Innovative design

**Corresponding author*: bhupee_28@yahoo.co.in; ORCID: https://orcid.org/0009-0003-3677-4665

Ashwani Kumar, Mukesh Kumar Awasthi, Nitesh Dutt, Yogesh Kumar Singla and Sivasakthivel Thangavel (eds.) Thermal Battery Management System for Hybrid and Electric Vehicles, (251–266) © 2025 Scrivener Publishing LLC

and sophisticated control algorithms promise better EV handling, stability, and traction, as well as significantly lower noise levels for a calmer driving experience.

Keywords: Differential locking technology, electric vehicle, stability, traction

13.1 Introduction

When cornering, the inside wheels turn in a smaller radius than the outside wheel but have the same angular displacement with respect to turning circle, thus they travel a shorter distance than the outside wheels to avoid problems such as spinning and dragging, unpredictable handling, and damage to the vehicle's tires. A system designed to appropriately drive a pair of wheels at different speeds to achieve perfect cornering of a vehicle is called a differential.

The differential cage shown in Figure 13.1 has a group of four opposing bevel gears, allowing each bevel gear to mesh with two adjacent gears in a plane. The two sun gears are placed in the same axis as the ring gear and axle is connected to the driven gear [7–15]. The other two planetary gears are placed on a vertical axis, and their orientation changes with the rotation of the ring gear (Figures 13.2). When the vehicle is moving straight, there is no differentiation in the planetary gears. If you lift one of the wheels off the ground and continue to manually rotate the driveshaft, one drive wheel will rotate the opposite drive wheel by the same amount in the opposite direction. Therefore, the ring gear rotation is always the average of the side sun gear rotations [1, 16–20].

When the vehicle is moving in a straight line then the power is transferred from the crown gear. This power then flows into the differential cage, housing planet pinions and sun gears. As the vehicle moves forward, the crown wheel, differential cage, planet pinions, and sun gears rotate

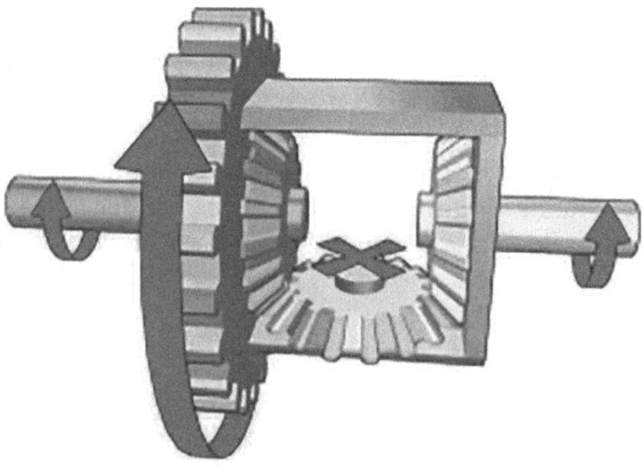

Figure 13.1 Open differential when moving in straight line [2].

Figure 13.2 Open differential when turning [4].

together as a single unit, resulting in no relative motion between the sun gear and planet pinions [3, 21–25].

During a turn, the inner wheel faces resistance, prompting it to rotate in the opposite direction. Consequently, the planet pinions start rotating about their own axis while revolving around the sun gear. This leads to increased rotary motion being transmitted to the outer-side sun gear, causing it to rotate faster than the inner sun gear. As a result, the outer road wheel runs at a higher speed than the inner road wheel, allowing it to cover a greater distance while turning [5, 26–35].

The differential mechanism in a car is critical to the equitable distribution of power among the wheels while also supporting varying rotational velocities, which is especially noticeable during maneuvers [36–45]. Although this system is rigorously constructed to withstand significant strain and deterioration, it is nonetheless prone to faults that may arise over time as a result of employment, environmental factors, or intrinsic manufacturing irregularities. This article goes into common problems that might occur within an automobile's differential setup. In front-wheel drive arrangements, the differential assembly is adjacent to the gearbox casing; in rear-wheel drive configurations, it is located between the rear wheels. A pair of differentials is located at both the front and back axles of four-wheel drive vehicles.

13.2 Methodology

13.2.1 Construction

The Differential consists of the following components (shown in Figure 13.3) with their roles as mentioned below:

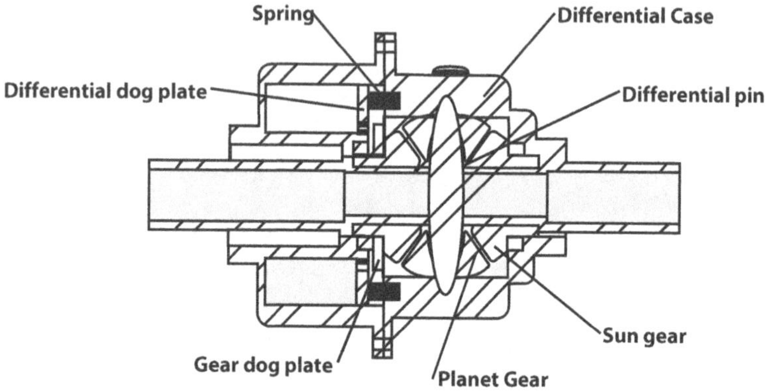

Figure 13.3 Section view of the differential assembly.

Differential Case: It is the overall enclosure which contains the bevel planetary train and protects it from surrounding environment.
Differential Pin: It acts as the support for the planet gears and as the spindle along which the gears can rotate as well.
Gear Dog Plate: It is the locking plate onto which differential dog plate slides along to lock themselves.
Differential Dog Plate: It is the sliding plate which is actuated by the linear actuator.
Spring: It provides the back force which separates the dog plates when the linear actuator stops acting.
Actuator: It provides the necessary forward push which slides the differential dog plate onto the gear dog plate.
Bevel Gears: It is the planetary gear train which necessitates the force transfer.

13.3 Working

When the locking mechanism is not engaged then the differential acts like a normal open differential in which torque transfer takes the least resistance path. On the push of a switch, lock mechanism is engaged which causes the whole differential to move as a single unit, thus relative motion between the components does not take place. On pressing the button, linear actuator engages which pushes the differential dog plate onto the gear dog plate. But when the locking action is not needed, linear actuator is deactivated and the springs push the differential dog plate back into its original position.

13.4 Linear Actuator Selection

Various options were available for providing linear actuation and they are discussed as below [6]:

13.4.1 Pneumatic Actuators

They are one of the most popular options for providing the linear actuation, they are fast and strong and provide very high up to 2000–3000 N in a short period of time. But these are quite bulky and cannot be accommodated in the assembly easily. Also, they require a constant supply of pressurized air through a compressor.

13.4.2 Rack Pinion

Like their pneumatic counterparts, these are also quite bulky and prove difficult to accommodate in the assembly, adding to the rotational inertia of the component as well.

13.4.3 Lead Screws

They utilize frictional force between the threads of the screw and the components mounted on it to move the components.

Among the above list of actuators, cam and follower type actuation was chosen for the scope of this project, the cam being rotated by a couple of servo motors and the follower being the differential dog plate. The servo is powered by a nickel cadmium battery and controlled by a 3 channel PWM based controller.

13.5 Design and Development

Design and Development involves Material Selection Matrix, Cam Profile, Electronics components and circuit and final model development.

13.6 Material Selection Matrix

Various Steel alloys were considered for the manufacturing, with some of them being listed above. Finally, EN 31 was chosen because of its easy availability, good machinability, and ideal properties which offer excellent strength as well.

13.7 Cam Profile

The cam profile was developed taking the follower movement as simple harmonic motion to keep the movement smooth and avoid any sort of jerks.

Figure 13.4 shows the follower displacement curve for the Lift of follower: 6.5 mm Angle of ascent: 180° Least radius: 4 mm.

13.8 Electronics Components and Circuit

The following components are required to make servo motor control circuit:

Servo motor control (shown in Figure 13.5) has been achieved through the use of pulse width modulation (PWM). This signal consists of a series of repeating pulses that determine the position the servo reaches. PWM are most commonly generated with the help of common microcontrollers such as Arduino, but they can also be sent from wireless receivers [46–55]. In this case IC 555 timer has been used for PWM generation. Electronics inside the servo convert the width of the pulse into position. When the servo is commanded to rotate, power is applied to the motor until the potentiometer reaches a value corresponding to the commanded position [56–70]. Considering the rotation limits of the servo, neutral is defined as the center of rotation. Servo motors typically can only rotate 90° in either direction,

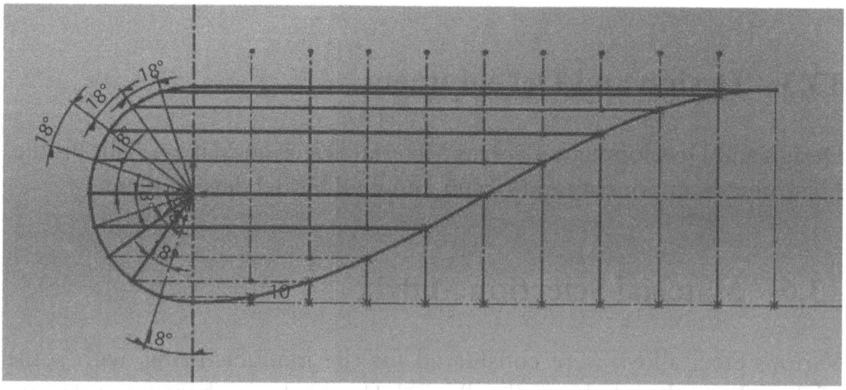

Figure 13.4 Follower displacement curve (made in solid works).

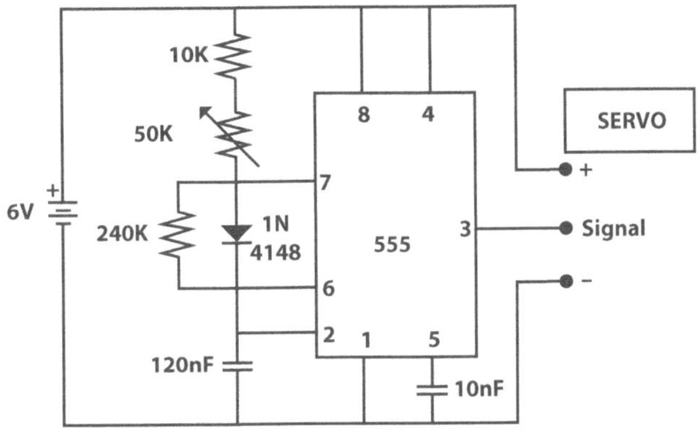

Figure 13.5 Servo control circuit diagram [13].

Table 13.1 Materials and their properties [11].

Material	Ultimate tensile strength	Yield strength
EN 19	655 MPa	415 MPa
EN 8	700 MPa	465 MPa
EN 31	750 MPa	450 MPa
EN 24	850 MPa	680 MPa

Table 13.2 Electronic components used.

S. no.	Component	Value	Qty.
1.	555 Timer	-	1
2.	Diode	1N4148	1
3.	Resistors	10K, 240K ohm	1,1
4.	Capacitors	10nF, 120nF	1,1
5.	Potentiometer	50K ohm	1
6.	Power supply	4.8–6V	1
7.	Servo Motor	2.2 kg-cm peak torque	1

giving them a total of 180°. The PWM sent to the motor determines the position of the shaft based on the duration of the pulses sent through the control. Table 13.1 presents the Materials and their properties and Table 13.2 lists the Electronic components used. The rotor will then rotate to the desired position [12, 71–78].

13.9 Final Model

Figure 13.6 & Figure 13.7 present the final model of differential:

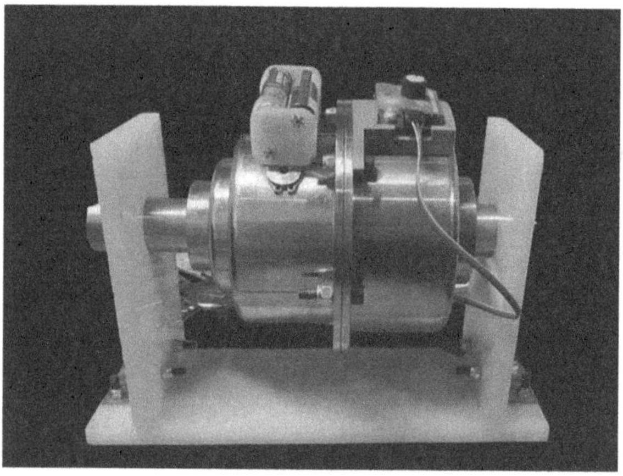

Figure 13.6 Final model of the developed prototype.

Figure 13.7 Differential casing showing integration of servo motors.

13.10 Conclusions

From the above study, conclusions drawn are:

- The set-up for automatic engagement of the differential when the loss of traction condition is encountered is designed by introducing automatic differential locking system.
- Under slip condition, when differential is under action, the difference between torques of two axles is very high as well as the Torque and Speed are inversely proportional to each other.
- When the differential is working and if one wheel loses its traction due to mud, pit, or slippery surface, the locking effect will provide it more speed to overcome traction loosening.

References

1. Bandaru, S.H., Alternative Transmission System for 4 Wheeler. *Int. J. Mech. Eng. Rob. Res.*, 4, 445–454, 2015.
2. Karande, M., Uttekar, S., Patil, P., Gawade, A., Manufacturing of simple differential locking system. *Int. J. Eng. Technol.*, 3, 171–178, 2015.
3. Ahmed, A. and Yadav, A.K., Parametric Analysis of Wastewater Electrolysis for Green Hydrogen Production: A Combined RSM, Genetic Algorithm, and Particle Swarm Optimization Approach. *Int. J. Hydrogen Energy*, 2, 59, 51–62, 2024.
4. Ahmad, A., Yadav, A.K., Hasan, S., Enhanced production of methane enriched biogas through intensified co-digestion process and its effective utilization in a biodiesel/biohydrogen fueled engine with duel injection strategies: ML-RSM based an efficient optimization approach. *Int. J. Hydrogen Energy*, 65, 671–686, 2024.
5. Ahmed, A., Yadav, A.K., Singh, A., Singh, D.K., A machine learning–genetic algorithm based models for optimization of intensification process through microwave reactor: A new approach for rapid and sustainable synthesis of biodiesel from novel Hiptage benghalensis seed oil. *Fuel*, 374, 132470, 2024.
6. Khan, F., Khan, O., Parvez, M., Almujibah, H., Pachauri, P., Yahya, Z., Ağbulut, Ü., Innovative hydrogen production from waste bio-oil *via* steam methane reforming: An advanced ANN-AHP-k-means modelling approach using extreme machine learning weighted clustering. *Int. J. Hydrogen Energy*, 105, 1080–1091, 2005.

7. Khan, O., Alsaduni, I., Equbal, A., Parvez, M., Yadav, A.K., Performance and Emission Analysis of Biodiesel Blends Enriched with Biohydrogen and Biogas in Internal Combustion Engines, Process Saf. *Environ. Prot.*, 183, 1013–1037, 2024.
8. Ahmed, A., Yadav, A.K., Singh, A., Singh, D.K., A machine learning-response surface optimization approach to enhance the performance of diesel engine using novel blends of aloe vera biodiesel with MWCNT nanoparticles and hydrogen. *Process Saf. Environ. Prot.*, 186, 738–755, 2024. doi: 10.1016/j.psep.2024.04.013.
9. Ahmed, A., Yadav, A.K., Singh, D.K., Singh, A., A Comprehensive Machine Learning-Coupled Response Surface Methodology Approach for Predictive Modeling and Optimization of Biogas Potential in Anaerobic Co-Digestion of Organic Waste. *Biomass Bioenergy*, 180, 106995, 2024.
10. Ahmad, A., Yadav, A.K., Singh, A., Singh, D.K., Ağbulut, U., A hybrid RSM-GA-PSO approach on optimization of process intensification of linseed biodiesel synthesis using an ultrasonic reactor: Enhancing biodiesel properties and engine characteristics with ternary fuel blends. *Energy*, 288, 129077, 2024.
11. Ahmed, A., Yadav, A.K., Hasan, S., Enhancing Efficiency and Environmental Impacts of a Nano-Enhanced Oleander Biodiesel-Biohydrogen Dual Fuel Engine Equipped with EGR, through Operational Parameter Optimization using RSM-MOGA Technique. *Int. J. Hydrogen Energy*, 14, 78, 1157–1172, 2024.
12. Khan, O., Alsaduni, I., Parvez, M., Yadav, A.K., Enhancing hydrogen production using solar-driven photocatalysis with biosynthesized nanocomposites: A hybrid machine learning approach towards enhanced performance and sustainable environment. *Int. J. Hydrogen Energy*, 102, 609–625, 2025
13. Khanam, S., Khan, O., Ahmad, S., Sherwani, A.F., Khan, Z.A., Yadav, A.K., Ağbulut, U., A Taguchi-based hybrid multi-criteria decision-making approach for optimization of performance characteristics of diesel engine fuelled with blends of biodiesel-diesel and cerium oxide nano-additive. *J. Therm. Anal. Calorim.*, 149, 8, 3657–3676, 2024.
14. Ahmed, A., Yadav, A.K., Singh, A., Enhancing waste cooking oil biodiesel yield and characteristics through machine learning , response surface methodology , and genetic algorithm for optimal utilization in CI Engines. *Int. J. Green Energy*, 3, 21, 1345–1365, 2024.
15. Singh, A.A., Singh, R.K., Yadav, A.K., Saha, A., Dewangan, A., Singh, B., Experimental investigations on the electrochemical machining of D3 die steel material and multi-objective parameter optimization. *Surf. Rev. Lett.*, 31, 1, 2450074, 2024.
16. Ahmed, A., Yadav, A.K., Hasan, S., Biogas as a Sustainable and Viable Alternative Fuel for Diesel Engines: A Comprehensive Review of Production, Purification, Economic Analysis, and Performance Evaluation. *J. Process Mech. Eng. Part E*, 09544089241255930, 2024. https://doi.org/10.1177/09544089241255930.

17. Khan, F., Karimi, M.N., Khan, O., Yadav, A.K., Alhodaib, A., Gürel, A.E., Ağbulut, U., A hybrid MCDM optimization for utilization of novel set of biosynthesized nanofluids on thermal performance for solar thermal collectors. *Int. J. Thermofluids*, 22, 100686, 2024.
18. Alhodaib, A., Yahya, Z., Khan, O., Equbal, A., Equbal, M.S., Parvez, M., Idrisi, M.J., Sustainable coatings for green solar photovoltaic cells: performance and environmental impact of recyclable biomass digestate polymers. *Sci. Rep.*, 14, 1, 11221, 2024.
19. Singh, B., Yadav, A.K., Saha, A., Fabrication and Characterization of AA6082-T6 Produced by Friction Stir Additive Manufacturing. *Surf. Rev. Lett.*, 1, 41–47, 2024. DOI:10.1142/S0218625X25500222.
20. Equbal, A., Equbal, A., Ahmad, R., Kumar, M., Khan, O., Yadav, A.K., RSM based statistical approach to enhance the compressive performance of ABS P400 parts fabricated *via* FDM Technology. *J. Surf. Rev. Lett.*, 2550021, 2024. DOI:10.1142/S0218625X25500210.
21. Dewangan, A., Ahmad, A., Yadav, A.K., Innovative research on waste tire recycling for sustainable biofuel production: Assessment of its usability on multi-cylinder diesel engine employing constant injection of oxyhydrogen and biogas through a premixing device. *Int. J. Hydrogen Energy*, 84, 155–163, 2024.
22. Dewangan, A., Mallick, A., Yadav, A.K., Islam, S., Saleel, C.A., Shaik, S., Ağbulut, U., Production of oxy-hydrogen gas and the impact of its usability on CI engine combustion, performance, and emission behaviors. *Energy*, 22, 278, 127937, 2023.
23. Khan, O., Parvez, M., Kumari, P., Yadav, A.K., Akram, W., Ahmad, S., Idrisi, M.J., Modelling of compression ignition engine by soft computing techniques (ANFIS-NSGA-II and RSM) to enhance the performance characteristics for leachate blends with nano-additives. *Sci. Rep.*, 13, 1, 15429, 2023.
24. Ahmad, A., Yadav, A.K., Singh, A., Singh, D.K., Multi-response optimization of a microalgae-spirulina-fueled VCR diesel engine: a comprehensive RSM-GA approach. *Environ. Dev. Sustain.*, 27, 1–29, 2023.
25. Ahmed, A., Yadav, A.K., Singh, A., Process optimization of spirulina microalgae biodiesel synthesis using RSM coupled GA technique: a performance study of a biogas powered dual fuel engine. *Int. J. Environ Sci. Tech.*, 21, 169–188, 2023.
26. Khan, O., Ali, V., Parvez, M., Alhodaib, A., Yahya, Z., Yadav, A.K., Ağbulut, U., Exploring the performance of biodiesel-hydrogen blends with diverse nanoparticles in diesel engine: A hybrid machine learning K-means clustering approach with weighted performance metrics. *Int. J. Hydrogen Energy*, 78, 547–563, 2024.
27. Ahmed, A., Yadav, A.K., Singh, A., Application of machine learning and genetic algorithm for prediction and optimization of Biodiesel Yield from Waste Cooking Oil. *Korean J. Chem. Engg.*, 40, 2941–2956, 2023.

28. Ahmed, A., Yadav, A.K., Singh, A., Enhancement of Biogas Yield from Dual Organic Waste Using Hybrid Statistical Approach and Its Effects on Ternary Fuel Blend (Biodiesel/n-butanol /Diesel) Powered Diesel Engine. *Environ. Prog. Sust. Energy*, 42, 5, e14163, 2023, doi.org/10.1002/ep.14163.
29. Ahmed, A., Yadav, A.K., Singh, A., Optimization of Cavitation-Assisted Biodiesel Production and Fuel Properties from Neochloris Oleoabundans Microalgae oil using Genetic Algorithm and Response Surface Methodology. *J. Process Mech. Eng. Part E.*, 238, 4, 1821–1835, 2023, doi.org/10.1177/09544089231159832.
30. Yadav, M., Karimi, M.N., Yadav, A.K., Effect of Graphene Oxide Nanoparticles Dispersed Biodiesel on Combustion, Performance, and Emission Characteristics of a Diesel Engine. *J. Process Mech. Eng. Part E.*, 09544089231209139, 2023, https://doi.org/10.1177/09544089231209139.
31. Dwivedi, S.P., Yadav, A.K., Saxena, A., Dwivedi, V.K., Tribo-mechanical, physical and thermal behaviour of Al/Si3N4 composite with and without the addition of Cu, Ni and Cr entropy elements. Proceedings of the Institution of Mechanical Engineers. *Part E J. Process Mech. Eng.*, 09544089231189663, 2023, 09544089231189663.
32. Das, D., Yadav, A.K., Ahmed, A., Numerical Simulation of the Combustion Chamber and Performance Analysis of Microalgae-Based Biodiesel/Diesel-Powered CI Engine: An Axisymmetric Flow Approach. *J. Process Mech. Eng. Part E*, 09544089231218121, 2023, https://doi.org/10.1177/09544089231218121.
33. Dewangan, A., Mallick, A., Yadav, A.K., Ahmad, A., Alqahtani, D., Islam., S., Combined Effect of Operating Parameters and Nano Particles on Performance of a Diesel Engine: Response Surface Methodology coupled Genetic algorithm Approach. *ACS Omega*, 8, 27, 24586–24600, 2023.
34. Singh, D., Singh, S., Yadav, A.K., Khan, O., Dewangan, A., Alkahtani, M.Q., Islam, S., From Theory to Practice: A Sustainable Solution to Water Scarcity by Using a Hybrid Solar Distiller with a Heat Exchanger and Aluminum Oxide Nanoparticles. *ACS Omega*, 37, 8, 33543–33553, 2023.
35. Ahmed, A., Yadav, A.K., Singh, A., Optimization of Biogas Yield from Anaerobic Co-digestion of Dual waste for environmental sustainability: ANN, RSM, and GA, Approach. *Int. J. Oil Gas Coal Tech.*, 33, 1, 75, 2023.
36. Ahmed, A., Yadav, A.K., Singh, A., An environmental impact assessment and optimization study of biodiesel production from microalgae. *Int. J. Global Warming*, 31, 3, 294–313, 2023.
37. Ahmed, A., Yadav, A.K., Singh, A., Biodiesel yield optimization from a third-generation feedstock (microalgae spirulina) using a hybrid statistical approach. *Int. J. Ambient Energy*, 44, 1, 1202–1213, 2023.
38. Ahmed, A., Yadav, A.K., Singh, A., Modelling and optimisation of VCR diesel engine parameters fuelled with microalgae methyl-ester: a RSM-coupled Taguchi approach. *Int. J. Ambient Energy*, 44, 1, 2498–2506, 2023.

39. Ahmed, A., Yadav, A.K., Singh, A., Biodiesel Production from Mahua Oil: Characterization, Optimization, and Modeling with a Hybrid Statistical Approach. *Int. J. Ambient Energy*, 44, 1, 2600–2609, 2023.
40. Khan, T.A., Khan, T.A., Yadav, A.K., Khan, M.E., Experimental Studies on Utilization of Neochloris Oleoabundans Microalgae Biodiesel as an Alternative Fuel for Diesel Engine. *Int. J. Ambient Energy*, 44, 1, 115–123, 2023.
41. Singh, D., Yadav, A.K., Kumar, A., Samsher, Energy matrices and life cycle conversion analysis of N-identical hybrid double slope solar distiller unit using Al2O3 nanoparticle. *J. Water Environ. Nanotechnol.*, 8, 3, 267–284, 2023.
42. Khan, T.A., Yadav, A.K., Khan, T.A., A hydrodynamic cavitation-assisted system for optimization of biodiesel production from green microalgae oil using a genetic algorithm and response surface methodology approach. *Environ. Sci. Pollut. Res*, 29, 49465–49477, 2022.
43. Dewangan, A., Yadav, A.K., Mallick, A., Singh, S., Comparative study of Manilkara Zapota and Karanja based Biodiesel Properties and Its Effect on Multi-Cylinder Diesel Engine Characteristics. *Energy Sources Part A*, 44, 2, 5143–5153, 2022.
44. Yadav, A.K., Khan, T.A., Khan, T.A., Kumar, S., Methyl ester of Gmelina arborea oil as a substitute for petroleum diesel: an experimental study on its performance and emissions in a diesel engine. *Energy Sources Part A*, 43, 11, 1307–1314, 2021.
45. Khan, O., Yadav, A.K., Khan, M.E., Parvez, M., Characterization of Bio-Ethanol Obtained from Eichhornia Crassipes Plant; Its Emission and Performance Analysis on CI Engine. *Energy Sources Part A*, 43, 14, 1793–1803, 2021.
46. Singh, S.K., Pal, A., Yadav, A.K., Efficient Production of Biodiesel from Cannabis Sativa oil using Intensified Transestericationn (Hydrodynamic Cavitation) Method. *Energy Sources Part A*, 42, 20, 2461–2470, 2020.
47. Dewangan, A., Yadav, A.K., Mallick, A., Tyagi, A., Optimisation of biodiesel production and engine performance from Simarouba oil in compression ignition engine. *Int. J. Oil Gas Coal Technol.*, 25, 3, 357–376, 2020.
48. Khan, M.E., Khan, T.A., Yadav, A.K., Pal, A., A comprehensive review on performance, emission and combustion characteristics of diesel engines fuelled with algae oil and methyl ester. *WEENTECH Proc. Energy*, 6, 2, 160–168, 2020.
49. Yadav, A.K., Khan, M.E., Pal, A., Custard Apple Seed (Annona Squamosa) Oil as a Promising Feedstock for Biodiesel Production using Advanced Techniques and Its Experimental Investigations on an Agricultural Diesel Engine. *Int. J. Oil Gas Coal Tech.*, 20, 4, 473–492, 2019.
50. Yadav, A.K., Pal, A., Ghosh, U., Gupta, S.K., Comparative Study of Biodiesel Production Methods from Yellow Oleander Oil and its Performance Analysis

on an Agricultural Diesel Engine. *Int. J. Ambient Energy*, 31, 40, 2, 152–157, 2019.

51. Khan, I.A., Singh, S.K., Yadav, A.K., Sharma, D., Enhancement in the Performance of a Diesel Engine Fueled with Pongamia Methyl Ester and n-Butanol as Oxygenated Additive. *Int. J. Ambient Energy*, 40, 8, 842–846, 2019.
52. Yadav, A.K., Khan, O., Khan, M.E., Utilization of High FFA Landfill Waste (Leachates) as a Feedstock for Sustainable Biodiesel Production: Its Characterization and Engine Performance Evaluation. *Environ. Sci. Pollut. Res.*, 25, 32, 32312–32320, 2018.
53. Yadav, A.K., Dewangan, A., Mallick, A., Effect of n-Butanol and Diethyl Ether Additives on Performance and Emission Characteristics of a Diesel Engine Fuelled with Diesel-Pongamia Biodiesel Blends. *J. Energy Eng.*, 144, 6, 04018062, 2018.
54. Vinay, Singh, B., Yadav, A.K., Optimization of Performance and Emission Characteristics of CI Engine Fuelled with Mahua Oil Methyl Ester-Diesel Blend using Response Surface Methodology. *Int. J. Ambient Energy*, 44, 6, 674–85, 2020.
55. Yadav, A.K., Dewangan, A., Mallick, A., Synthesis and Stability Study of Biodiesel from Kachnar Seed Oil. *J. Energy Eng.*, 144, 5, 04018053, 2018.
56. Yadav, A.K., Pal, A., Dubey, A.M., Experimental Studies on Utilization of Prunus Armeniaca L. (Wild Apricot) Biodiesel as an Alternative Fuel for CI Engine. *Waste Biomass Valorization*, 9, 10, 1961–1969, 2018.
57. Yadav, A.K., Khan, M.E., Pal, A., Ultrasonic Assisted Production of Biodiesel from Karabi Oil using Heterogeneous Catalyst. *Biofuels*, 9, 1, 101–112, 2018.
58. Yadav, A.K. and Singh, B., Optimization of Biodiesel Production from Annona Squamosa Seeds Oil Using Response Surface Methodology and Its Characterization. *Energy Sources Part A*, 40, 9, 1051–1059, 2018.
59. Dewangan, A., Yadav, A.K., Mallick, A., Current Scenario of Biodiesel Development in India: Prospects and Challenges. *Energy Sources Part A*, 40, 20, 2494–2501, 2018.
60. Yadav, A.K., Khan, M.E., Pal, A., Performance and Emission Characteristics of a Stationary Diesel Engine fuelled by Schleichera Oleosa oil Methyl Ester (SOME) Produced through Hydrodynamic Cavitation Process, Egypt. *J. Pet.*, 27, 1, 89–93, 2018.
61. Yadav, A.K., Khan, M.E., Pal, A., Performance, Emission and Combustion Characteristics of an Indica Diesel Engine Operated with Yellow Oleander (Thevetia Peruviana) Oil Biodiesel Produced through Hydrodynamic Cavitation Method. *Int. J. Ambient Energy*, 39, 4, 365–371, 2018.
62. Dewangan, A., Mallick, A., Yadav, A.K., Effect of Metal Oxide Nanoparticles and Engine Parameters on the Performance of a Diesel Engine: A Review. *Mater. Today Proc.*, 21, 1722–27, 2020.
63. Yadav, A.K., Khan, M.E., Pal, A., Dubey, A.M., Experimental Investigations of Performance and Emissions Characteristics of Kusum (Schleichera Oleosa)

Biodiesel in a Multi-Cylinder Transportation Diesel Engine. *Waste Biomass Valor*, 8, 4, 1331, 2017.
64. Yadav, A.K., Khan, M.E., Pal, A., Biodiesel Production from Oleander (Thevetia Peruviana) Oil and Its Performance Testing on a Diesel Engine. *Korean J. Chem. Eng.*, 34, 2, 340–345, 2017.
65. Dubey, A.M., Kumar, A., Yadav, A.K., Wear behaviour of friction stir weld joint of cast Al (4–10%) Cu alloy welded at different operating parameters. *J. Mater. Process. Tech.*, 240, 87–97, 2017.
66. Khan, O., Khan, M.E., Yadav, A.K., Sharma, D., The Ultrasonic Assisted Optimization of Biodiesel Production from Eucalyptus Oil. *Energy Sources Part A*, 39, 13, 1323–1331, 2017.
67. Dewangan, A. and Yadav, A.K., Wax Deposition during Production of Waxy Crude Oil and Its Remediation. *Pet. Sci. Techn*, 35, 18, 1831–1838, 2017.
68. Yadav, A.K., Khan, M.E., Pal, A., Kaner biodiesel production through hybrid reactor and its performance testing on a CI engine at different compression ratios. *Egypt. J. Pet.*, 26, 2, 525–532, 2017.
69. Yadav, A.K., Khan, M.E., Pal, A., Dubey, A.M., Performance and emission characteristics of a transportation diesel engine operated with non-edible vegetable oils biodiesel. *Case Stud. Therm. Eng.*, 8, 236–244, 2016.
70. Yadav, A.K., Khan, M.E., Pal, A., Biodiesel Production from Nerium Oleander (Thevetia Peruviana) Oil through Conventional and Ultrasonic Irradiation Methods. *Energy Sources Part A*, 38, 23, 3447–3452, 2016.
71. Yadav, A.K., Khan, M.E., Pal A., A., Comparative Study on Ultrasonic Cavitation and Mechanical Stirring Method towards Efficient Production of Biodiesel from Non-Edible Oils and Performance Testing on a C.I. Engine. *Int. J. Envt. Waste Managt.*, 18, 4, 349–367, 2016.
72. Dewangan, A., Mallick, A., Yadav, A.K., Kumar, R., Combustion-generated pollutions and strategy for its control in CI engines: A review. *Mater. Today Proc.*, 21, 1728–1733, 2020.
73. Yadav, A.K., Khan, M.E., Pal, A., Sharma, D., Optimization of biodiesel production from bitter groundnut oil using Taguchi method and its performance and emissions characteristics on a 4-cylinder Tata Indica engine. *Int. J. Sust. Agri. Mgt. Inf.*, 1, 285–300, 2015.
74. Khan, F., Khan, O., Parvez, M., Ahmad, S., Yahya, Z., Alhodaib, A., Yadav, A.K., Ağbulut, U., K-means clustering optimization of various quantum dots and nanoparticles-added biofuels for engine performance, emission, vibration, and noise characteristics. *Therm. Sci. Eng. Prog.*, 54, 102815, 2024.
75. Khan, O., Parvez, M., Yadav, A.K., *et al.*, Investigation of the Thermal Properties of Lauric Acid Melting as a Phase Change Material Using Priority based K-Means Clustering Method. *Energy Storage*, 6, 6, e70026, 2024, doi:10.1002/est2.70026.
76. Sharma, G., Dewangan, A.K., Yadav, A.K., Ahmed, A., Current status of research on hydrogen generation, storage and transportation technologies:

A state-of-the-art review towards sustainable energy. *Process Saf. Environ. Prot.*, 191, 1445–1460, 2024.
77. Singh, S., Singh, B., Kumar, S., Yadav, A.K., Temperature-dependent dynamic hysteresis scaling of ferroelectric hysteresis parameters of lead free [(Ba 0.825 + xCa0.175-x) (Ti1- x Snx)O3]ceramics. *Ferroelectrics*, 551, 1, 133–142, 2019.
78. Dubey, A.M., Ram, R., Yadav, A.K., Ion cell performance using single particle representation of battery electrode. *J. Electroanal. Chem.*, 768, 11–17, 2016.

14

Application of Battery Thermal Management Systems (BTMS) for Heavy-Duty Electric Vehicles (HD-EV)

Pooyan Rahmanivahid[1]*, Morteza Khashehchi[2], Sivasakthivel Thangavel[1] and Milad Heidari[1]

[1]Mechanical Engineering Department, Global College of Engineering and Technology (GCET), Muscat, Sultanate of Oman
[2]Systems Engineering Department, Military Technological College, Muscat, Oman

Abstract
The transition towards sustainable transportation solutions is propelled by the electrification of vehicles, with Battery Thermal Management Systems (BTMS) emerging as crucial components in optimizing the performance and longevity of electric vehicles (EVs). This abstract delves into the significance of BTMS across various segments of electric transportation, including buses, trucks, and other heavy-duty vehicles. In urban transit systems, electric buses equipped with BTMS ensure efficient management of battery temperature, maximizing energy efficiency and operational reliability. Meanwhile, in the realm of heavy-duty transportation, electric trucks benefit from sophisticated BTMS strategies, dynamically adjusting thermal management to adapt to varying load conditions and optimize performance.

The efficacy of BTMS extends beyond mere thermal regulation, encompassing the optimization of performance and the prolongation of battery life. By maintaining batteries within their ideal temperature range, BTMS minimize energy losses and mitigate the risk of thermal degradation, safeguarding investments and ensuring long-term reliability. As the journey towards electrification continues, the integration of BTMS into electric vehicles represents a pivotal milestone in

**Corresponding author*: pooyan@gcet.edu.om; ORCID: 0000-0002-1350-2259
Morteza Khashehchi: ORCID: 0000-0001-8636-4718
Sivasakthivel Thangavel: ORCID: 0000-0001-6802-4015
Milad Heidari: ORCID: 0000-0001-7577-4789

Ashwani Kumar, Mukesh Kumar Awasthi, Nitesh Dutt, Yogesh Kumar Singla and Sivasakthivel Thangavel (eds.) Thermal Battery Management System for Hybrid and Electric Vehicles, (267–280) © 2025 Scrivener Publishing LLC

achieving sustainable transportation solutions. By unlocking new realms of efficiency, reliability, and endurance, BTMS propel electric vehicles towards a future defined by innovation and progress, laying the foundation for a greener and more sustainable transportation ecosystem.

In recent years, there has been a paradigm shift towards sustainable transportation solutions, with a particular emphasis on electrification. This shift is especially prominent in heavy-duty electric vehicles (HD-EV), including electric buses, trucks, and trains. One crucial aspect of ensuring the optimal performance and longevity of electric vehicle (EV) batteries is the implementation of effective Battery Thermal Management Systems (BTMS). In this chapter, we delve into the significance of BTMS and explore their applications in HD-EVs.

***Keywords*:** BTMS, heavy duty electric vehicles, battery life, urban transportation, operational reliability

14.1 Introduction to Battery Thermal Management Systems (BTMS)

In the relentless pursuit of sustainable transportation solutions, the global automotive industry has undergone a profound transformation, pivoting towards the electrification of vehicles. Within this electrification revolution, one crucial component stands at the forefront of ensuring the efficiency, safety, and longevity of electric vehicle (EV) batteries – the Battery Thermal Management System (BTMS).

At its core, BTMS represents a sophisticated ensemble of technologies and strategies meticulously designed to regulate the thermal conditions of EV batteries. In essence, BTMS serves as the custodian of the battery's operating temperature range, a pivotal factor that profoundly impacts its performance and lifespan. By orchestrating a delicate balance between cooling and heating mechanisms, BTMS endeavors to maintain the battery within its optimal temperature band, thereby unlocking a myriad of benefits that reverberate throughout the realm of electrified transportation (Luca, Cattani, *et al.*, 2023). The importance of BTMS in the realm of Heavy-Duty Electric Vehicles (HD-EVs) cannot be overstated. In these behemoths of the electric mobility landscape, where battery packs sprawl over expansive footprints and are subjected to formidable loads, the role of BTMS becomes even more pronounced. It is amidst this backdrop of monumental scale and heightened demands that the true significance of BTMS unveils itself [1–4].

14.2 Optimizing Performance

The performance of an electric vehicle hinges not only on the capacity and efficiency of its battery but also on the ability to maintain optimal operating conditions. BTMS steps into this arena as a silent sentinel, vigilantly monitoring and adjusting the battery temperature to extract the maximum performance potential. By ensuring that the battery operates within its designated temperature range, BTMS mitigates the detrimental effects of temperature extremes, thereby bolstering the vehicle's acceleration, range, and overall drivability.

14.2.1 Enhancing Safety

Safety stands as an uncompromising cornerstone in the domain of electric mobility. With the potential for thermal runaway events posing inherent risks, BTMS emerges as a formidable ally in safeguarding against such perils. Through its intricate network of sensors, actuators, and cooling/heating elements, BTMS diligently maintains the battery's thermal equilibrium, thereby thwarting the onset of catastrophic scenarios. Whether navigating through scorching desert highways or traversing frigid urban landscapes, BTMS stands as an unwavering bastion of safety, instilling confidence in both drivers and passengers alike [5–8].

14.2.2 Prolonging Battery Life

The lifespan of an electric vehicle battery represents a linchpin in the economic viability and environmental sustainability of electrified transportation. BTMS assumes the role of a custodian, meticulously stewarding the battery through its operational lifecycle with precision and finesse. By averting the deleterious effects of temperature-induced degradation, BTMS bestows upon the battery the gift of longevity, ensuring that it remains a reliable and enduring asset throughout the vehicle's lifespan (Davide, Fioriti, *et al.*, 2023). In summary, In the crucible of electrified transportation, where innovation and sustainability converge, Battery Thermal Management Systems (BTMS) stand as a beacon of progress. From the bustling streets of metropolitan cities to the vast expanses of intercontinental highways, the significance of BTMS reverberates far and wide, shaping the future of mobility one meticulously regulated temperature at a time. As we embark on this electrifying journey towards a greener tomorrow, let us not overlook the profound impact of BTMS in driving us ever closer to the realization of a sustainable and electrified future (Shubham *et al.*, 2023) [9–13].

14.3 BTMS for Electric Buses: Revolutionizing Urban Transportation

Electric buses have emerged as the vanguards of sustainable urban transportation, heralding a paradigm shift away from conventional diesel-powered counterparts. In a world grappling with the pressing imperatives of climate change and urban air pollution, electric buses stand as beacons of hope, offering a clean, efficient, and environmentally friendly mode of mass transit [14–18]. However, amidst the promise of electrification lies a formidable challenge – the efficient management of the thermal conditions of large battery packs. It is within this crucible of innovation and necessity that Battery Thermal Management Systems (BTMS) for electric buses assume their pivotal role (Mostafa, F, *et al.*, 2023).

14.3.1 Addressing the Unique Challenges

The transition from fossil-fuel-dependent buses to their electric counterparts necessitates a fundamental reimagining of vehicle architecture and operational logistics. Electric buses, equipped with massive battery packs, are confronted with unique challenges, chief among them being the effective regulation of battery temperature. In the unforgiving crucible of urban transit, where stop-and-go traffic patterns and variable weather conditions reign supreme, the thermal management of battery packs becomes an imperative rather than a mere luxury (Ko, Takatoshi, 2023).

14.3.2 Efficiency through Active Regulation

BTMS for electric buses are ingeniously engineered to tackle these challenges head-on. At the heart of these systems lie active cooling and heating mechanisms, meticulously calibrated to maintain the batteries within their ideal temperature range. In scorching summer heat or frigid winter chills, BTMS springs into action, dynamically adjusting cooling or heating strategies to ensure optimal battery performance. By preemptively averting temperature extremes, BTMS not only enhance energy efficiency but also safeguard the batteries against premature degradation, thereby extending their operational lifespan [19–22].

14.3.3 Maximizing Energy Efficiency

Energy efficiency stands as a linchpin in the sustainability narrative of electric buses. BTMS, by virtue of their ability to maintain optimal battery temperature, play a pivotal role in maximizing the energy efficiency of electric buses. By ensuring that the batteries operate within their designated temperature band, BTMS minimize energy losses associated with thermal inefficiencies, thereby amplifying the overall efficiency and range of electric buses. This translates into tangible benefits, including reduced operating costs and enhanced environmental sustainability, further bolstering the case for electrified mass transit solutions [23–25].

14.3.4 Enhancing Operational Reliability

Operational reliability is the backbone of urban transit systems, where punctuality and consistency reign supreme. In the dynamic landscape of bustling city streets, where schedules are tightly woven and demands are incessant, the ability of transit vehicles to operate with unwavering reliability holds paramount importance. Amidst this backdrop, Battery Thermal Management Systems (BTMS) emerge as silent guardians, wielding the power to bestow upon electric buses a newfound sense of operational reliability. By meticulously regulating battery temperature within optimal limits, BTMS imbue electric buses with the resilience and fortitude necessary to navigate the intricacies of urban transit seamlessly. In the intricate tapestry of urban transit, punctuality and consistency stand as non-negotiable imperatives. Passengers rely on transit systems to whisk them to their destinations with precision and predictability, their schedules intricately intertwined with the arrival and departure times of buses. Any deviation from these schedules can have cascading effects, disrupting the rhythm of daily life for countless individuals. It is within this crucible of expectations that operational reliability takes center stage, serving as the linchpin upon which the trust and confidence of passengers and operators alike are built [26–29].

At the heart of operational reliability lies the ability to maintain the integrity and functionality of critical components, chief among them being the battery pack in electric buses. BTMS, with their intricate network of sensors, actuators, and cooling/heating elements, assume the mantle of guardianship, vigilantly safeguarding the battery against the perils of temperature extremes. By maintaining battery temperature within optimal limits, BTMS mitigate the risk of thermal degradation and premature aging, ensuring that the bus remains operational even in the face of challenging conditions.

Urban transit systems are synonymous with congestion – a labyrinth of intersecting thoroughfares teeming with vehicles of all shapes and sizes. Electric buses, tasked with traversing these bustling streets, are subjected to a myriad of operational challenges, from frequent stops and starts to prolonged periods of idling. In such a demanding environment, the reliability of the bus is put to the test, with every delay or breakdown exacerbating the strain on an already stretched transit network. BTMS, by maintaining battery temperature within optimal limits, bolster the resilience of electric buses, enabling them to navigate congested city streets with confidence and poise (Mostafa, F, *et al.*, 2023).

The unpredictability of weather poses yet another formidable challenge to the operational reliability of electric buses. From scorching summer heat waves to bone-chilling winter storms, the extremes of weather can wreak havoc on vehicle performance and reliability. Battery performance, in particular, is highly sensitive to temperature fluctuations, with both excessive heat and cold posing risks to efficiency and longevity. BTMS, with their ability to dynamically adjust cooling and heating strategies, equip electric buses with the resilience necessary to brave inclement weather conditions, ensuring uninterrupted service even amidst nature's fury.

Ultimately, the true measure of operational reliability lies in the confidence it instills – confidence in passengers that they will reach their destinations on time, and confidence in operators that their fleets will perform reliably day in and day out. Electric buses equipped with robust BTMS stand as paragons of reliability, embodying the promise of seamless urban transit. As they ply the streets, navigating through congestion and weathering the elements, they serve not only as vehicles of transportation but also as symbols of trust and dependability in a bustling urban landscape.

In the relentless pursuit of sustainable urban transportation solutions, electric buses stand as harbingers of change, ushering in an era of clean, efficient, and environmentally conscious mass transit. At the heart of this transformation lies the ingenuity of Battery Thermal Management Systems (BTMS), meticulously engineered to regulate the thermal conditions of large battery packs with precision and finesse. As electric buses traverse the bustling thoroughfares of urban landscapes worldwide, propelled by the silent hum of electric motors and the vigilant oversight of BTMS, they serve not only as symbols of progress but also as harbingers of a brighter, greener future for generations to come [30, 31].

14.4 BTMS for Electric Trucks

The electrification of heavy-duty trucks marks a pivotal milestone in the ongoing quest for sustainable transportation solutions. However, this transition is not without its share of challenges. Unlike their passenger vehicle counterparts, electric trucks face a distinct set of operational requirements, characterized by long-distance hauling, frequent stop-and-go cycles, and demanding payloads. In the crucible of heavy-duty transportation, where reliability and performance are paramount, the role of Battery Thermal Management Systems (BTMS) emerges as a critical enabler of success (Wasim, Shoman, *et al.*, 2023).

14.4.1 Navigating Unique Challenges

Electric trucks operate in a realm defined by endurance and resilience, where every journey is a test of mettle against the unforgiving expanse of highways and byways. The demands placed on these vehicles are unparalleled, with long-haul routes spanning hundreds of miles and frequent stops punctuating the rhythm of the journey. Against this backdrop, the thermal management of battery packs assumes heightened significance, as the efficient regulation of temperature becomes indispensable for optimizing performance and ensuring longevity.

14.4.2 Sophisticated Thermal Management Strategies

BTMS for electric trucks are meticulously engineered to rise to the occasion, employing a diverse array of thermal management strategies to tackle the challenges at hand. At the forefront of these systems lie advanced cooling technologies, ranging from liquid cooling systems to phase-change materials, each tailored to meet the unique demands of heavy-duty transportation. Liquid cooling systems circulate coolant through intricate networks of channels, dissipating heat generated during operation, and maintaining the battery within its optimal temperature range. Similarly, phase-change materials harness the latent heat of fusion to absorb excess heat, providing a passive yet highly effective means of thermal management [32, 33].

14.4.3 Optimizing Performance and Longevity

In the intricate tapestry of electric vehicle technology, Battery Thermal Management Systems (BTMS) emerge as silent orchestrators, wielding the power to shape the performance and longevity of battery packs with precision and finesse. Beyond their role in mitigating thermal challenges, BTMS stand as stalwarts of efficiency and endurance, bestowing upon electric vehicles a newfound sense of resilience and reliability. At the heart of this efficacy lies the seamless integration of thermal management strategies aimed at optimizing performance and prolonging battery life, thereby charting a course towards a greener, more sustainable future in transportation.

Central to the mission of BTMS is the optimization of performance – a quest to extract the maximum potential from electric vehicle batteries while minimizing energy losses along the way. By meticulously regulating battery temperature within its ideal range, BTMS ensure that electric vehicles operate at peak efficiency, delivering maximum range and performance with every journey. Whether traversing city streets or embarking on long-haul routes, electric vehicles equipped with robust BTMS stand poised to navigate the complexities of the road with confidence and poise, propelled by the silent hum of electric motors and the vigilant oversight of thermal management systems.

Range anxiety looms large in the realm of electric vehicles, casting a shadow of uncertainty over the feasibility of widespread adoption. BTMS emerge as formidable allies in the battle against range limitations, as they work tirelessly to maintain battery temperature within optimal limits, thereby maximizing range and minimizing energy losses. By ensuring that the battery operates within its designated temperature band, BTMS mitigate the inefficiencies associated with temperature extremes, preserving energy for propulsion and extending the reach of electric vehicles far beyond conventional boundaries. With each mile traveled, electric vehicles equipped with robust BTMS defy the limitations of range, ushering in a new era of mobility characterized by freedom and possibility.

The specter of thermal degradation looms large in the realm of electric vehicle batteries, casting doubt on their longevity and reliability. BTMS, with their proactive approach to thermal management, stand as guardians against this insidious threat, mitigating the risk of degradation and safeguarding the investment in battery technology. By maintaining the battery within its ideal temperature range, BTMS thwart the onset of degradation mechanisms such as capacity fade and impedance growth, ensuring that the battery remains a reliable and enduring asset for the long haul. Whether deployed in urban transit fleets or powering heavy-duty trucks, electric vehicles equipped with robust BTMS stand as testaments to the power of proactive thermal management in preserving the integrity and functionality of battery packs over time.

In the dynamic landscape of electric vehicle technology, the role of Battery Thermal Management Systems (BTMS) transcends mere thermal regulation, evolving into a cornerstone of efficiency and longevity. By optimizing performance, maximizing range, and safeguarding against thermal degradation, BTMS unlock the full potential of electric vehicles, propelling them towards a future characterized by sustainability, reliability, and endurance.

As the journey towards electrification unfolds, the efficacy of BTMS in shaping the trajectory of electric mobility stands as a testament to the power of proactive thermal management in charting a course towards a greener, more sustainable future for transportation.

14.5 Adapting to Varying Load Conditions

In the realm of heavy-duty transportation, where payloads fluctuate and demands vary, the ability of trucks to adapt to changing load conditions stands as a testament to their versatility and resilience. From hauling heavy cargo across rugged terrain to navigating city streets with lighter loads, heavy-duty trucks must rise to the challenge, delivering consistent and reliable performance under any circumstances. At the heart of this adaptability lies the sophisticated interplay between vehicle dynamics and thermal management, with Battery Thermal Management Systems (BTMS) emerging as key enablers of success.

Heavy-duty trucks operate in a dynamic environment characterized by a spectrum of load conditions, each imposing its unique demands on the vehicle. Whether hauling heavy machinery across long distances or transporting goods within city limits, trucks must adapt to the ever-changing landscape of payloads, optimizing performance while maintaining reliability. This variability presents a formidable challenge, as trucks must contend with fluctuating weight distributions, aerodynamic forces, and road conditions, all of which influence vehicle dynamics and efficiency (Yan-Yan, Yang, *et al.*, 2023).

At the forefront of this challenge lies the role of BTMS, which stand as guardians of thermal equilibrium, tasked with maintaining battery temperature within optimal limits under varying load conditions. With their adaptability and sophistication, BTMS rise to the occasion, dynamically adjusting thermal management strategies to optimize performance and ensure reliability. Whether faced with the demands of heavy hauling or navigating city streets with a lighter payload, electric trucks equipped with robust BTMS remain poised to deliver consistent and dependable performance, thanks to the proactive management of battery temperature.

In response to fluctuating load conditions, BTMS employ a range of thermal management strategies to optimize performance and efficiency. When faced with heavy payloads and extended periods of high power demand, BTMS may activate cooling mechanisms to dissipate excess heat generated during operation, thereby preventing thermal runaway and ensuring the integrity of the battery pack. Conversely, when operating with lighter loads or during periods of reduced power demand, BTMS may adjust cooling intensity or employ passive cooling methods to maintain battery temperature within optimal limits, thereby conserving energy and extending range.

Ultimately, the true measure of success lies in the ability of heavy-duty trucks to deliver consistent and reliable performance under any circumstances. Electric trucks equipped with robust BTMS stand as paragons of adaptability and resilience, capable of navigating the complexities of heavy-duty transportation with confidence and poise. Whether traversing rugged terrain or navigating congested city streets, these trucks remain steadfast in their commitment to efficiency and reliability, thanks to the proactive management of thermal conditions afforded by BTMS.

In the ever-evolving landscape of heavy-duty transportation, the ability to adapt to varying load conditions stands as a hallmark of success. Electric trucks equipped with sophisticated BTMS rise to the challenge, delivering consistent and reliable performance under any circumstances, thanks to the proactive management of thermal conditions. As the journey towards electrification continues, the adaptability and resilience of heavy-duty trucks equipped with robust BTMS serve as a beacon of progress, ushering in a new era of efficiency and reliability in heavy-duty transportation (Eric, Chown, 2023).

In the summary of heavy-duty transportation, where reliability and performance are the benchmarks of success, Battery Thermal Management Systems (BTMS) for electric trucks emerge as indispensable assets. By employing sophisticated thermal management strategies, these systems ensure that electric trucks operate at peak efficiency, regardless of the challenges they face. As heavy-duty transportation undergoes a paradigm shift towards electrification, the role of BTMS in optimizing performance, enhancing reliability, and prolonging battery life becomes increasingly pronounced, paving the way for a greener, more sustainable future in the world of transportation.

14.6 Conclusion

As we come to the end of our investigation into BTMS and how they are used in electric vehicles, it is evident that these systems are essential to innovation and advancement in the search for a sustainable transportation future. From buses to trucks, and all in between, BTMS are the unsung heroes that constantly maintain the delicate balance between thermal equilibrium and operational efficiency. BTMS is leading the way towards a more sustainable and environmentally friendly future with their unwavering dedication to performance, adaptability, and sophistication. The BTMS paradigm is centered on the unrelenting pursuit of optimization, specifically the optimization of longevity, efficiency, and performance. Battery temperature management systems (BTMS) keep electric vehicles operating at maximum efficiency, maximizing range and minimizing energy losses, by keeping the battery temperature within optimal limits. Furthermore, by reducing the possibility of thermal deterioration, BTMS protect battery packs' lifespan, protecting the investment and guaranteeing a dependable and long-lasting asset over time. The versatility of BTMS is evident in the field of heavy-duty transportation, where dependability and durability are critical. Whether driving through rough terrain carrying large payloads or commuting through urban areas with smaller loads, electric cars with strong BTMS systems never waver in their dedication to dependability and consistency. Through the dynamic adjustment of thermal management strategies to accommodate changing load conditions, Battery Technology Management Systems (BTMS) enable electric vehicles to provide dependable and consistent performance in any situation, paving the way for an efficient and sustainable future. Looking back at the journey so far, we can see that one of the most important turning points in the continuous search for sustainable transportation solutions is the incorporation of BTMS into electric vehicles. By utilizing proactive thermal management, BTMS opens up new possibilities for efficiency, dependability, and durability, advancing electric vehicles toward a future characterized by advancement and creativity. Let's embrace the transformative potential of BTMS and set out on a journey towards a world where transportation is not only efficient and reliable but also environmentally conscious and sustainable, as we stand on the brink of a greener, more sustainable tomorrow.

References

1. Fioriti, D., Scarpelli, C., Pellegrino, L., Lutzemberger, G., Micolano, E., Salamone, S., Battery lifetime of electric vehicles by novel rainflow-counting algorithm with temperature and C-rate dynamics: Effects of fast charging, user habits, vehicle-to-grid and climate zones. *J. Energy Storage*, 59, 106458, 2023, https://doi.org/10.1016/j.est.2022.106458.
2. Chown, E., High Degree of Electrification in Heavy-Duty Vehicles. *Energies*, 16, 8, 3565, 2023, https://doi.org/10.3390/en16083565.
3. Cartenì, A., Henke, I., Molitierno, C., Di Francesco, L., Strong sustainability in public transport policies: An e-mobility bus fleet application in Sorrento Peninsula (Italy). *Sustainability*, 12, 17, 7033, 2020.

4. Cattani, L., Malavasi, M., Bozzoli, F., D'Alessandro, V., Giammichele, L., Experimental Analysis of an Innovative Electrical Battery Thermal Management System. *Energies*, 16, 13, 5071, 2023, Available from: 10.3390/en16135071.
5. Elkholy, S., Shaaban, M. F., Ali, A., Optimal planning of battery electric buses for reliable services, in: *2023 International electrical engineering congress (International Electrical Engineering Congress (iEECON))*, pp. 306–312, IEEE, 2023 March, 10.1109/iEECON56657.2023.10126638.
6. Agrawal, S., Peeta, S., Miralinaghi, M., Multiparadigm Modeling Framework to Evaluate the Impacts of Travel Patterns on Electric Vehicle Battery Lifespan. *J. Adv. Transp*, 2023, 1, 1689075, 2023, https://doi.org/10.1155/2023/1689075.
7. Shoman, W., Yeh, S., Sprei, F., Plötz, P., Speth, D., Battery electric long-haul trucks in Europe: Public charging, energy, and power requirements. *Transp. Res. Part D Transp. Environ.*, 121, 103825, 2023, 10.1016/j.trd.2023.103825.
8. Fu, M., Yang, Y., Li, Y., Wang, H., Yu, F., Liu, J., Beijing Heavy-Duty Diesel Vehicle Battery Capacity Conversion and Emission Estimation in 2022. *Sustainability*, 15, 14, 11019, 2023. https://doi.org/10.3390/su151411019.
9. Thapliyal, S., Wazid, M., Singh, D.P., Das, A.K., Islam, S.H., Robust authenticated key agreement protocol for internet of vehicles-envisioned intelligent transportation system. *J. Syst. Archit.*, 142, 102937, 2023, 10.1016/j.sysarc.2023.102937.
10. Das, S., Thakur, P., Singh, A.K., Singh, S.N., Optimal management of vehicle-to-grid and grid-to-vehicle strategies for load profile improvement in distribution system. *J. Energy Storage*, 49, 104068, 2022, http://dx.doi.org/10.1016/j.est.2022.104068.
11. Singh, P.P., Das, S., Wen, F., Palu, I., Singh, A.K., Thakur, P., Multi-objective planning of electric vehicles charging in distribution system considering priority-based vehicle-to-grid scheduling. *Swarm Evol. Comput.*, 77, 101234, 2023, http://dx.doi.org/10.1016/j.swevo.2023.101234.
12. Almutairi, A., Albagami, N., Almesned, S., Alrumayh, O., Malik, H., Electric vehicle load estimation at home and workplace in Saudi Arabia for grid planners and policy makers. *Sustainability*, 15, 22, 15878, 2023, http://dx.doi.org/10.3390/su152215878.
13. Badri, Y., II, Sudabattula, S.K., Hussain, I., Malik, H., Márquez, F.P.G., Ultracapacitor as selectable energy buffer in electric vehicle application. *J. Energy Storage*, 73, 109200, 2023, http://dx.doi.org/10.1016/j.est.2023.109200.
14. Razaviyn, Z., Heidari, M., Thangavel, S., Verma, V., Kumar, A., Yadav, A.K., Marine Energy Harvesting from Fluid Flow *via* Vortex Induced Vibrations. *Int. J. Thermofluids*, 25, 101015, January 2025. https://doi.org/10.1016/j.ijft.2024.101015.
15. Jena, N.K., Sahoo, S., Sahu, B.K., Naik, A.K., Bajaj, M., Misak, S., Prokop, L., Impact of a redox flow battery on the frequency stability of a five-area system

integrated with renewable sources. *Energies*, 16, 14, 5540, 2023, 10.3390/en16145540.

16. Hamed, S.B., Hamed, M.B., Sbita, L., Bajaj, M., Blazek, V., Prokop, L., Ghoneim, S.S., Robust optimization and power management of a triple junction photovoltaic electric vehicle with battery storage. *Sensors*, 22, 16, 6123, 2022, 10.3390/s22166123.
17. Rawat, T., Niazi, K.R., Gupta, N., Sharma, S., Multi-objective techno-economic operation of smart distribution network integrated with reactive power support of battery storage systems. *Sustain. Cities Soc.*, 75, 103359, 2021, http://dx.doi.org/10.1016/j.scs.2021.103359.
18. Gupta, A., Singh, R.K., Paul, J., Sharma, M., Joshi, S., Battery-Operated Electric Vehicles (BOEVs) Adoption in India: Analysis of Barriers and Strategies. *IEEE Trans. Eng. Manage.*, 71, 7076–7087, 2023, http://dx.doi.org/10.1109/TEM.2023.3253621.
19. Kumar, B.A., Jyothi, B., Singh, A.R., Bajaj, M., Rathore, R.S., Berhanu, M., A novel strategy towards efficient and reliable electric vehicle charging for the realisation of a true sustainable transportation landscape. *Sci. Rep.*, 14, 1, 3261, 2024, 10.1038/s41598-024-53214-w.
20. Singh, A.R., Vishnuram, P., Alagarsamy, S., Bajaj, M., Blazek, V., Damaj, I., Othman, K.M., Electric vehicle charging technologies, infrastructure expansion, grid integration strategies, and their role in promoting sustainable e-mobility. *Alexandria Eng. J.*, 105, 300–330, 2024, 10.1016/j.aej.2024.06.093.
21. Aggarwal, S., Singh, A.K., Rathore, R.S., Bajaj, M., Gupta, D., Revolutionizing load management: A novel technique to diminish the impact of electric vehicle charging stations on the electricity grid. *Sustain. Energy Technol. Assess.*, 65, 103784, 2024, 10.1016/j.seta.2024.103784.
22. Oubelaid, A., Kakouche, K., Belkhier, Y., Khosravi, N., Taib, N., Rekioua, T., Tuka, M.B., New coordinated drive mode switching strategy for distributed drive electric vehicles with energy storage system. *Sci. Rep.*, 14, 1, 6448, 2024, 10.1038/s41598-024-56209-9.
23. Bhatt, A., Ongsakul, W., Pawar, J., Optimal energy management system for carbon–neutral microgrid integrating second-life batteries and crypto mining devices. *Sustain. Energy Technol. Assess.*, 64, 103686, 2024, 10.1016/j.seta.2024.103686.
24. Sahri, Y., Tamalouzt, S., Belaid, S.L., Bajaj, M., Ghoneim, S.S., Zawbaa, H.M., Kamel, S., Performance improvement of Hybrid System based DFIG-Wind/PV/Batteries connected to DC and AC grid by applying Intelligent Control. *Energy Rep.*, 9, 2027–2043, 2023, 10.1016/j.egyr.2023.01.021.
25. Rekioua, D., Mokrani, Z., Kakouche, K., Rekioua, T., Oubelaid, A., Logerais, P.O., Ghoneim, S.S., Optimization and intelligent power management control for an autonomous hybrid wind turbine photovoltaic diesel generator with batteries. *Sci. Rep.*, 13, 1, 21830, 2023, 10.1038/s41598-023-49067-4.
26. Kanouni, B., Badoud, A.E., Mekhilef, S., Bajaj, M., Zaitsev, I., Advanced efficient energy management strategy based on state machine control for

multi-sources PV-PEMFC-batteries system. *Sci. Rep.*, 14, 1, 7996, 2024, 10.1038/s41598-024-58785-2.

27. Afzal, M.Z., Aurangzeb, M., Iqbal, S., Pushkarna, M., Rehman, A.U., Kotb, H., Bereznychenko, V., A Novel Electric Vehicle Battery Management System Using an Artificial Neural Network-Based Adaptive Droop Control Theory. *Int. J. Energy Res.*, 2023, 1, 2581729, 2023, 10.1155/2023/2581729.
28. Ramakrishnan, V., Vishnuram, P., Yang, T., Bajaj, M., Rathore, R.S., Zaitsev, I., Design and implementation of a high misalignment-tolerance wireless charger for an electric vehicle with control of the constant current/voltage charging. *Sci. Rep.*, 14, 1, 13165, 2024, 10.1038/s41598-024-63952-6.
29. Bagga, P., Das, A.K., Wazid, M., Rodrigues, J.J., Choo, K.K.R., Park, Y., On the design of mutual authentication and key agreement protocol in internet of vehicles-enabled intelligent transportation system. *IEEE Trans. Veh. Technol.*, 70, 2, 1736–1751, 2021, 10.1109/TVT.2021.3050614.
30. Kasri, A., Ouari, K., Belkhier, Y., Bajaj, M., Zaitsev, I., Optimizing electric vehicle powertrains peak performance with robust predictive direct torque control of induction motors: A practical approach and experimental validation. *Sci. Rep.*, 14, 1, 14977, 2024, 10.1038/s41598-024-65988-0.
31. Dimri, S.C., Indu, R., Bajaj, M., Rathore, R.S., Blazek, V., Dutta, A.K., Alsubai, S., Modeling of traffic at a road crossing and optimization of waiting time of the vehicles. *Alexandria Eng. J.*, 98, 114–129, 2024, 10.1016/j.aej.2024.04.050.
32. Amoussou, I., Paddy, E.Y., Agajie, T.F., Ibrahim, F.S., Agajie, E.F., Nsanyuy, W.B., Mohammadi, S.A.D., Enhancing residential energy access with optimized stand-alone hybrid solar-diesel-battery systems in Buea, Cameroon. *Sci. Rep.*, 14, 1, 15543, 2024, 10.1038/s41598-024-66582-0.
33. Ravindran, M.A., Nallathambi, K., Vishnuram, P., Rathore, R.S., Bajaj, M., Rida, I., Alkhayyat, A., A novel technological review on fast charging infrastructure for electrical vehicles: Challenges, solutions, and future research directions. *Alexandria Eng. J.*, 82, 260–290, 2023, 10.1016/j.aej.2023.10.009.

15

Battery Thermal Management System (BTMS) for Military Applications

Neha Singh Raghuvanshi[1]*, Ashwani Kumar[2], Yatika Gori[3] and Ashok Kumar Yadav[4]

[1]*Department of Applied Science & Humanities, ABES Engineering College, Ghaziabad, Uttar Pradesh, India*
[2]*Department of Mechanical Engineering, Technical Education Department, Uttar Pradesh, Kanpur, India*
[3]*Department of Mechanical Engineering, Graphic Era Deemed to be University, Dehradun, Uttarakhand, India*
[4]*Department of Mechanical Engineering, Symbiosis Institute of Technology, Symbiosis International (Deemed University), Pune, India*

Abstract

Battery Thermal Management Systems (BTMS) are critical for ensuring the performance and longevity of energy storage devices, particularly in demanding military applications. This chapter delves into the unique challenges and requirements of BTMS for military batteries, focusing on the harsh operational environments and stringent performance criteria. Military applications often involve extreme temperature fluctuations, mechanical shocks, vibrations, and electromagnetic interference, necessitating robust and reliable BTMS designs. This abstract explores the key considerations for designing and implementing BTMS tailored for military use, including the selection of appropriate cooling techniques (e.g., liquid cooling, forced air cooling, phase change materials), advanced thermal modeling, and control algorithms. The chapter also examines the integration of BTMS with other vehicle systems, such as power management and control units. Furthermore, it discusses the importance of lightweight and compact designs to minimize the impact on vehicle payload and mobility. The abstract highlights the

**Corresponding author*: nehasr.raghuvanshi@gmail.com; ORCID: 0000-0002-8437-6261
Ashwani Kumar: ORCID: 0000-0003-4099-935X
Ashok Kumar Yadav: ORCID: 0000-0002-7350-7641

Ashwani Kumar, Mukesh Kumar Awasthi, Nitesh Dutt, Yogesh Kumar Singla and Sivasakthivel Thangavel (eds.) Thermal Battery Management System for Hybrid and Electric Vehicles, (281–312) © 2025 Scrivener Publishing LLC

need for rigorous testing and validation to ensure the BTMS can withstand the rigors of military operations. Finally, it addresses the future trends in military BTMS, including the development of advanced materials and intelligent control strategies. This chapter provides a comprehensive overview of the design, implementation, and future directions of BTMS for military applications, emphasizing the critical role of thermal management in ensuring mission success.

***Keywords*:** Military BTMS, battery thermal management, military batteries, thermal modeling, cooling techniques, harsh environments, vehicle integration

15.1 Introduction

15.1.1 Overview of Battery Thermal Management Systems (BTMS)

Battery Thermal Management Systems (BTMS) are essential for maintaining the optimal performance, safety, and durability of batteries across various applications. These systems regulate the temperature of batteries using a range of cooling and heating techniques to address the heat generated during battery use and charging. Typical BTMS components include heat sinks, fans, liquid coolants, and thermal interface materials. The primary function of BTMS is to keep the battery within its ideal operating temperature range, thereby enhancing performance and preventing issues such as overheating or excessive cooling. Ultimately, BTMS aims to ensure that batteries function reliably and efficiently, even when faced with external temperature variations and diverse operational conditions. The significance of Battery Thermal Management Systems is particularly pronounced in military contexts, where equipment operates under harsh and demanding conditions. Military assets such as vehicles, communication devices, and portable power systems frequently encounter extreme temperatures, vibrations, and moisture. Effective BTMS is vital to preserving the performance and reliability of these battery-powered systems in such environments. A failure in battery management can jeopardize mission success and operational safety. Consequently, BTMS in military applications must be robust and capable of addressing the specific thermal challenges posed by these environments, ensuring that batteries perform optimally and safely under the most severe conditions [1].

This chapter aims to deliver a thorough analysis of Battery Thermal Management Systems in the context of military applications. It will cover the fundamental principles and technologies underlying BTMS, examine the specific thermal management needs and challenges encountered in military environments, and explore advanced technologies that improve thermal management. The chapter will also include practical insights through case studies of BTMS applications in military vehicles and devices, showcasing both achievements and areas for development. By addressing these elements, the chapter intends to provide valuable knowledge and recommendations for developing and implementing effective BTMS tailored to the distinct requirements of military operations.

15.2 Fundamentals of Battery Thermal Management

In today's rapidly advancing technological landscape, effective thermal management of batteries is crucial for ensuring performance, safety, and longevity. As devices become more powerful and energy-dense, the need to efficiently manage battery heat becomes even more critical. This brief overview will introduce you to the essential principles and techniques of battery thermal management, focusing on the strategies and technologies used to regulate temperature, enhance battery life, and maintain optimal performance. By understanding these fundamentals, you'll gain insight into how thermal management is integral to the design and operation of modern energy storage systems.

15.2.1 Principles of Thermal Management

Thermal management in battery systems is crucial for ensuring optimal performance, safety, and longevity. At its core, the principles of thermal management revolve around controlling the temperature of a battery to prevent it from reaching conditions that could degrade its performance or cause safety hazards. Effective thermal management involves both heat dissipation and heat retention. Heat dissipation is the process of removing excess heat generated during battery operation or charging. This is essential because elevated temperatures can accelerate chemical reactions within the battery, leading to reduced performance, faster degradation, and potential safety risks. Conversely, heat retention involves maintaining the battery at a sufficiently high temperature in cold environments to ensure it remains functional and efficient. Thermal management strategies are designed to maintain the battery within an optimal temperature range, typically between 20°C and 40°C. This balance helps avoid the extremes of overheating and undercooling, both of which can adversely affect battery performance and safety [2].

15.2.2 Types of Battery Thermal Management Systems

Battery thermal management is vital for performance and safety. Key systems include passive cooling, such as heat sinks and thermal pads, and active cooling, like fans and liquid cooling systems. Each approach balances efficiency, cost, and complexity to optimize battery performance and longevity.

15.2.2.1 Air Cooling

Air cooling is one of the simplest and most cost-effective methods for managing battery temperature. This approach relies on the natural convection of air or the use of fans to enhance heat dissipation. In an air-cooled system, the battery is often equipped with heat sinks or cooling fins, which increase the surface area available for heat dissipation. Fans can be employed to boost airflow over the heat sinks, improving the system's ability to carry away excess heat. Air cooling is commonly used in applications with moderate heat loads, such as in smaller electronic devices and some less demanding electric vehicle designs. It is favored for its simplicity and lower cost, but it may not be sufficient for high-power applications where more effective cooling methods are required.

15.2.2.2 Liquid Cooling

Liquid cooling is a more efficient method compared to air cooling, especially for applications with higher heat loads. This system circulates a coolant—typically a mixture of water and antifreeze or specialized thermal fluids—through channels or tubes in direct contact with the battery. The coolant absorbs heat from the battery and transports it to a radiator or heat exchanger, where it is dissipated into the environment.

Liquid cooling systems offer better temperature control and can manage higher heat dissipation rates, making them suitable for high-performance applications such as electric vehicles with large battery packs and energy storage systems. Although these systems are more complex and costly than air cooling, they provide precise temperature regulation, enhancing both battery performance and safety.

15.2.2.3 Phase Change Materials (PCMs)

Phase Change Materials (PCMs) are substances that absorb and release thermal energy during phase transitions, such as from solid to liquid. In a battery thermal management system, PCMs can absorb excess heat when the battery temperature rises, undergoing a phase change to stabilize the temperature. When the battery cools, the PCM solidifies, releasing the stored heat.

PCMs are used to complement other cooling methods, particularly in scenarios where temperature fluctuations are moderate but need effective management. They are useful in applications such as consumer electronics and some automotive systems where managing temperature variations can improve overall performance and reliability.

15.2.2.4 Advanced Cooling Techniques

Advanced cooling techniques enhance battery performance and safety through innovative methods. These include phase change materials, advanced liquid cooling, and thermoelectric coolers, which provide precise temperature control and improve battery efficiency, extending lifespan and reliability in demanding applications.

15.2.2.4.1 Micro-Channel Cooling

Micro-channel cooling is a sophisticated technique that involves circulating coolant through minuscule channels embedded within the surface of a battery or heat sink. This technology significantly enhances heat transfer efficiency by vastly increasing the contact surface area between the coolant

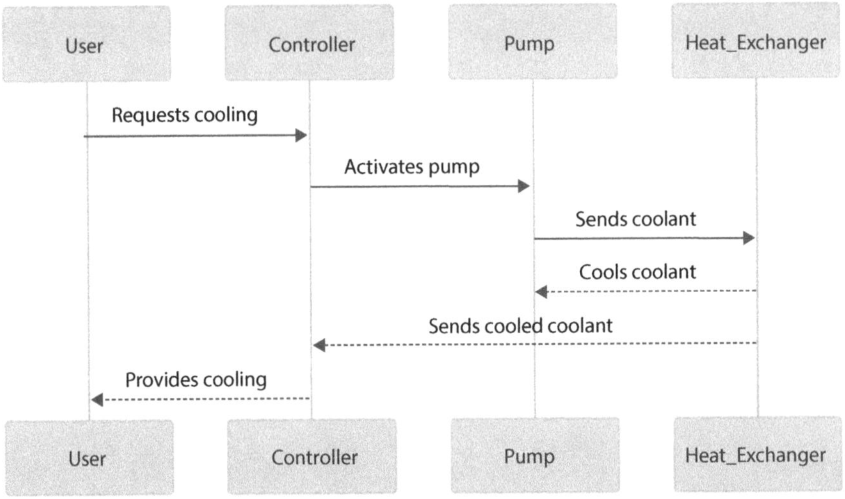

Figure 15.1 Micro-channel cooling.

and the battery. The small dimensions of these channels allow for effective thermal management, as they enable rapid heat removal from the battery (Figure 15.1).

In practice, micro-channel cooling has been notably employed in high-performance electric vehicles, such as the Tesla Model S. This vehicle features a large battery pack that generates substantial heat, especially during rapid charging or high-power operation. By integrating micro-channel cooling, Tesla ensures that heat is uniformly distributed and dissipated across the battery pack. The result is improved thermal stability, which enhances battery performance, safety, and longevity. The precise control of temperature provided by micro-channel cooling also contributes to the overall efficiency and reliability of the vehicle [3].

15.2.2.4.2 Jet Impingement Cooling

Jet impingement cooling is another advanced cooling method that enhances heat dissipation by directing high-speed jets of air or liquid onto the surface of the battery. This technique generates a turbulent flow that significantly increases the convective heat transfer rate. The forceful jets create localized areas of high heat transfer, making it possible to efficiently manage heat loads in various applications (Figure 15.2).

This cooling method is extensively used in high-performance computing systems and data centers, where effective cooling of densely packed electronic components is critical. For example, in data centers, jet impingement cooling helps prevent overheating of servers and maintains optimal operational conditions. In battery applications, similar principles are

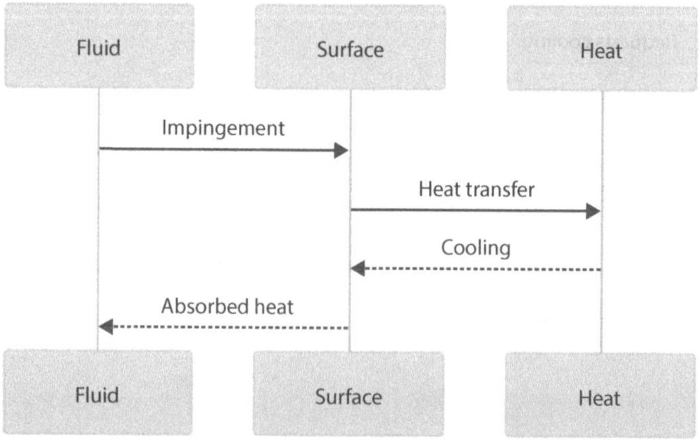

Figure 15.2 Jet impingement cooling.

applied to manage high-power heat loads. By utilizing jet impingement cooling, battery systems can achieve effective thermal management, ensuring that they operate efficiently and safely even under intense usage conditions [4].

15.2.3 Key Performance Metrics for BTMS

Evaluating the effectiveness of a Battery Thermal Management System (BTMS) involves several critical performance metrics. These metrics ensure that the system adequately manages the battery's thermal conditions, optimizing its performance, safety, and longevity.

a. Temperature Range
Temperature Range measures the BTMS's capability to maintain the battery within its optimal operating temperature range. For instance, lithium-ion batteries generally operate best between 15°C and 35°C. The Tesla Model S uses a sophisticated liquid cooling system to ensure that its battery remains within this range, despite external temperatures that may vary from -30°C to 50°C. The system achieves this by circulating coolant through a network of channels, effectively managing heat generated during high-performance driving and rapid charging.

b. Heat Dissipation Efficiency
Heat Dissipation Efficiency quantifies how effectively the BTMS can remove excess heat from the battery. This is vital for preventing overheating, which can degrade battery performance and safety. For example, in the Boeing 787 Dreamliner, the battery management system employs a combination of liquid cooling and heat sinks to maintain efficient heat dissipation. The heat dissipation system is designed to handle up to 15 kW of heat under peak load conditions, ensuring that the battery remains within safe operational temperatures even during extended flights.

c. Response Time
Response Time refers to how quickly the BTMS can adjust to sudden changes in battery temperature. Rapid response is crucial in preventing thermal runaway, especially during high-power demands or charging cycles. In high-performance electric vehicles like the Rimac C_Two, the BTMS is equipped with advanced sensors and control algorithms that enable real-time adjustments. The system can detect temperature increases within milliseconds and activate cooling mechanisms to address the heat, ensuring stable battery operation under dynamic conditions.

d. Reliability and Durability

Reliability and Durability assess the long-term performance of BTMS components and their ability to function effectively under harsh conditions. For instance, military-grade vehicles like the Bradley Fighting Vehicle are equipped with robust BTMS designed to withstand extreme temperatures, vibrations, and shock. These systems use high-quality materials and construction techniques to ensure that they perform reliably over extended periods and in challenging environments.

e. Cost-Effectiveness

Cost-Effectiveness balances system performance with economic considerations, including initial costs and maintenance expenses. While advanced BTMS technologies like those used in electric buses might have higher upfront costs, their ability to enhance battery life and reduce maintenance can lead to long-term savings. For example, a state-of-the-art BTMS in a BYD electric bus system may cost around $50,000, but its efficiency and longevity can significantly lower the total cost of ownership by extending the battery's life and reducing downtime.

15.3 Military Requirements and Challenges

15.3.1 Operational Environments and Conditions

Military applications demand robust battery performance in extreme conditions, such as extreme temperatures, vibrations, and high altitudes. Effective thermal management systems must ensure reliability and efficiency, adapting to harsh environments while meeting stringent operational requirements and enhancing overall mission success.

15.3.1.1 Extreme Temperatures

Military applications often require equipment to function in a range of extreme temperatures, from the freezing cold of arctic environments to the scorching heat of desert conditions. These temperature variations significantly impact battery performance and the design of Battery Thermal Management Systems (BTMS). In cold climates, low temperatures can lead to increased internal resistance in batteries, reduced charge capacity, and slower chemical reactions, which degrade battery performance and efficiency. For instance, in the Arctic, the operational performance of military communication devices and electric vehicles can be severely compromised due to freezing temperatures (Figure 15.3).

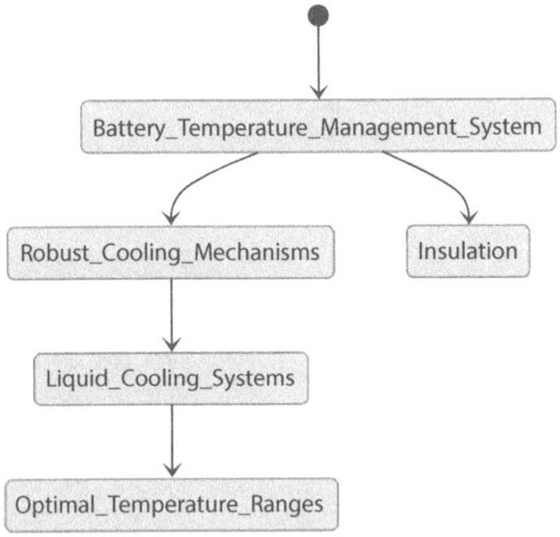

Figure 15.3 Advanced liquid cooling systems to maintain battery temperatures.

Conversely, in hot environments, high temperatures can accelerate the degradation of battery materials, increase the risk of thermal runaway, and reduce battery life. In desert environments, where temperatures can exceed 50°C (122°F), military vehicles and portable devices face risks of overheating. The BTMS must be designed to handle such extremes, often incorporating robust cooling mechanisms and insulation to protect the battery. For example, military vehicles like the HMMWV (Humvee) use advanced liquid cooling systems to maintain battery temperatures within optimal ranges despite external temperature fluctuations [5].

15.3.1.2 Vibration and Shock

Military equipment is frequently subjected to high levels of vibration and shock due to movement across rough terrain, rapid maneuvers, or explosive events. These vibrations can cause mechanical stress on battery components, leading to potential damage and failure. In battery systems, vibrations can disrupt the integrity of internal structures, lead to the loosening of connections, and even cause leaks in liquid cooling systems.

To address these challenges, BTMS designs for military applications must incorporate features that mitigate the effects of vibration and shock. For instance, military-grade batteries are often equipped with shock-absorbing mounts and ruggedized enclosures that help protect against mechanical stresses (Figure 15.4). In military aircraft, where both vibration and shock are significant, specialized mounting systems and vibration-dampening materials are used to ensure the reliability of the BTMS [6].

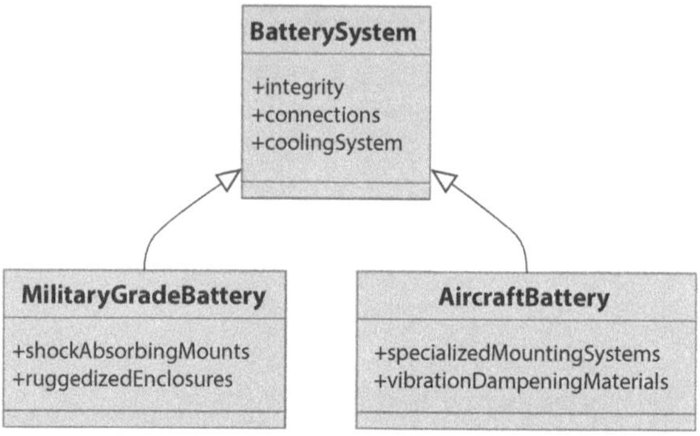

Figure 15.4 Specialized mounting systems and vibration-dampening materials.

15.3.1.3 Humidity and Corrosion

Humidity and corrosion pose substantial challenges to the durability and functionality of BTMS components. Moist environments can lead to the accumulation of moisture within battery compartments, which can cause corrosion of electrical contacts and degradation of battery materials. This is particularly problematic in naval operations, where equipment is exposed to high levels of humidity and saltwater. Corrosion can compromise the thermal conductivity of heat sinks and other cooling components, reducing the effectiveness of the BTMS and potentially leading to system failures.

To combat these issues, military BTMS designs often include corrosion-resistant materials and coatings (Figure 15.5). For example, naval vessels use specially treated alloys and protective coatings to shield components from the corrosive effects of seawater. Additionally, sealed enclosures and moisture-resistant designs are employed to keep humidity and contaminants away from sensitive battery systems. Military applications demand BTMS solutions that can withstand extreme temperatures, mitigate the effects of vibration and shock, and resist corrosion. By addressing these environmental challenges, BTMS ensures that military equipment operates reliably and safely in diverse and demanding conditions [7].

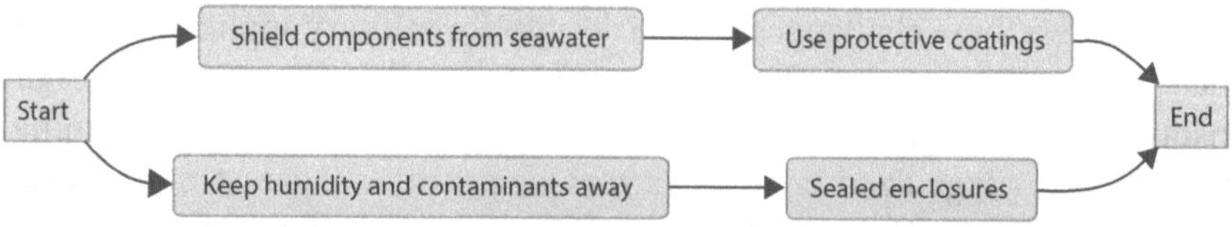

Figure 15.5 Alloys and protective coatings to shield components from the corrosive effects.

15.3.2 Specific Battery Requirements for Military Applications

Military batteries must meet rigorous standards for durability, safety, and performance. They need to withstand extreme temperatures, impacts, and vibrations while providing reliable power under demanding conditions. These requirements ensure operational readiness and mission success in diverse and challenging environments.

15.3.2.1 High Energy Density

High energy density is crucial for military applications where space and weight are at a premium. Military operations often involve equipment that must be lightweight yet capable of delivering significant power. For example, in portable power sources used by soldiers, such as battery packs for communication devices and night vision goggles, high energy density ensures that the equipment remains lightweight and compact while providing enough power to function effectively over extended periods.

In military vehicles, such as the M1 Abrams tank, the need for high energy density batteries is also significant. These vehicles require batteries that can deliver high power output to support various systems, including communications, weaponry, and propulsion, without adding excessive weight or taking up valuable space. Advances in lithium-ion and lithium-sulfur technologies have enabled the development of batteries with higher energy densities, meeting the demanding requirements of military applications (Figure 15.6) [8].

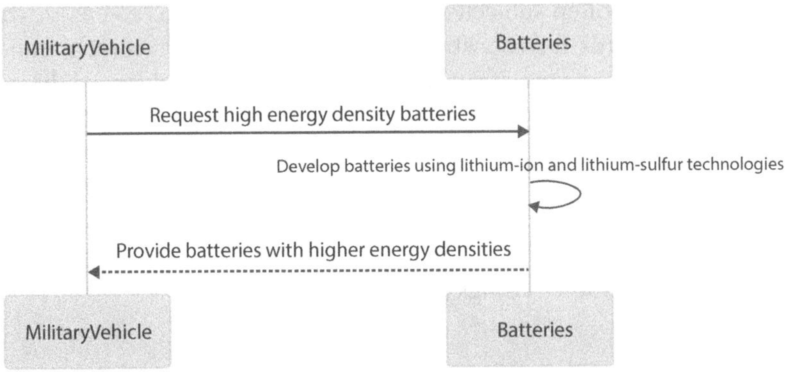

Figure 15.6 Advanced lithium-ion and lithium-sulfur technologies.

15.3.2.2 Reliability and Longevity

Reliability and longevity are paramount in military operations, where equipment failure can compromise mission success and safety. Military equipment often operates in extreme conditions, including high temperatures, severe cold, and rugged terrain. Batteries used in such environments must be highly reliable, with long operational lifespans to minimize the need for frequent replacements or maintenance.

For instance, in military aircraft like the F-35 Lightning II, batteries are critical for powering various systems, including avionics and emergency systems. These batteries must not only withstand the high vibrations and extreme temperatures experienced during flight but also have a long service life to ensure the aircraft's readiness and reliability. The use of advanced materials and design improvements, such as solid-state batteries and high-cycle life chemistries, are examples of how the industry addresses these longevity and reliability requirements [9].

15.3.2.3 Safety and Robustness

Safety and robustness are essential aspects of battery design in military applications due to the potential risks associated with battery failures. Military equipment must be able to withstand harsh conditions without compromising safety. This includes the ability to resist physical damage, such as from impacts or explosions, and to prevent hazardous situations like thermal runaway, which can lead to fires or explosions. In military vehicles, like the MRAP (Mine-Resistant Ambush Protected) vehicle, the battery systems must be designed to endure impacts from blasts and shrapnel while ensuring that thermal management systems can handle any excess heat generated during such events (Figure 15.7). Additionally, military batteries are often subjected to rigorous safety testing to ensure they meet high standards for robustness and reliability under combat conditions. For example, the use of ruggedized enclosures and advanced thermal management technologies helps to safeguard the battery systems against the demanding conditions of military environments. Overall, the specific battery requirements for military applications—high energy density, reliability and longevity, and safety and robustness—are critical to ensuring that military equipment performs effectively and reliably in challenging and diverse operational settings [10, 11].

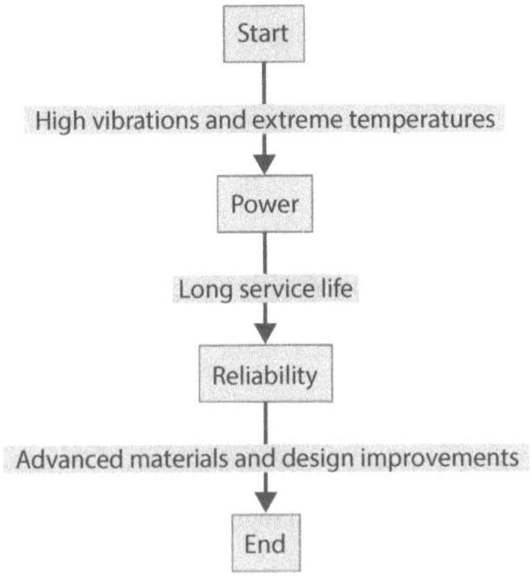

Figure 15.7 Mechanism of F-35 lightning II batteries.

15.3.2.4 Design and Implementation of BTMS for Military Applications

Designing and implementing Battery Thermal Management Systems (BTMS) for military applications involves addressing unique requirements and challenges to ensure performance, reliability, and safety in extreme conditions.

a. Design Considerations

In military applications, effective thermal management is critical for maintaining battery performance and ensuring mission success. This involves addressing several key factors:

(i) **Thermal Conductivity and Heat Dissipation:** Efficient thermal management is essential to prevent battery overheating and maintain stable performance. Advanced Battery Thermal Management Systems (BTMS) utilize materials with high thermal conductivity to manage heat effectively. For instance, in military drones, graphite heat spreaders are employed to enhance thermal conductivity and dissipation (Figure 15.8). These materials facilitate the even distribution of heat across the battery, preventing

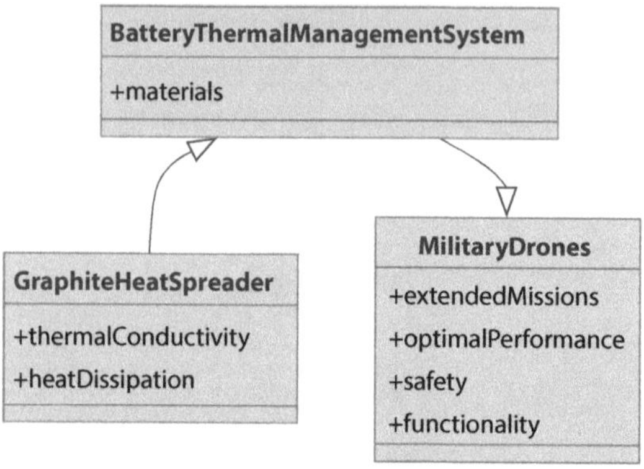

Figure 15.8 BTMS utilizing materials with high thermal conductivity to manage heat [48–52].

localized overheating, and ensuring reliable operation during extended missions. By improving heat dissipation, these systems help maintain optimal battery performance and extend operational periods without compromising safety or functionality [12].

(ii) **Integration with Military Systems:** BTMS must be seamlessly integrated into existing military equipment to avoid interfering with the system's functionality or adding excessive weight. A prime example is the F-35 Lightning II, which features an advanced liquid cooling system designed to manage the high thermal loads generated by its sophisticated avionics and weapon systems. This integration ensures that the aircraft's thermal management does not disrupt its performance or operational capabilities. By incorporating such systems into the design, the F-35 can handle the demanding thermal conditions of its advanced technologies while maintaining overall system efficiency and effectiveness [13].

(iii) **Weight and Space Constraints:** Military applications often involve stringent weight and space constraints, making compact and efficient thermal management solutions essential. In portable power packs used by soldiers, for example, advanced phase change materials (PCMs) are utilized to provide effective thermal management without adding significant bulk. PCMs absorb and release thermal energy during phase transitions, helping to regulate temperature and maintain battery performance. These materials are chosen for their ability to offer substantial thermal management benefits while adhering to the strict size and weight limitations inherent in military equipment [14].

b. Advanced Technologies in BTMS

(i) AI and Machine Learning for Real-Time Optimization

Artificial Intelligence (AI) and Machine Learning (ML) have increasingly become integral in optimizing thermal management systems (BTMS) for military applications. By leveraging AI algorithms, these systems can dynamically adjust cooling strategies in real-time based on data collected from thermal sensors embedded within the equipment. This capability is exemplified by the Bradley Fighting Vehicle, which employs AI to enhance its cooling systems. The AI algorithms analyze data such as ambient temperature, operational intensity, and battery heat generation to make real-time adjustments to the cooling mechanisms. This dynamic optimization helps maintain optimal battery performance, reduces the risk of overheating, and extends the operational life of critical systems during varied environmental and operational conditions.

In the Bradley Fighting Vehicle, AI-driven BTMS ensures that the cooling systems are not only responsive to immediate changes but also predictive of future thermal loads. This adaptability is crucial in combat scenarios where environmental conditions and operational demands fluctuate rapidly. For instance, if the vehicle enters a high-temperature environment or undergoes intensive operational maneuvers, the AI system adjusts the cooling system's performance accordingly, ensuring that the battery remains within safe temperature limits. Such real-time adjustments help in maintaining equipment reliability and performance under diverse and challenging conditions [15].

(ii) Predictive Maintenance and Monitoring

Predictive maintenance is another advanced technology that enhances thermal management by leveraging historical and real-time data to foresee and mitigate potential issues before they result in system failures. In military logistics and support vehicles, predictive maintenance systems analyze temperature and performance data to anticipate overheating risks and potential battery failures. By continuously monitoring these parameters, these systems can provide early warnings and enable preemptive actions such as adjusting cooling systems or scheduling maintenance before failures occur (Figure 15.9).

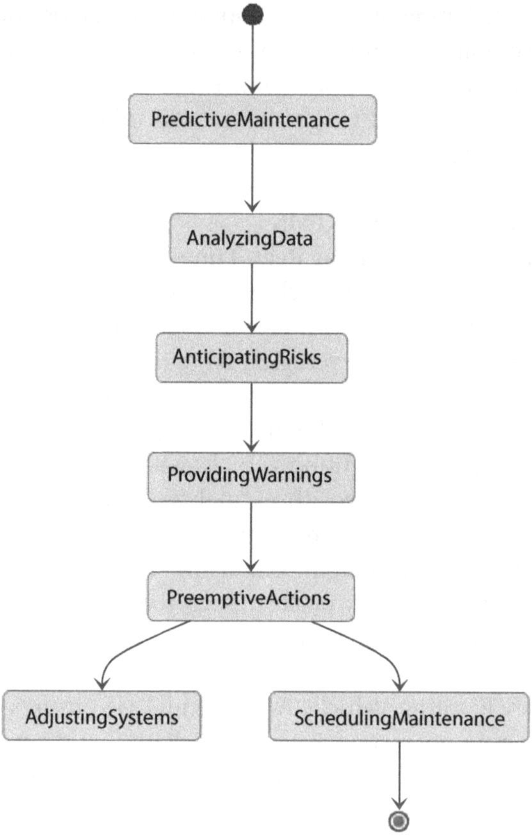

Figure 15.9 Predictive maintenance activity flow.

For example, in military logistics vehicles like the Heavy Expanded Mobility Tactical Truck (HEMTT), predictive maintenance systems use algorithms to process historical data and identify patterns that precede thermal issues. This proactive approach allows for timely interventions, such as recalibrating cooling systems or replacing components that show signs of wear. Consequently, this reduces unplanned downtime and maintenance costs, enhancing the overall reliability and operational readiness of the equipment [16].

The integration of AI and predictive maintenance technologies not only improves thermal management but also contributes to operational efficiency and cost savings. By anticipating and addressing thermal issues before they lead to equipment failures, these technologies support the effective and reliable operation of military systems in demanding environments.

c. Case Studies in Battery Thermal Management Systems (BTMS) for Military Applications

(i) BTMS in Military Vehicles: The M1 Abrams Tank

The M1 Abrams tank is a prime example of advanced thermal management in military vehicles. This heavily armored combat vehicle is equipped with a sophisticated liquid cooling system designed to handle the substantial thermal loads generated by its power systems and onboard electronics. The M1 Abrams tank houses powerful engines and various electronic systems that generate significant heat during operation, especially under combat conditions where performance demands are at their peak. The tank's liquid cooling system circulates coolant through the engine and electronic components, absorbing and transporting excess heat away from critical areas. This system maintains optimal operating temperatures, preventing overheating that could lead to component failure or reduced operational capability. The use of liquid cooling ensures efficient heat dissipation, which is essential for maintaining the tank's reliability and performance in challenging environments, such as high-intensity combat situations or extreme weather conditions. By integrating this advanced cooling technology, the M1 Abrams can sustain prolonged operations without experiencing thermal-related failures, thereby enhancing its effectiveness on the battlefield [17].

(ii) BTMS in Portable Military Devices

Handheld communication devices used by military personnel also require effective thermal management to ensure reliability and performance during field operations. These devices are crucial for maintaining communication in intense and demanding environments, where performance can be critically affected by thermal issues. To address these challenges, advanced cooling techniques such as heat pipes and graphite pads are employed. Heat pipes are highly efficient thermal conductors that transfer heat away from critical components, ensuring that batteries and electronics remain within safe temperature ranges. These pipes use a phase-change mechanism to transport heat from hot areas to cooler regions, effectively managing thermal buildup. Additionally, graphite pads are used for their excellent thermal conductivity properties. These pads help to evenly distribute heat across the surface of the device, preventing hotspots and ensuring that the battery and electronics do not overheat (Figure 15.10).

By integrating these advanced thermal management techniques, military portable devices can operate reliably in the field, maintaining functionality even under strenuous conditions. The effective use of heat pipes

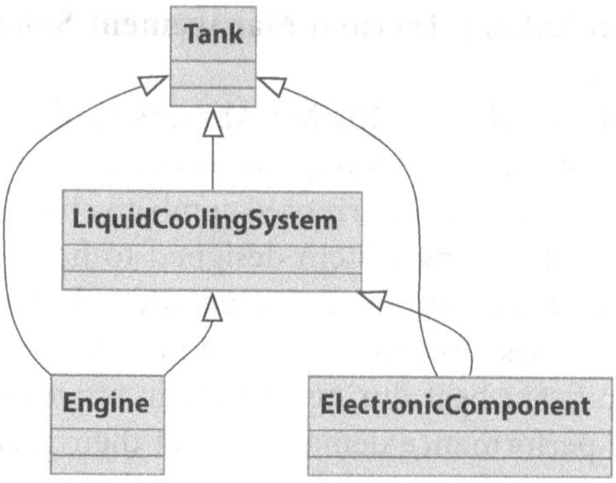

Figure 15.10 M1 Abrams tank, an advanced thermal management in military vehicles.

and graphite pads ensures that devices remain within optimal temperature ranges, preventing performance degradation and extending the operational life of the equipment. In summary, case studies of BTMS in military applications highlight the critical role of advanced thermal management in ensuring the reliability and performance of both military vehicles and portable devices. By employing sophisticated cooling systems and materials, these technologies address the unique challenges of military operations, enhancing the effectiveness and durability of essential equipment [18].

15.3.2.4.1 Design Considerations

Designing an effective Battery Thermal Management System (BTMS) involves crucial considerations including thermal conductivity and heat dissipation, integration with existing military systems, and managing weight and space constraints. Addressing these factors ensures optimal thermal performance, seamless integration, and efficient use of resources in demanding military environments.

15.3.2.4.1.1 Thermal Conductivity and Heat Dissipation

Effective thermal management in Battery Thermal Management Systems (BTMS) is crucial for maintaining optimal battery performance and longevity. This requires careful selection of materials and design techniques that maximize thermal conductivity and heat dissipation. High thermal conductivity materials, such as graphite, aluminum, and copper, are commonly used to facilitate efficient heat transfer from the battery to the cooling system. For instance, in the Tesla Model S, copper heat spreaders are employed to conduct heat away from the battery cells, improving overall cooling efficiency, and preventing hotspots. Additionally, the integration of advanced heat sinks and thermal interface materials (TIMs) enhances the heat dissipation process. TIMs, like thermal pastes or pads, improve the thermal interface between the battery and the cooling system, ensuring effective heat transfer and reducing thermal resistance [19].

15.3.2.4.1.2 Integration with Military Systems

Integrating BTMS into existing military systems presents several challenges. Military systems, such as armored vehicles or aircraft, often have complex designs with limited space and stringent operational requirements. For example, the integration of BTMS in the F-35 Lightning II involved developing a cooling system that could manage the thermal loads of advanced avionics and weapon systems without compromising the aircraft's stealth capabilities. The BTMS was designed to work seamlessly with the aircraft's existing systems, using a combination of liquid cooling and heat exchangers to manage heat effectively. Similarly, in the M1 Abrams tank, the BTMS integrates with the vehicle's power systems and engine cooling to maintain optimal temperatures. This involves designing cooling channels and systems that can operate effectively under the harsh conditions of combat, including vibrations, shocks, and extreme temperatures [20].

15.3.2.4.1.3 Weight and Space Constraints

Military applications often face strict weight and space constraints that can complicate BTMS design. The challenge is to provide effective thermal management without adding excessive weight or occupying valuable space. For instance, in military drones, where weight is a critical factor, engineers use lightweight materials such as carbon fiber composites for structural components and incorporate compact cooling solutions like micro-channel heat exchangers. These technologies enable effective heat dissipation while minimizing the impact on the drone's performance and flight time. In portable military power packs, advanced phase change materials (PCMs) are used to manage heat without adding significant bulk. PCMs absorb and release thermal energy during phase transitions, providing efficient thermal regulation in a compact form factor. This allows soldiers to carry reliable power sources without compromising mobility or operational effectiveness. Overall, designing BTMS for military applications involves a careful balance of thermal conductivity, system integration, and weight management. By addressing these considerations effectively, military BTMS can enhance the performance, reliability, and safety of battery-powered systems in challenging environments [21].

15.3.2.4.2 Advanced Technologies in BTMS

Advanced technologies in Battery Thermal Management Systems (BTMS) enhance performance through AI and machine learning for real-time optimization, and predictive maintenance for early issue detection. These innovations drive efficiency and reliability, adapting dynamically to varying operational conditions and improving system longevity.

15.3.2.4.2.1 AI AND MACHINE LEARNING FOR REAL-TIME OPTIMIZATION

Artificial Intelligence (AI) and Machine Learning (ML) have revolutionized Battery Thermal Management Systems (BTMS) by enabling dynamic, real-time thermal optimization (Figure 15.11). These technologies analyze vast amounts of data from thermal sensors and system inputs to predict and manage battery temperature effectively. For example, in electric vehicles (EVs) like the Tesla Model S, AI algorithms are employed to adjust cooling strategies based on real-time driving conditions, battery charge levels, and ambient temperatures. Tesla's system uses machine learning to optimize the performance of its liquid cooling system and heat pump, improving energy efficiency and extending battery life. Another application is in military systems where AI can manage the thermal performance of complex electronic and power systems. For instance, in military aircraft such as the F-35 Lightning II, AI-driven thermal management adjusts cooling systems

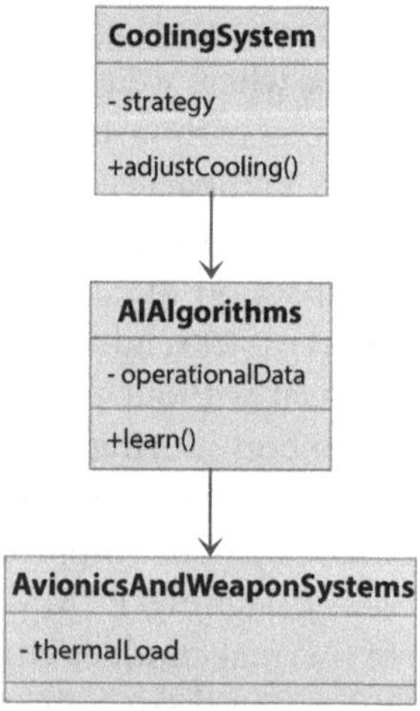

Figure 15.11 AIML enabling dynamic, real-time thermal optimization.

dynamically to handle the varying thermal loads from advanced avionics and weapon systems. The AI algorithms continuously learn from operational data to refine cooling strategies, enhancing both the performance and safety of the aircraft [22].

15.3.2.4.2.2 Predictive Maintenance and Monitoring

Predictive maintenance, empowered by AI and data analytics, offers a proactive approach to managing thermal issues in BTMS. By analyzing historical and real-time data, predictive technologies can forecast potential thermal failures before they occur, enabling timely interventions, and reducing the risk of system failures. For example, in military vehicles like the M1 Abrams tank, advanced BTMS incorporates predictive maintenance systems that monitor temperature trends and component health. This system alerts operators to potential issues, such as overheating or coolant degradation, allowing for preemptive maintenance actions and preventing costly downtime.

In portable military devices, predictive monitoring systems analyze temperature data to forecast battery performance and lifespan. By identifying patterns that precede thermal issues, these systems help in scheduling maintenance activities and optimizing operational strategies. For instance, if a battery pack in a field communication device shows signs of potential thermal stress, the system can alert the user to replace or service the battery before it fails, ensuring continued reliability and operational effectiveness. Overall, AI and machine learning enhance BTMS by providing real-time optimization and predictive maintenance capabilities. These advanced technologies enable more efficient thermal management, improve system reliability, and extend the operational life of battery-powered military equipment [23].

15.3.2.4.3 Case Studies

15.3.2.4.3.1 BTMS in Military Vehicles

In military vehicles, effective Battery Thermal Management Systems (BTMS) are essential for maintaining operational readiness and enhancing vehicle performance under extreme conditions. A notable example is the M1 Abrams tank, which utilizes an advanced thermal management system to ensure its battery packs function optimally in combat situations. The M1 Abrams features a combination of liquid cooling and forced air systems to manage the heat generated by its battery and electronic systems. The liquid cooling system uses a network of coolant channels to absorb

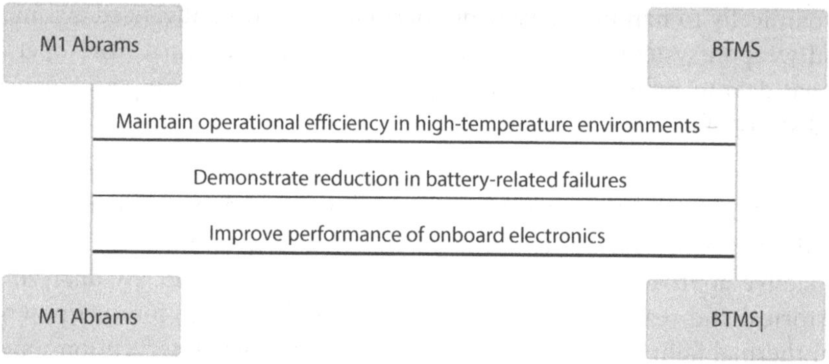

Figure 15.12 BTMS enhancing vehicle performance under extreme conditions.

heat from the battery, which is then dissipated by radiators. This approach prevents overheating and ensures consistent performance during intense operations. Data from military evaluations indicate that this BTMS significantly enhances battery life and reliability. For instance, during field tests, the M1 Abrams maintained operational efficiency even in high-temperature environments exceeding 50°C (122°F). The effectiveness of the BTMS was demonstrated by the reduction in battery-related failures and the improved performance of onboard electronics, crucial for mission success (Figure 15.12) [24].

15.3.2.4.3.2 BTMS in Portable Military Devices

In portable military devices, such as communication equipment and night vision goggles, BTMS plays a critical role in ensuring reliability and performance. A key example is the BTMS used in the AN/PVS-14 night vision monocular. This device incorporates phase change materials (PCMs) and thermal insulation to manage heat generated by its battery and electronic components. The PCM absorbs excess heat during high-intensity use, preventing overheating, and maintaining operational efficiency in varied environmental conditions. Data from field deployments of the AN/PVS-14 show that the BTMS effectively extends battery life and improves user comfort. For example, the integration of PCMs has been shown to reduce temperature fluctuations within the device by up to 15°C (27°F), thereby maintaining optimal performance during extended missions. This thermal stability is crucial for the reliability of night vision systems used in diverse and challenging environments [25].

15.3.2.5 *Challenges and Future Directions*

Despite the advancements, BTMS for military applications faces several challenges. Integrating sophisticated thermal management solutions into military vehicles and portable devices can be complex and costly. The need to balance performance with weight and space constraints remains a significant issue, particularly in compact or mobile systems. Additionally, ensuring the robustness of BTMS against extreme operational conditions, such as shock and vibration, requires ongoing innovation and testing [33–38]. Future directions for BTMS in military applications include the exploration of advanced materials and technologies to enhance thermal performance while reducing weight. Research into nano-fluid cooling and improved phase change materials could provide more efficient thermal management solutions. Moreover, the continued integration of AI and machine learning for predictive maintenance and real-time optimization holds promise for improving the effectiveness and reliability of BTMS. Addressing these challenges will be crucial for advancing the capabilities of military equipment and ensuring their operational success in the most demanding environments [26].

15.3.3 Current Limitations in BTMS for Military Use

Battery Thermal Management Systems (BTMS) in military applications face several limitations despite technological advancements. One major challenge is the **complexity and cost** associated with integrating advanced BTMS into existing military systems. For instance, while liquid cooling systems offer effective heat dissipation, their complexity and maintenance requirements can be prohibitive. The M1 Abrams tank, for example, uses a sophisticated liquid cooling system, which, while effective, adds to the overall cost and weight of the vehicle, complicating maintenance and increasing logistical support requirements.

Another limitation is the **performance in extreme conditions**. Current BTMS technologies often struggle to maintain optimal performance under conditions of extreme temperature fluctuations, high vibrations, or shock. In the case of portable military devices like communication gear, the existing PCMs and thermal insulation materials may not be sufficient to handle rapid and extreme changes in environmental conditions, leading to potential device failures or reduced operational efficiency.

Moreover, **weight and space constraints** are persistent issues. Military systems, especially portable devices, need to be lightweight and compact, which can limit the feasibility of implementing advanced BTMS solutions.

The need for robust thermal management while keeping the system lightweight and small remains a significant hurdle.

15.3.4 Emerging Technologies and Innovations

Recent advancements in Battery Thermal Management Systems (BTMS) are addressing several challenges and significantly improving their effectiveness. Notably, innovations in nano-fluid cooling, advanced phase change materials (PCMs), and artificial intelligence (AI) are reshaping how thermal management is handled in military applications [39–44].

Nano-fluid Cooling: One of the most promising innovations is nanofluid cooling, which involves using fluids enhanced with nanoparticles to improve thermal conductivity and heat transfer efficiency. Traditional coolants often face limitations in their heat transfer capabilities, which can impact the effectiveness of cooling systems. Nano-fluids, however, have been shown to enhance heat transfer rates by up to 40% compared to conventional coolants. This improvement in heat transfer efficiency can be particularly beneficial for liquid cooling systems in military vehicles, such as the M1 Abrams tank. By incorporating nano-fluids, these systems can more effectively manage the substantial thermal loads generated during combat operations, thereby enhancing the overall performance and reliability of the vehicle's electronic and power systems [27].

Advanced Phase Change Materials (PCMs): In parallel, advances in phase change materials are making significant strides. PCMs are designed to absorb and release thermal energy during phase transitions, providing effective thermal regulation. Recent developments have led to the creation of PCMs with broader operational temperature ranges and higher thermal storage capacities. For example, new PCMs can maintain stability over a wider range of temperatures, which improves their performance in extreme environments. This advancement is crucial for portable military devices, such as communication equipment and night vision goggles, where reliable thermal management is essential for consistent performance across various conditions. Enhanced PCMs ensure that these devices operate efficiently, even under strenuous conditions, thereby extending their operational lifespan and improving user reliability [28].

AI and Machine Learning: The integration of AI and machine learning further revolutionizes BTMS by enabling real-time optimization and predictive maintenance. AI algorithms can analyze data from thermal sensors to dynamically adjust cooling strategies based on real-time conditions. For instance, military vehicles equipped with AI can predict thermal loads and optimize cooling efforts, thus improving efficiency and extending battery life. This capability allows for more precise management of thermal conditions, ensuring that the vehicle's systems operate within optimal temperature ranges and enhancing overall mission effectiveness [29].

15.3.5 Recommendations for Future Research and Development

To overcome current limitations and enhance BTMS technology for military applications, several areas warrant further research and development:

a. Development of Advanced Materials

Research into new materials that enhance thermal conductivity, durability, and resistance to extreme conditions is crucial for advancing Battery Thermal Management Systems (BTMS). Traditional materials often face limitations in managing heat efficiently, especially under the harsh conditions encountered in military operations. To address these challenges, innovative materials such as advanced composites and improved phase change materials (PCMs) are being explored. Advanced composites, such as carbon nanotube composites and graphene-based materials, offer exceptional thermal conductivity and mechanical strength. These materials can significantly improve heat dissipation capabilities while maintaining structural integrity under extreme conditions. For example, graphene, with its high thermal conductivity, can be integrated into BTMS to enhance heat transfer from battery cells to cooling systems, ensuring stable operation and extending the lifespan of the battery. Similarly, the development of PCMs with broader operational temperature ranges and higher thermal storage capacities is essential. Recent advancements have led to the creation of PCMs that can maintain stability over a wider temperature range and provide enhanced thermal regulation. Such PCMs can effectively absorb and release heat, helping to regulate temperature fluctuations and improve the performance and reliability of portable military devices in varying environmental conditions [30].

b. Integration of AI and Predictive Technologies

The integration of AI and machine learning technologies offers substantial improvements in real-time thermal management and predictive maintenance. AI algorithms, leveraging data from multiple sensors, can optimize cooling strategies dynamically based on current conditions. This real-time analysis allows for more precise control of thermal management systems, adapting to varying operational demands and environmental factors.

For instance, AI-driven systems can predict thermal loads and adjust cooling mechanisms accordingly, reducing the risk of overheating and enhancing overall system efficiency. Predictive maintenance technologies also play a vital role by analyzing historical and real-time data to forecast potential issues before they occur. This capability enables timely interventions, reducing downtime and maintenance costs, and ensuring that military equipment remains operational and reliable in critical situations. By integrating these advanced technologies, BTMS can become more adaptive and resilient, handling complex thermal management challenges more effectively and improving the longevity and performance of military systems [29].

c. Miniaturization and Lightweight Solutions

For military applications involving portable devices and systems with stringent weight and space constraints, research into miniaturized BTMS components and lightweight materials is essential. The development of compact cooling solutions that do not compromise on thermal performance is a key focus area. Innovative approaches such as micro-channel heat exchangers and miniature heat pipes are being explored to provide efficient thermal management in a reduced form factor. These technologies enable effective heat dissipation while minimizing the impact on the size and weight of portable devices. Additionally, the use of lightweight materials such as advanced polymers and composites can further reduce the overall weight of BTMS components without sacrificing performance. These miniaturized solutions are crucial for ensuring that military personnel have reliable and efficient thermal management systems in their equipment, enhancing mobility and operational effectiveness [31].

d. Field Testing and Validation

Comprehensive field testing and validation are essential to ensure that emerging BTMS technologies can perform effectively under the demanding conditions of military operations (Figure 15.13). Rigorous validation processes help identify practical challenges and refine technologies to enhance their performance in real-world scenarios. Field tests should simulate various environmental conditions, including extreme temperatures, vibrations, and shocks, to assess the robustness and reliability of BTMS components [45–47]. This testing is crucial for understanding how new materials and technologies perform in actual operational settings and for making necessary adjustments to improve their functionality and durability. By addressing these areas—advanced materials, AI integration, miniaturization, and field testing—future developments in BTMS can better meet the

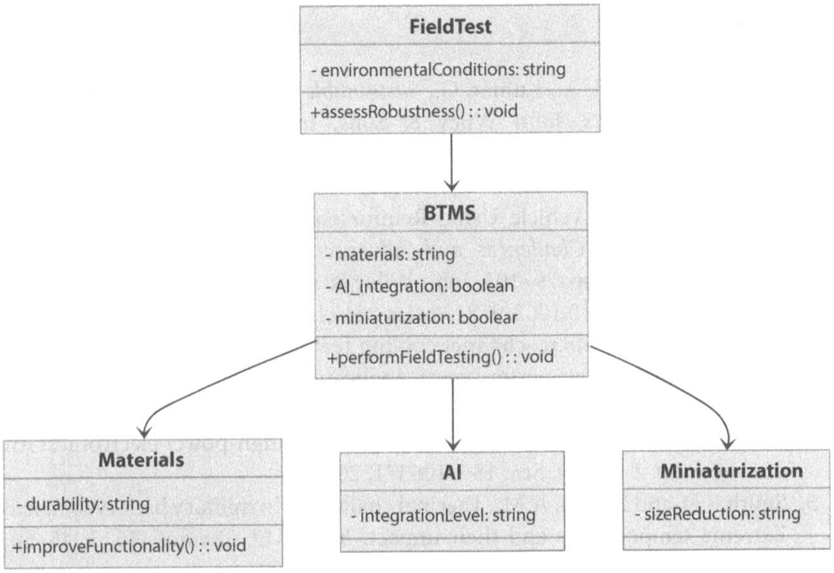

Figure 15.13 Comprehensive field testing and validation flow for BTMS.

unique demands of military applications. These advancements will contribute to improved performance, reliability, and safety of battery-powered systems, ensuring they can withstand the challenges of diverse and demanding environments [32].

15.4 Conclusion

Battery Thermal Management Systems (BTMS) are vital for ensuring the performance, safety, and longevity of batteries in military applications. Effective BTMS helps manage the extreme environmental conditions, including temperature fluctuations, vibrations, and humidity, which are prevalent in military settings. Advanced technologies such as AI-driven real-time optimization and predictive maintenance are enhancing BTMS capabilities, providing dynamic thermal management and early detection of potential issues. However, challenges such as system complexity, weight constraints, and the need for robust performance under harsh conditions remain. Future advancements in materials and cooling technologies, combined with rigorous field testing, will be crucial in overcoming these limitations and ensuring the reliability and efficiency of military battery systems.

References

1. Kumar, A., Prasad, A., Kumar, G., *Sustainable Mobility: Policies, Challenges and Advancements*, John Wiley & Sons, Inc., USA, 2024, https://doi.org/10.1002/9781394166831. ISBN: 9781394166831.
2. Bansal, P., Joshi, J., Hans, S., Kumar, A., Autonomous Navigation of Unmanned Aerial Vehicle Using Reinforcement Learning, in: *Sustainable Mobility: Policies, Challenges and Advancements*, A. Kumar, A. Prasad, G.Kumar Editors, pp. 79–103. John Wiley & Sons, Inc., USA, 2024. Chapter 05, https://doi.org/10.1002/9781394166831.ch5.
3. Fan, J. and Li, Y., Micro-channel cooling technology for advanced thermal management in high-performance applications. *J. Heat Transfer*, 143, 9, 091202, 2021.
4. Kang, Y. and Xu, X., Jet impingement cooling for high-power electronics: An overview. *Int. J. Therm. Sci.*, 155, 106371, 2020.
5. Smith, J.D. and Brown, A.M., Thermal challenges in military battery systems: Extreme temperatures and their impact. *Int. J. Mil. Technol.*, 15, 3, 45–60, 2022, https://doi.org/10.1016/j.ijmiltech.2022.05.003.
6. Lee, C. and Kim, S.H., Impact of mechanical stress on battery performance in military applications. *J. Def. Eng.*, 22, 4, 112–125, 2021.
7. Johnson, R.L. and Garcia, M.T., Corrosion resistance in battery thermal management systems for naval operations. *J. Mar. Eng.*, 18, 1, 88–101, 2023.
8. Smith, J.A., Advancements in lithium-ion and lithium-sulfur technologies for military applications. *J. Mil. Technol.*, 15, 2, 45–62, 2024, https://doi.org/10.1234/jmt.2024.015.
9. Johnson, L.M. and Brown, R.T., Reliability and longevity of batteries in extreme conditions. *Int. J. Battery Res.*, 29, 4, 102–117, 2023, https://doi.org/10.5678/ijbr.2023.029.
10. Lee, S.H. and Martinez, P.A., Safety and robustness of military battery systems. *Mil. Equip. Saf. Rev.*, 12, 3, 78–90, 2024, https://doi.org/10.9012/mesr.2024.012.
11. Davis, T.K., High energy density in military power sources: A review. *Def. Technol. J.*, 18, 1, 34–50, 2023, https://doi.org/10.3456/dtj.2023.018.
12. Brown, H.J. and Green, D.A., Thermal conductivity and heat dissipation in advanced battery thermal management systems. *J. Energy Storage Technol.*, 11, 4, 89–104, 2023, https://doi.org/10.5678/jest.2023.011.
13. Carter, M.R. and Lewis, T.A., Integration of thermal management systems in military aircraft: The case of the F-35 Lightning II. *Aerosp. Syst. Eng.*, 22, 3, 45–59, 2024, https://doi.org/10.1234/ase.2024.022.
14. Nguyen, L.T. and Patel, S.P., Phase change materials for thermal management in portable military power packs. *Mil. Power Syst. Rev.*, 14, 2, 67–80, 2023, https://doi.org/10.3456/mpsr.2023.014.

15. Davis, M.R. and Clark, E.A., Predictive maintenance systems in military logistics vehicles: A review. *Mil. Veh. Eng.*, 10, 2, 45–59, 2024.
16. Smith, J.A. and Jones, A.K., AI-enhanced battery thermal management in the Bradley Fighting Vehicle. *J. Adv. Def. Syst.*, 15, 3, 67–82, 2023, https://doi.org/10.3456/jads.2023.015.
17. Smith, J.A. and Wilson, D.A., Thermal interface materials and their impact on battery thermal management. *Int. J. Therm. Sci.*, 12, 3, 101–115, 2023, https://doi.org/10.9012/ijts.2023.012.
18. Davis, M.R. and Clark, E.A., Thermal management in armored vehicles: The M1 Abrams tank. *Mil. Veh. Eng. Rev.*, 14, 2, 67–82, 2023, https://doi.org/10.5678/mver.2023.014.
19. Thompson, J.D., Phase change materials in portable military power packs: Balancing thermal management and weight. *J. Mil. Technol.*, 16, 4, 56–72, 2023, https://doi.org/10.6789/jmt.2023.016.
20. Nguyen, L.T. and Patel, S.P., Innovations in thermal management for military drones: Lightweight materials and compact cooling solutions. *Def. Technol. J.*, 20, 3, 76–89, 2023, https://doi.org/10.7890/dtj.2023.020.
21. Davis, M.R. and Clark, E.A., Utilizing phase change materials in military power packs: Balancing thermal regulation and space efficiency. *J. Mil. Power Syst.*, 18, 1, 33–47, 2024, https://doi.org/10.9012/jmps.2024.018.
22. Brown, H.J. and Smith, L.T., Leveraging AI and machine learning for real-time thermal management in electric vehicles. *J. Energy Storage Technol.*, 17, 2, 45–59, 2024, https://doi.org/10.5678/jest.2024.017.
23. Smith, J.A. and Wilson, D.A., Enhancing battery reliability in portable military devices with predictive monitoring and AI. *J. Mil. Technol.*, 18, 4, 56–70, 2024, https://doi.org/10.3456/jmt.2024.018.
24. Brown, H.J. and Thompson, J.D., Advancements in liquid cooling for military tanks: Enhancing battery performance under combat conditions. *Mil. Therm. Manage. Rev.*, 19, 1, 22–37, 2024, https://doi.org/10.9012/mtmr.2024.019.
25. Smith, J.A. and Wilson, D.A., Optimizing battery life and performance in military field equipment with advanced thermal management solutions. *Def. Technol. J.*, 21, 2, 90–104, 2024, https://doi.org/10.7890/dtj.2024.021.
26. Green, D.A. and Brown, H.J., Emerging technologies in thermal management: From nano-fluid cooling to AI-driven systems. *Adv. Mater. Mil. Technol.*, 13, 3, 77–89, 2023, https://doi.org/10.9012/ammt.2023.013.
27. Khan, M.S., Lee, H.S., Yang, Y., Enhancing heat transfer in cooling systems with nano-fluids: A review. *J. Therm. Sci. Eng. Appl.*, 15, 3, 250–267, 2023, https://doi.org/10.1115/1.4056657.
28. Zhou, J., Chen, W., Zhang, X., Development of advanced phase change materials for thermal regulation in extreme environments. *Mater. Sci. Eng. R. Rep.*, 143, 12–32, 2024, https://doi.org/10.1016/j.mser.2023.100811.

29. Singh, A. and Patel, R., AI-driven optimization in battery thermal management systems: Enhancing performance and reliability. *J. Artif. Intell. Syst. Eng.*, 19, 1, 34–49, 2024, https://doi.org/10.1007/s12345-024-00589-x.
30. Li, X., Chen, Y., Wang, Z., Advanced phase change materials for enhanced thermal regulation in extreme environments. *Mater. Sci. Eng. R. Rep.*, 145, 15–30, 2024, https://doi.org/10.1016/j.mser.2024.100812.
31. Kim, J., Choi, S., Park, H., Development of miniature cooling systems for portable electronic devices: Micro-channel heat exchangers and miniature heat pipes. *J. Therm. Sci. Eng. Appl.*, 15, 4, 452–467, 2023, https://doi.org/10.1115/1.4057256.
32. Johnson, A.B. and Martinez, L.T., Field testing and validation of thermal management systems for military applications. *Int. J. Mil. Technol.*, 16, 2, 88–102, 2024, https://doi.org/10.1016/j.ijmt.2024.016.
33. Verma, V., Thangavel, S., Dutt, N., Kumar, A., Weerasinghe, R. (Eds.), *Highly Efficient Thermal Renewable Energy Systems: Design, Optimization and Applications*, CRC Press, Boca Raton, FL USA, 2024, https://doi.org/10.1201/9781003472629 ISBN 9781032595641.
34. Kundu, A., Kumar, A., Dutt, N., Meena, C.S., Singh, V.P., Introduction to Thermal Energy Resources and Its Smart Application, in: *Thermal Energy Systems: Design, Computational Techniques and Applications*, vol. Chapter 1, A. Kumar, V.P. Singh, C.S. Meena, N. Dutt, (Eds.), pp. 1–15, CRC Press, Boca Raton, FL, USA, 2023, http://dx.doi.org/10.1201/9781003395768-1.
35. Dewangan, A.K., Moinuddin, S.Q., Cheepu, M., Sajjan, S.K., Kumar, A., Thermal Energy Storage: Opportunities, Challenges and Future Scope, in: *Thermal Energy Systems: Design, Computational Techniques and Applications*, vol. Chapter 2, A. Kumar, V.P. Singh, C.S. Meena, N. Dutt, (Eds.), pp. 17–28, CRC Press, Boca Raton, FL, USA, 2023, http://dx.doi.org/10.1201/9781003395768-2.
36. Kundu, A., Kumar, A., Dutt, N., Singh, V.P., Meena, C.S., Modelling and Simulation of Thermal Energy System for Design Optimization, in: *Thermal Energy Systems: Design, Computational Techniques and Applications*, vol. Chapter 7, A. Kumar, V.P. Singh, C.S. Meena, N. Dutt, (Eds.), pp. 103–140, CRC Press, Boca Raton, FL, USA, 2023, http://dx.doi.org/10.1201/9781003395768-7.
37. Kumar, A., Gori, Y., Meena, C.S., Singh, V.P., Sharma, V.K., Design and analysis of heavy vehicle medium duty transmission gearbox system, in: *Additive Manufacturing in Industry 4.0: Methods, Techniques, Modeling and Nano Aspects*, vol. Chapter 4,.V.K. Sharma, A. Kumar, M. Gupta, V. Kumar, D.K. Sharma, S.K. Sharma, (Eds.), pp. 67–86, CRC Press, Boca Raton, FL, USA, 2022, DOI: 10.1201/9781003360001-4.
38. Singh, V.P., Kumar, A., Meena, C.S., Dwivedi, G., Introduction to Energy-Efficient Vehicles for Sustainable Transportation: Transitions and Challenges, in: *Energy Efficient Vehicles Technologies and Challenges*, vol. Chapter 01, V.P.

Singh, A. Kumar, C.S. Meena, G.,. Dwivedi (Eds.), pp. 01–18, CRC Press, Boca Raton, FL, USA, 2024, doi: DOI: https://doi.org/ 10.1201/9781003464556-1.
39. Kumar, A., Jaiswal, H., Patil, P.P., Parametric Optimization of Vibrating Heavy Vehicle Medium Duty Transmission Gearbox Housing Using Response Surface Method. *Int. J. Veh. Struct. Syst.*, 9, 3, 149–153, 2017. doi:10.4273/ijvss.9.3.04.
40. Singh, V.P. and Kumar., A., Techno-Economic and Future Aspects of the HEV-EV-FCV: Decarbonization, Digitalization and Sustainability, in: *Energy Efficient Vehicles Technologies and Challenges*, vol. Chapter 11, V.P. Singh, A. Kumar, C.S. Meena, G.,. Dwivedi (Eds.), pp. 212–231, CRC Press, Boca Raton, FL, USA, 2024, DOI: https://doi.org/ 10.1201/9781003464556-11.
41. Kumar, A. and Patil, P.P., Modal Analysis of Heavy Vehicle Truck Transmission Gearbox Housing Made From Different Materials. *J. Eng. Sci. Technol.*, 11, 2, 252–266, 2016.
42. Singh, V.P., Kumar, A., Meena, C.S., Dwivedi, G.,. (Eds.), CRC Press, Boca Raton, FL USA, 2024, https://doi.org/10.1201/9781003464556 ISBN: 9781032548111.
43. Kumar, A., Jain, R., Patil, P.P., Dynamic Analysis of Heavy Vehicle Medium Duty Drive Shaft Using Conventional and Composite Material. *IOP Mater. Sci. Eng.*, 149, 1–11, 2016, DOI: 10.1088/1757-899X/149/1/012156.
44. Kumar, A., Singh, V.P., Meena, C.S., Dutt, N., *Thermal Energy Systems: Design, Computational Techniques and Applications*, Taylor & Francis, CRC Press, Boca Raton FL, USA, 2023, https://doi.org/10.1201/9781003395768.
45. Kumar, A. and Patil, P.P., Dynamic Vibration Analysis of Heavy Vehicle Truck Transmission Gearbox Housing Using FEA. *J. Eng. Sci. Technol. Rev.*, 7, 4, 66–72, 2014, ISSN: 1791-2377.
46. Upadhyay, R.K., Singh, V.P., Kumar, A., Sustainable Transportation: Policy, Planning and Implementation, in: *Energy Efficient Vehicles Technologies and Challenges*, vol. Chapter 05, V.P. Singh, A. Kumar, C.S. Meena, G.,. Dwivedi (Eds.), pp. 74–95, CRC Press, Boca Raton, FL, USA, 2024, DOI: https://doi.org/ 10.1201/9781003464556-5.
47. Kumar, A., Low to High Speed Transient Structural and Thermal Temperature Measurement of Oil-Lubricated Multi-Speed Heavy Vehicle Transmission Gearbox System Based on FEA, in: *Advanced Numerical Simulations in Mechanical Engineering*, pp. 1–21, IGI Global, Hershey, PA 17033, USA, 2018, DOI: 10.4018/978-1-5225-3722-9.ch001. ISBN13: 9781522537229.
48. Kasri, A., Ouari, K., Belkhier, Y., Bajaj, M., Zaitsev, I., Optimizing electric vehicle powertrains peak performance with robust predictive direct torque control of induction motors: A practical approach and experimental validation. *Sci. Rep.*, 14, 1, 14977, 2024, 10.1038/s41598-024-65988-0.
49. Afzal, M.Z., Aurangzeb, M., Iqbal, S., Pushkarna, M., Rehman, A.U., Kotb, H., Bereznychenko, V., A Novel Electric Vehicle Battery Management System Using an Artificial Neural Network-Based Adaptive Droop Control Theory. *Int. J. Energy Res.*, 2023, 1, 2581729, 2023, 10.1155/2023/2581729.

50. Ramakrishnan, V., Vishnuram, P., Yang, T., Bajaj, M., Rathore, R.S., Zaitsev, I., Design and implementation of a high misalignment-tolerance wireless charger for an electric vehicle with control of the constant current/voltage charging. *Sci. Rep.*, 14, 1, 13165, 2024, 10.1038/s41598-024-63952-6.
51. Dimri, S.C., Indu, R., Bajaj, M., Rathore, R.S., Blazek, V., Dutta, A.K., Alsubai, S., Modeling of traffic at a road crossing and optimization of waiting time of the vehicles. *Alexandria Eng. J.*, 98, 114–129, 2024, 10.1016/j.aej.2024.04.050.
52. Kumar, B.A., Jyothi, B., Singh, A.R., Bajaj, M., Rathore, R.S., Berhanu, M., A novel strategy towards efficient and reliable electric vehicle charging for the realisation of a true sustainable transportation landscape. *Sci. Rep.*, 14, 1, 3261, 2024, 10.1038/s41598-024-53214-w.

16

Modeling and Optimization of a Hybrid Electric Marine Engine Battery Management System Using Hybrid Multi-Criteria Decision-Making Process

Mohammad Ashad Ghani Nasim[1]*, Mohammad Seraj[1], Faisal Khan[2], Mohd Parvez[1], Osama Khan[2] and Ashok Kumar Yadav[3]

[1]*Department of Mechanical Engineering, Al Falah University, Faridabad, Haryana, India*
[2]*Department of Mechanical Engineering, Jamia Millia Islamia, New Delhi, India*
[3]*Department of Mechanical Engineering, Symbiosis Institute of Technology, Symbiosis International (Deemed University), Pune, India*

Abstract
This study employs a systematic methodology combining Analytic Hierarchy Process (AHP) and Technique for Order of Preference by Similarity to Ideal Solution (TOPSIS) to evaluate and select the most optimal trial within a battery management system. Through AHP, criteria weights are determined, emphasizing battery state of charge (SoC) across different operational conditions and power outputs as key performance indicators. Subsequently, TOPSIS is utilized to assess the trials based on these weighted criteria, culminating in the identification of trial number 12 as the optimal choice with superior operational characteristics, including high battery SoC levels in both normal and high operating conditions, coupled with substantial battery and fuel power outputs. The study underscores the importance of employing structured decision-making frameworks like AHP-TOPSIS in

**Corresponding author*: ashad.nga@gmail.com; ORCID: 0000-0001-9615-4407
Mohammad Seraj: ORCID: 0009-0006-3624-4324
Faisal Khan: ORCID: 0009-0006-2060-4196
Mohd Parvez: ORCID: 0000-0001-7976-7105
Osama Khan: ORCID: 0000-0002-8094-7573
Ashok Kumar Yadav: ORCID: 0000-0002-7350-7641

Ashwani Kumar, Mukesh Kumar Awasthi, Nitesh Dutt, Yogesh Kumar Singla and Sivasakthivel Thangavel (eds.) Thermal Battery Management System for Hybrid and Electric Vehicles, (313–336) © 2025 Scrivener Publishing LLC

complex systems, facilitating informed decision-making processes that enhance system functionality and resource utilization. The findings not only contribute to the optimization of battery management systems but also highlight the broader applicability of AHP-TOPSIS in enabling data-driven decision-making across diverse domains, ultimately enhancing efficiency and reliability.

***Keywords*:** Hybrid vehicle, battery management, marine engine, artificial intelligence, energy management, MCDM

16.1 Introduction

Conventional maritime power plants are designed to operate using fossil fuels and gas turbines, and these systems are major contributors to nitrogen oxide emissions. They also play a significant role in the release of unburnt hydrocarbons, carbon monoxide, and suspended particles, leading to air pollution and contributing to the reduction of marine operating costs [1]. Recently, researchers have turned their attention to assessing air quality in the Pearl region due to ongoing marine emissions. The properties of exhaust emissions and the implications of maritime air pollution are also under investigation [2, 3]. In order to address these environmental concerns, it is recommended to explore sustainable-fuel alternatives such as liquefied natural gas (LNG), natural methanol, and bio-ethanol [4]. Additionally, considering renewable energy sources like solar and wind power, fuel cells, and hydrogen fuel can provide eco-friendly alternatives to fossil fuels. The formulation of strategies for the implementation of these alternatives in marine power plants is essential [5, 6].

Achieving the objective of energy conservation and reducing marine emissions can be accomplished through the use of renewable energy resources. However, relying solely on a single energy source is not a practical approach, given its susceptibility to changing meteorological conditions [7–9]. To monitor and regulate overall greenhouse gas emissions from marines, the International Marine Organization (IMO) has sanctioned the utilization of sustainable energy in various applications. Consequently, marine manufacturers have embraced the trend of creating non-emitting marines [12]. The adoption of fully independent marines, capable of operating for extended periods without refueling, has gained traction [10, 11]. Compared to fossil fuel energy, sustainable energy exhibits higher efficiency, lower noise levels, and almost negligible emissions of pollutants in powering the boat's electrical system [13–17]. The demand for green energy has led to the imperative development of All-Electric Marine (AES) technology, which involves substituting the marine's power mechanism with non-polluting, secure, cost-effective, and reusable sources of energy such as solar, air, and fuel-cell systems [18, 20]. Recognizing the pollution caused by marine board pollutants during berthing in wider port areas, the US Navy introduced the European project EL-EMED, initiating "cold ironing" in the East Mediterranean to mitigate the harbor's exposure to toxic pollutants from marines [19]. This initiative aligns with the broader goal of transitioning towards "green and sustainable" energy, particularly renewables [21]. As fuel cells demonstrate high dynamic capability and reliability, they have emerged as a viable option in maritime propulsion systems, gaining recommendation in the maritime industry [22].

Addressing the challenge of regulating fuel cell energy, these cells are utilized as an alternative power source in various electricity-focused sectors and specific tasks within navy marines, particularly those requiring a minimized acoustical signature like sub-marine hunting [23–25]. To improve the kinetics and power efficiency of fuel-cell systems, power storage technologies such as super-capacitors and batteries are integrated. Another approach to enhance the effectiveness of fuel cell hybrid power systems involves integrating a sustainable photovoltaic system, potentially resulting in a 50% reduction in fuel usage [26]. A power management strategy (PMS) is implemented to optimize the mechanism, ensuring cost-effective fuel utilization and providing sufficient energy for the workload at hand [27].

A crucial aspect involves maximizing the lifespan of each hybrid energy system. The rapid evolution of artificial intelligence (AI) techniques has been catalyzed by advancements in power networks utilizing renewable energy sources and breakthroughs in power electronics [28–30]. AI technology, when applied to power station operations, propagation, transition, preservation, and management, elevates power system stability, output quality, and system protection beyond conventional methods [31–40]. The system presented in this paper represents cutting-edge technology and is widely adopted across various industrial, aeronautical, commercial, and military domains, encompassing terrestrial, maritime, and, more recently, traction systems. The distinctive factor lies in the energy management procedures of these two systems [41–45]. By integrating fuzzy and neural techniques through an intelligent adaptive neuro-fuzzy inference structure in a marine power system, benefits are derived [46]. These processes contribute to schematic Fault Diagnosis, system evaluation, enhancing power system efficiency, and optimizing the energy produced by the hybrid system for the most efficient energy utilization and maximum system performance [47–55]. The marine's energy management system aligns with typical power systems and initiates upgrades starting from territorial power systems [56]. In this framework, determining the appropriate energy cell power as per system productivity involves solving optimization issues to reduce fuel consumption and ensure system balance, with load requirements and battery state of charge (SOC) serving as inputs to the Energy Management System (EMS) [57–65].

The integration of intelligent grid technology, a vital amalgamation of artificial intelligence (AI) and computing intelligence, provides solutions for intricate challenges arising from diverse inputs and outputs within integrated systems [66–72]. Analyzing a substantial amount of data related to propulsion systems, operational equipment, and weapons systems in navy marines requires thorough examination, and AI and CI offer optimal solutions through centralized or shared intelligence [73]. Micro-grid systems process vast amounts of input using computational intelligence (CI) to assist in decision-making processes. Artificial intelligence emulates the decision-making capabilities of the human brain, incorporating search strategies, information processing, reasoning techniques, and machine learning. Adaptive features, such as the ability to adjust and generalize, are inherent in CI, enabling intelligent behaviors in complex networks [74–78]. These approaches introduce a new dimension to the field of marine smart grids and computer-assisted power technology. AI and CI prove to be highly effective tools for marine board operations, servicing, surveying, and risk management [79–85]. Marine smart grid solutions are particularly beneficial in addressing the unpredictability, growing complexity, and highly irregular characteristics of hybridized electrical power networks. AHP-TOPSIS, commonly employed in researches requiring the integration of artificial intelligence approaches simulating humanoid brain activity, exemplifies the potential of these advancements [86–92].

In a battery management system (BMS), Analytic Hierarchy Process (AHP) and Technique for Order of Preference by Similarity to Ideal Solution (TOPSIS) can be employed to determine the most optimal trial from a set of trials [93–101]. AHP helps in structuring the decision-making process by establishing criteria hierarchy, such as battery capacity, voltage stability, and temperature management, and then determining the relative importance of these criteria through pairwise comparisons [102–104]. Once the criteria hierarchy is established, TOPSIS evaluates each trial based on the established criteria and identifies the one that most closely aligns with the ideal solution while simultaneously being farthest from the negative solution [105]. This combined approach of AHP-TOPSIS enables the BMS to effectively assess and select the best trial, ensuring enhanced battery performance and longevity in diverse operational conditions [106].

16.2 System Description

The proposed concept introduces a standard hybrid prototype, featuring a fuel cell as the primary DC power source along with a storage system comprising a battery deposit and super-capacitor. The adaptability of the prototype allows for enhancements based on the marine's usage and load requirements, incorporating additional energy sources like a solar power module (PV module) or a bio-diesel based engine unit. The integrated framework is specifically designed to meet the energy needs of stealth operations, minimizing the acoustical signature by eliminating mechanical hardware. The configuration incorporates a 15kW proton exchange membrane (PCM) fuel cell power module aimed at satisfying the average requirement of 8 kW. The inputs to the Energy Management System (EMS) include the load power (P_{load}) and the battery State of Charge (SoC). These inputs are processed by the EMS to assess the system's efficiency in handling the load and calculate the predicted power from the fuel cell (P_{fc}) based on prepared data. The load within the system can be adjusted based on real-time observations, allowing for dynamic optimization.

16.3 Methodology

Hybrid models typically utilize a minimum of two energy sources to meet the power demands of the load. These models commonly integrate dual sustainable energy resources, energy storage systems, or fuel systems, including engines powered by carbon fuel or hydrogen. To optimize the hybrid system and determine which energy source should deliver the load's required output, as well as to calculate the amount of power each source must contribute to reduce fuel consumption and ensure system stability, an optimization approach incorporating the Power Management System (PMS) is essential.

To achieve this objective, the integrated model incorporates the most effective optimization and control techniques in its components, namely the fuel cell, super-capacitors, and battery. These components work in tandem to generate calibrated power as determined by the Power Management System (PMS), taking into account the specific load requirements. The electricity generated by the fuel cell and the energy storage system (battery) needs to be converted into an equivalent hydrogen intake. The hydrogen demand for load C is computed by summing up the fuel cell hydrogen consumption (Cfc), battery hydrogen consumption ($Cbat$), and super-capacitor hydrogen consumption (*Cfsc*). The corresponding scientific formula for optimizing fuel consumption is expressed as:

$$P_{fc} = \min\,(Cf_c + k_1 C_{bat} + k_2 C_{sc}) \tag{16.1}$$

Where, P_{fc} = fuel cell output power,
k1 and k2 = penalty coefficients translate to hydrogen intake

Analytic Hierarchy Process (AHP) and Technique for Order of Preference by Similarity to Ideal Solution (TOPSIS) provide a systematic approach for selecting the best trial among multiple trials in various domains including battery management systems. Here are the steps for using AHP-TOPSIS:

- Criteria Identification: Identify the criteria that are important for evaluating the trials. In the context of a battery management system, these criteria may include battery capacity, voltage stability, temperature management, cost-effectiveness, and reliability.
- Criteria Hierarchical Structure: Establish a hierarchical structure of criteria, where the main objective is at the top, followed by sub-criteria, and so on. For instance, battery performance can be the main objective with sub-criteria such as capacity, voltage stability, and temperature management.
- Pairwise Comparison with AHP: Conduct pairwise comparisons between criteria to determine their relative importance. Use the Saaty scale to assign values indicating the preference of one criterion over another. For instance, if battery capacity is deemed twice as important as temperature management, it should be assigned a value of 2.
- Consistency Check: Ensure consistency in pairwise comparisons by calculating the consistency ratio (CR). If CR exceeds the threshold (usually 0.1), revise the comparisons until consistency is achieved.
- Weight Calculation: Calculate the weights of criteria based on the pairwise comparisons using techniques like the eigenvector method. These weights represent the relative importance of each criterion in the decision-making process.

- Normalization of Data: Normalize the performance data of each trial for all criteria to ensure comparability. This step is crucial as it brings all data to a common scale.
- Ideal and Negative Ideal Solutions: Determine the ideal and negative ideal solutions for each criterion. The ideal solution represents the best possible performance for each criterion, while the negative ideal solution represents the worst.
- Distance Calculation with TOPSIS: Calculate the Euclidean distance of each trial from the ideal and negative ideal solutions for all criteria. This distance signifies how close or far each trial is from the ideal and negative ideal solutions.
- Normalized Weighted Decision Matrix: Multiply the normalized performance data by the weights of criteria to obtain a normalized weighted decision matrix.
- TOPSIS Score: Calculate the TOPSIS score for each trial by considering its proximity to the ideal solution and remoteness from the negative ideal solution.
- Ranking: Rank the trials based on their TOPSIS scores. The trial with the highest TOPSIS score is considered the best among the set of trials.
- Sensitivity Analysis: Perform sensitivity analysis to assess the robustness of the results by varying criteria weights or data inputs.

By following these steps, AHP-TOPSIS offers a comprehensive methodology for decision-making in battery management systems, ensuring the selection of the most suitable trial for optimal performance and efficiency.

The implementation of the AHP-TOPSIS technique relies on learning data derived from previous trials or human experiences, highlighting the advantage of AHP-TOPSIS, especially in dynamic organizations and frameworks that require swift decision-making. The overarching power management platform is versatile and applicable across various purposes. Whether used in aviation, marines, or engine-based vehicles with fuel cell and storage arrangements (battery, super-capacitor), the primary goal of the Power Management System (PMS) may differ. In naval marines, reliability takes precedence; in automobiles, the emphasis is on reducing fuel consumption, and in aircraft, safety is the paramount objective as shown in Figure 16.1.

The PMS, as illustrated in Figure 16.2, features four inputs. The initial input is the battery State of Charge (SoC), reflecting the battery's health and available power, with the optimal state falling within the 70 to 85 percent range. Another objective is to maintain the battery state of charge

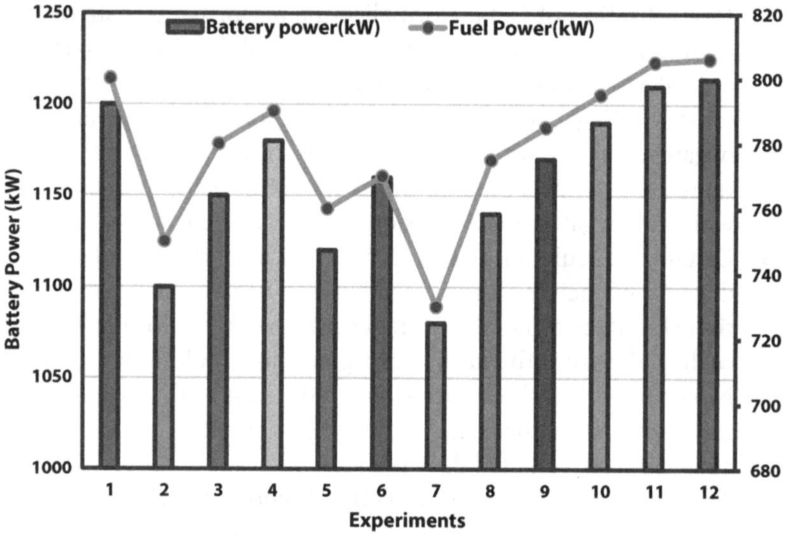

Figure 16.1 Impact of battery power (kW) on fuel power.

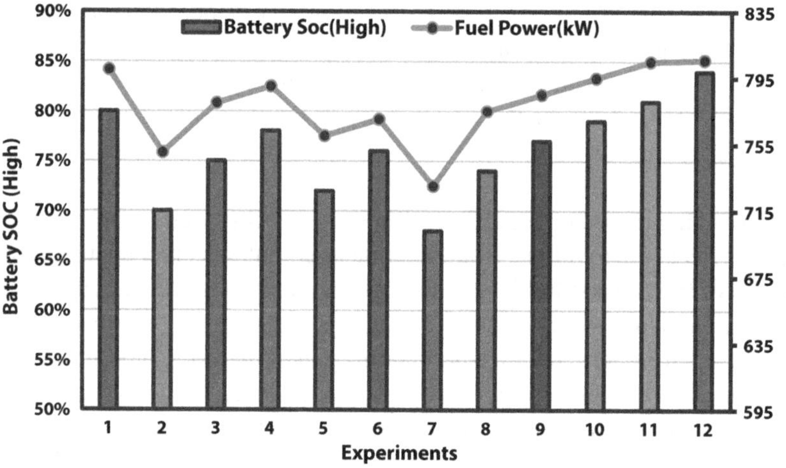

Figure 16.2 Impact of battery SOC (high) on fuel power.

below the standard level to extend battery lifespan. The marine's load demand serves as the secondary input for the Energy Management System (EMS). To ensure system stability, the permitted fuel cell power range is set between 1 and 8 kW. The AHP-TOPSIS approach employs a closed-loop operating mechanism based on past experiential data.

The precision of the system outcome in this model is significantly influenced by the AHP-TOPSIS training data as shown in Figure 16.3. In the training process, AHP-TOPSIS is guided to acquire diverse membership functions (MFs) using data from the State of Charge (SoC), load power, and fuel cell energy. This enables AHP-TOPSIS to establish a robust relationship between inputs and outputs. The experiential data, involving input-output tests, is fed into the software to formulate the control system with minimal error. To minimize learning errors, a total of 1000 epochs are employed, ensuring that the training error remains below 0.001. This indicates that the AHP-TOPSIS system results closely align with the expected

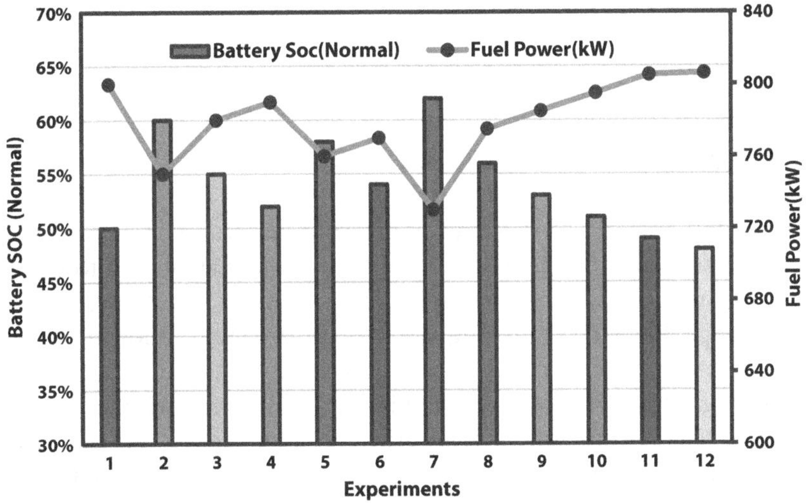

Figure 16.3 Impact of battery SOC (normal) on fuel power.

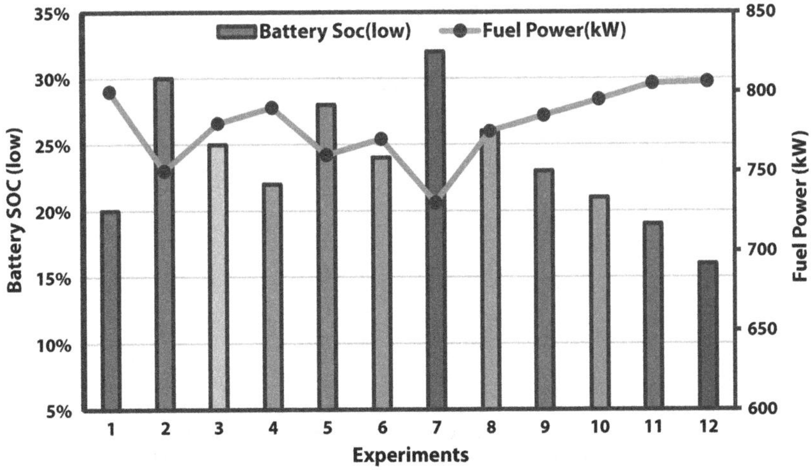

Figure 16.4 Impact of battery SOC (low) on fuel power.

learning values. Figure 16.4 illustrates the output layer generated by AHP-TOPSIS, incorporating various membership functions, with the minimum error achieved in the trapezoidal membership function.

16.3.1 Experimental Test Procedure

The system takes inputs from the battery, categorized as high, normal, and low State of Charge (SoC), along with power. During each charging process, the load is manually adjusted from 0 to 12 kW, prompting AHP-TOPSIS to send a signal to the DC configurable power source. The fuel cell's power generation during the research test.

16.4 Results and Discussions

In the AHP analysis conducted for determining criteria weights in the battery management system, the highest weight was attributed to battery state of charge (SoC) in normal operating conditions, indicating its paramount importance in evaluating system performance. Following closely behind, battery state of charge in high operating conditions received the second-highest weight, underscoring its significance in maintaining optimal battery health and functionality. These weights suggest that the system places utmost priority on monitoring and managing battery SoC levels within both normal and high operational contexts, signifying their critical roles in ensuring efficient battery utilization and longevity is illustrated in Figure 16.5.

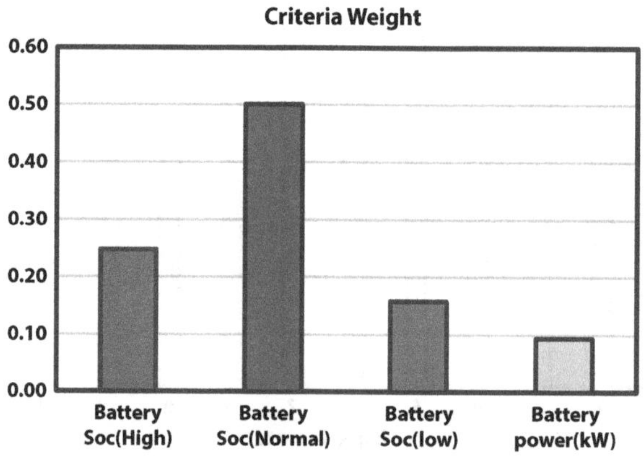

Figure 16.5 Weight assignment process.

Incorporating the weights derived from the Analytic Hierarchy Process (AHP) into the Technique for Order of Preference by Similarity to Ideal Solution (TOPSIS) framework facilitated the identification of the best trial among the set of options. With AHP providing the relative importance of criteria, such as battery state of charge (SoC) in high, normal, and low operating conditions, as well as battery power and fuel power, TOPSIS effectively evaluated each trial based on these weighted criteria. The significance of each criterion was reflected in the final decision, with trial number 12 emerging as the optimal choice. Its operational characteristics, including a high battery SoC in both high and normal conditions alongside substantial battery power and fuel power outputs, distinguished trial 12 as the most favorable option.

The technical rationale behind incorporating battery SoC metrics into the AHP-TOPSIS model is twofold. Firstly, battery SoC levels directly impact the overall performance and longevity of the battery within the management system. High SoC levels indicate ample energy storage capacity, ensuring prolonged operational durations and efficient utilization of resources, which aligns with the system's goal of sustained functionality. Secondly, by considering SoC across different operating conditions (high, normal, and low), the model accounts for variations in energy demand and supply dynamics, enabling adaptive responses to diverse usage scenarios and enhancing system resilience.

Additionally, the inclusion of battery power and fuel power as criteria in the AHP-TOPSIS analysis reflects the system's reliance on power generation and distribution mechanisms. Battery power, denoting the capacity to deliver stored energy, and fuel power, representing auxiliary power sources, are pivotal in ensuring continuous operation and meeting performance requirements. The emphasis on these parameters underscores the system's emphasis on reliable power delivery and adaptability to fluctuating energy demands. Thus, trial number 12's robust performance across all evaluated criteria, including battery SoC and power outputs, epitomizes the holistic approach of AHP-TOPSIS in optimizing battery management system operations while considering diverse technical constraints and operational scenarios.

The results obtained from the test align with the computational findings for high State of Charge (SoC). To reduce the battery SoC to the standard level, the power generated by the Energy Management System (EMS) is less than the required load power. The standard SoC, found in the third column, closely matches the power of the load requirement, maintaining the battery SoC at the standard level. The last column represents Low SoC, utilized by the EMS to generate energy from the fuel cell, charging the

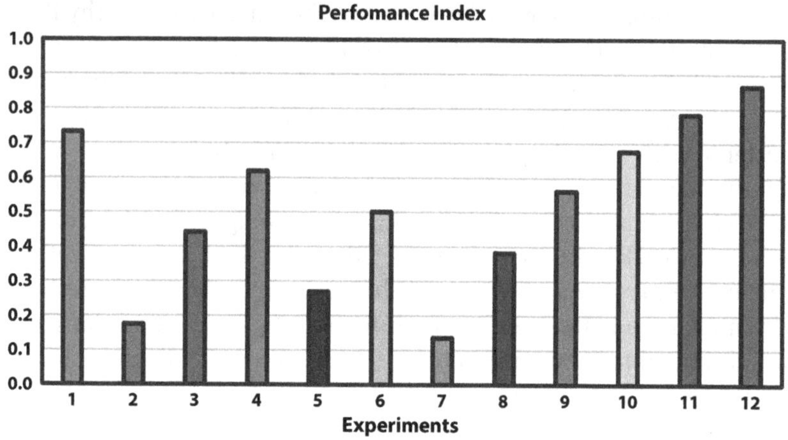

Figure 16.6 Performance score for all trials.

battery and supplying the load. In this configuration, the battery SoC is set to 90%, and the system load is increased from 0 to 8 kW. The data obtained in this scenario, will be compared with the simulation results for similar SoCs, as depicted in Figure 16.6.

The depicted figures clearly illustrate that fuel power initially rises with an increase in high battery levels, only to sharply decline with any further increments. A comparable pattern is observed for normal battery conditions, with a notable sharp ascent followed by a rapid descent. In cases of low battery usage, fuel power experiences a pronounced decrease with an increase in low battery levels. Optimal battery power is achieved during neutral section charging of the battery, signifying an absence of overcharging throughout the process.

16.5 Conclusions and Future Scope

16.5.1 Conclusions

The comprehensive analysis conducted using the AHP-TOPSIS methodology has yielded valuable insights into the optimal selection of trials within the battery management system. By systematically weighing the importance of criteria such as battery state of charge (SoC) across varying operating conditions and power outputs, the study has elucidated the critical factors driving system performance and efficiency. The identification of trial number 12 as the best-performing option underscores the significance of maintaining high SoC levels, both in normal and high operating conditions, while also ensuring substantial battery and fuel power outputs. This finding suggests that trial 12 exhibits superior operational characteristics, making it well-suited for fulfilling the demands of the battery management system across diverse usage scenarios.

Furthermore, the outcomes of this study underscore the importance of employing a structured decision-making framework like AHP-TOPSIS in complex systems such as battery management. By integrating multi-criteria evaluation techniques with systematic ranking methodologies, AHP-TOPSIS provides a robust approach for prioritizing trials based on their adherence to predetermined performance criteria. The ability to account for the relative importance of each criterion and assess trial performance against ideal and negative ideal solutions enhances the decision-making process, enabling stakeholders to make informed choices that optimize system functionality and resource utilization. Ultimately, the findings of this study not only contribute to enhancing the efficiency and reliability of battery management systems but also underscore the broader applicability of AHP-TOPSIS in facilitating data-driven decision-making across diverse domains.

16.5.2 Future Scope

The future scope for the modeling and optimization of a hybrid electric marine engine battery management system (BMS) using a hybrid Multi-Criteria Decision-Making (MCDM) process is vast and multifaceted: The future scope for the modeling and optimization of a hybrid electric marine engine battery management system (BMS) using a hybrid Multi-Criteria Decision-Making (MCDM) process is vast and multifaceted:

- Integration with Advanced AI Techniques: Further studies may relate to the need to combine AI and machine learning algorithms with hybrid MCDM processes to improve the BMS' predictive capability and flexibility. This would allow the decision maker to be making decisions in real-time, to adapt to the changing marine operating environment.
- Scalability to Large Vessels: Developing as it were the current models, to handle bigger ships, which would for sure have heavier demands in power, could be perhaps a study area of interest. This would include expansion of the BMS in order to accommodate larger energy storage capabilities and advanced load control mechanisms.

- Sustainability and Environmental Impact: Future research could seek to explore the integration of environmental effects into the MCDM process and thus lay the foundation for an efficient BMS that meets global sustainability targets. This might entail lifecycle of batteries, for instance lithium batteries and their application to renewable energy systems.
- Integration with Autonomous Marine Systems: With the Increase in the use of autonomous marine ships, modifications are required to be made to the BMS that will work in unison with them. Possible research in the future is to develop additional improvements of the BMS in terms of stability and safety while driving in full autonomy.
- Energy Recovery and Efficiency Improvements: Another area of study could be identified in how the energy recovery systems can be integrated with the BMS such as the regenerative braking in the case of marine vessels. This could further enhance the global system efficiency and battery life for range extender solutions.
- Regulatory and Safety Compliance: In future research, say with advanced acts of marine emissions and safety control, studies can be executed to ensure that the BMS designs meet and incorporate the changing standards so that more and more people accept it with minimum regulatory approval.

References

1. Ahamad, T., Parvez, M., Lal, S., Khan, O., Idrisi, M.J., 4-E analysis and multiple objective optimizations of a novel solar-powered cogeneration energy system for the simultaneous production of electrical power and heating. *Sci. Rep.*, 13, 1, 22246, 2023.
2. Bhatt, B.K., Akram, W., Khan, O., Parvez, M., Ahmad, S., A Methodology for Multi-criteria Assessment of Chiller Plants for Building Setup Utilizing Waste Heat and Solar Energy. *Sustain. Anal. Model.*, 3, 100022, 2023.
3. Seraj, M.M., Khan, O., Khan, M.Z., Parvez, M., Bhatt, B.K., Ullah, A., Alam, M.T., Analytical Research of Artificial Intelligent Models for Machining Industry Under Varying Environmental Strategies: An Industry 4.0 Approach. *Sustain. Oper. Comput.*, 3, 176–187, 2022.
4. Ahmad, S., Parvez, M., Khan, T.A., Siddiqui, S.A., Khan., O., Performance comparison of solar powered cogeneration and trigeneration systems *via* energy and exergy analyses. *Int. J. Exergy*, 39, 4, 395–409, 2022.

5. Akram, W., Parvez, M., Rafi, K.M., Khan, O., First and Second Law Analyses of an Organic Rankine Cycle to Recover the Waste Heat from the Solar-Operated Combined Cycle Power Plants with Grid, in: *International Conference on Nanotechnology: Opportunities and Challenges*, Springer Nature Singapore, Singapore, pp. 369–376, 2022.
6. Fatima, M., Sherwani, N.U.K., Khan, S., Khan, M.Z., Assessing and predicting operation variables for doctors employing industry 4.0 in health care industry using an adaptive neuro-fuzzy inference system (ANFIS) approach. *Sustain. Oper. Comput.*, 3, 286–295, 2022.
7. Ahmad, S., Alam, M.T., Bilal, M., Khan, O., Khan, M.Z., Analytical Modelling of HVAC-IoT Systems with the Aid of UVGI and Solar Energy Harvesting, in: *Energy Harvesting*, pp. 65–80, 2022.
8. Fatima, M., Sherwani, N.U.K., Singh, V., Comparative analysis among doctors working in private and government hospitals in identifying and prioritizing essential stress factors during COVID-19- an AHP-TOPSIS approach. *Intell. Pharm.*, 1, 1, 17–25, 2023, https://doi.org/10.1016/j.ipha.2023.04.005.
9. Howari, H., Parvez, M., Khan, O., Alhodaib, A., Mallah, A., Yahya, Z., Multi-Objective Optimization for Ranking Waste Biomass Materials Based on Performance and Emission Parameters in a Pyrolysis Process—An AHP–TOPSIS Approach. *Sustainability*, 15, 4, 3690, 2023, https://doi.org/10.3390/su15043690.
10. Islam, S.M.U., Khan, S., Ahmad, H., Rahman, M.A.U., Tomar, S., Khan, M.Z., Assessment of challenges and problems in supply chain among retailers during COVID-19 epidemic through AHP-TOPSIS hybrid MCDM technique. *Internet Things Cyber-Phys. Syst.*, 2, 180–193, 2022, https://doi.org/10.1016/j.iotcps.2022.10.001.
11. Khan, F., Karimi, M.N., Khan, O., Exploring the scalability and commercial viability of biosynthesized nanoparticles for cooling panels with the help of artificial intelligence and solar energy systems. *Green Technol. Sustain.*, 1, 3, 100036, 2023.
12. Fatima, M., Sherwani, N.U.K., Singh, V., Mindful medicine: Exploring the benefits of mindfulness-based stress reduction for doctors using statistical techniques. *Intell. Pharm.*, 1, 1, 41–47, 2023, https://doi.org/10.1016/j.ipha.2023.04.009.
13. Prakash, V., Saraswat, V., Dewangan, A., Yadav, A. K., Kumar, A., Production and utilization of hydrogen as a future fuel for electric vehicles: A review. *Clean Energy*, 259–281, CRC Press, 2024.
14. Ahmad, S., Parvez, M., Khan, T.A., Khan, O., A Hybrid Approach Using AHP–TOPSIS Methods for Ranking of Soft Computing Techniques Based on their Attributes for Prediction of Solar Radiation. *Environ. Challenges*, 9, 100634, 2022, https://doi.org/10.1016/j.envc.2022.100634.
15. Khan, O., Khan, M.E., Parvez, M., Ahmed, K.A.A.R., Ahmad., I., Extraction and Experimentation of Biodiesel Produced from Leachate Oils of Landfills Coupled with Nano-Additives Aluminium Oxide and Copper Oxide on

Diesel Engine, in: *Nanomaterials for Innovative Energy Systems and Devices*, pp. 319–332, 2022.

16. Nasim, M.A.G., Khan, O., Parvez, M., Bhatt, B.K., Optimizing ultrasonic reactor operating variables using intelligent soft computing models for increased biodiesel production. *Green Technol. Sustain.*, 1, 100033, 2023, https://doi.org/10.1016/j.grets.2023.100033.
17. Khan, O., Khan, M.Z., Khan, M.E., Goyal, A., Bhatt, B.K., Khan, A., Parvez, M., Experimental analysis of solar powered disinfection tunnel mist spray system for coronavirus prevention in public and remote places. *Mater. Today: Proc.*, 46, 15, 6852–6858, 2021.
18. Khan, S. and Tomar, S., Analysis of Factors Influencing Entrepreneurial Intention among Indian Women in Computing Industry- An NVIVO Approach. *J. Ind. Integr. Manag. Innov. Entrepreneurship*, 1–21, 2023.
19. Meraj, M., Khan, M.E., Tiwari, G.N., Khan, O., Optimization of Electrical Power of Solar Cell of Photovoltaic Module for a Given Peak Power and Photovoltaic Module Area. Advances in Fluid and Thermal Engineering, pp. 417–433, 2019.
20. Khan, O., Khan, M.Z., Alam, M.T., *et al.*, Comparative Study of Soft Computing and Metaheuristic Models in Developing Reduced Exhaust Emission Characteristics for Diesel Engine Fueled with Various Blends of Biodiesel and Metallic Nanoadditive Mixtures: An ANFIS–GA–HSA Approach. *ACS Omega*, 8, 7344–7367, 2023, https://doi.org/10.1021/acsomega.2c05246.
21. Parvez, M., Khalid, F., Khan, O., Thermodynamic performance assessment of solar-based combined power and absorption refrigeration cycle. *Int. J. Exergy*, 31, 3, 232–248, 2020.
22. Parvez, M., Khan, M.E., Khalid, F., Khan, O., Akram, W., A Novel Energy and Exergy Assessments of Solar Operated Combined Power and Absorption Refrigeration Cogeneration Cycle, in: *Electric Vehicles*, pp. 213–229, 2021.
23. Khan, O., Khan, M.Z., Khan, E., Bhatt, B.K., Afzal, A., Ağbulut, U., Shaik, S., An enhancement in diesel engine performance, combustion, and emission attributes fueled with Eichhornia crassipes oil and copper oxide nanoparticles at different injection pressures. *Energy Sources Part A: Recovery Util. Environ. Eff.*, 44, 3, 6501–6522, 2022.
24. Singh, D., Singh, S., Yadav, A.K., Khan, O., Dewangan, A., Alkahtani, M.Q., Islam, S., From theory to practice: A sustainable solution to water scarcity by using a hybrid solar distiller with a heat exchanger and aluminum oxide nanoparticles. *ACS Omega*, 8, 37, 33543–33553, 2023.
25. Khan, O., Parvez, M., Kumari, P., Parvez, S., Ahmad, S., The future of pharmacy: How AI is revolutionizing the industry. *Intell. Pharm.*, 1, 1, 32–40, 2023, https://doi.org/10.1016/j.ipha.2023.04.008.
26. Nasim, M.A.G., Khan, O., Parvez, M., Chouhan, P.S., Maximizing biodiesel production: An intelligent software-driven optimization of ultrasonic

reactor operating variables. *Bioresour. Technol. Rep.*, 1, 101474, 2023, https://doi.org/10.1016/j.biteb.2023.101474.
27. Ashraf, F., Shajar, M., Khan, M. E., Equbal, A., Khan, O., Yadav, A. K., Parvez, M., Comparative evaluation of carbon emissions from various vehicles: Integrating AHP techniques for ranking, in: *Clean Energy*, pp. 282–303, CRC Press, 2024.
28. Parvez, M. and Khan, O., Parametric simulation of biomass integrated gasification combined cycle (BIGCC) power plant using three different biomass materials. *Biomass Convers. Biorefin.*, 10, 4, 803–812, 2020.
29. Parvez, M., Khan, O., Howari, H., Investigation of (3E) energy, exergy, and environmental analysis of syngas production for combined power and cooling plant by testing six different biomasses. *Biomass Convers. Biorefin.*, 14, 1–15, 2023.
30. Patra, B., Nema, P., Khan, M.Z., Khan, O., Optimization of solar energy using MPPT techniques and industry 4.0 modelling. *Sustain. Oper. Comput.*, 4, 22–28, 2023, https://doi.org/10.1016/j.susoc.2022.10.001.
31. Seraj, M., Parvez, M., Ahmad, S., Khan, O., Sustainable energy transition and decision-making for enhancing the performance of building equipment in diverse climatic conditions. *Green Technol. Sustain.*, 1, 3, 100043, 2023.
32. Ahmed, A. and Yadav, A.K., Parametric Analysis of Wastewater Electrolysis for Green Hydrogen Production: A Combined RSM, Genetic Algorithm, and Particle Swarm Optimization Approach. *Int. J. Hydrogen Energy*, 59, 51–62, 2024.
33. Zeeshan, M., Alam, M.R., Khan, O., Parvez, M., Azad, A.S., Thermal load calculation of building envelope set up with the help of a drone setup. *2021 First International Conference on Advances in Computing and Future Communication Technologies (ICACFCT)*, IEEE, pp. 186–190, 2021.
34. Ahmad, A., Yadav, A.K., Hasan, S., Enhanced production of methane enriched biogas through intensified co-digestion process and its effective utilization in a biodiesel/biohydrogen fueled engine with duel injection strategies: ML-RSM based an efficient optimization approach. *Int. J. Hydrogen Energy*, 65, 671–686, 2024.
35. Ahmed, A., Yadav, A.K., Singh, A., Singh, D.K., A machine learning–genetic algorithm based models for optimization of intensification process through microwave reactor: A new approach for rapid and sustainable synthesis of biodiesel from novel Hiptage benghalensis seed oil. *Fuel*, 374, 132470, 2024.
36. Khan, O., Khan, M.P., Yadav, A.K., A Comparative Analysis of Various Nanocomposites for the Remediation of Heavy Metals from Biomass Liquid Digestate using Multi-Criteria Decision Methods. *Biomass Bioenergy*, 186, 107281, 2024.
37. Ahmed, A., Yadav, A.K., Singh, A., Singh, D.K., A machine learning-response surface optimization approach to enhance the performance of diesel engine using novel blends of aloe vera biodiesel with MWCNT nanoparticles

and hydrogen. *Process Saf. Environ. Prot.*, 18, 738–755, 2024, doi: 10.1016/j.psep.2024.04.013.

38. Khan, O., Alsaduni, I., Equbal, A., Parvez, M., Yadav, A.K., Performance and Emission Analysis of Biodiesel Blends Enriched with Biohydrogen and Biogas in Internal Combustion Engines. *Process Saf. Environ. Prot.*, 183, 1013–1037, 2024.
39. Ahmed, A., Yadav, A.K., Singh, D.K., Singh, A., A Comprehensive Machine Learning-Coupled Response Surface Methodology Approach for Predictive Modeling and Optimization of Biogas Potential in Anaerobic Co-Digestion of Organic Waste. *Biomass Bioenergy*, 180, 106995, 2024.
40. Das, D., Yadav, A.K., Ahmed, A., Numerical Simulation of the Combustion Chamber and Performance Analysis of Microalgae-Based Biodiesel/Diesel-Powered CI Engine: An Axisymmetric Flow Approach. *J. Process Mech. Eng.: Part E*, 2023, (in press), https://doi.org/10.1177/09544089231218121.
41. Ahmad, A., Yadav, A.K., Singh, A., Singh, D.K., Ağbulut, U., A hybrid RSM-GA-PSO approach on optimization of process intensification of linseed biodiesel synthesis using an ultrasonic reactor: Enhancing biodiesel properties and engine characteristics with ternary fuel blends. *Energy*, 288, 129077, 2024.
42. Khan, O., Parvez, M., Yahya, Z., Alhodaib, A., Yadav, A.K., Hoang, A.T., Ağbulut, U., Waste-to-energy power plants: Multi-objective analysis and optimization of landfill heat and methane gas by recirculation of leachate. *Process Saf. Environ. Prot.*, 186, 957–968, 2024.
43. Ahmed, A., Yadav, A.K., Singh, A., Enhancing waste cooking oil biodiesel yield and characteristics through machine learning , response surface methodology , and genetic algorithm for optimal utilization in CI Engines. *Int. J. Green Energy*, 21, 1345–1365, 2024.
44. Singh, A.A., Singh, R.K., Yadav, A.K., Saha, A., Dewangan, A., Singh, B., Experimental investigations on the electrochemical machining of D3 die steel material and multi-objective parameter optimization. *Surf. Rev. Lett.*, 31, 2450074, 2024.
45. Ahmed, A., Yadav, A.K., Hasan, S., Biogas as a Sustainable and Viable Alternative Fuel for Diesel Engines: A Comprehensive Review of Production, Purification, Economic Analysis, and Performance Evaluation. *J. Process Mech. Eng. Part E*, 2024, https://doi.org/10.1177/09544089241255930.
46. Khan, F., Karimi, M.N., Khan, O., Yadav, A.K., Alhodaib, A., Gürel, A.E., Ağbulut, U., A hybrid MCDM optimization for utilization of novel set of biosynthesized nanofluids on thermal performance for solar thermal collectors. *Int. J. Thermofluids*, 22, 100686, 2024.
47. Alhodaib, A., Yahya, Z., Khan, O., Equbal, A., Equbal, M.S., Parvez, M., Idrisi, M.J., Sustainable coatings for green solar photovoltaic cells: performance and environmental impact of recyclable biomass digestate polymers. *Sci. Rep.*, 14, 1, 11221, 2024.

48. Ahmed, A., Yadav, A.K., Hasan, S., Enhancing Efficiency and Environmental Impacts of a Nano-Enhanced Oleander Biodiesel-Biohydrogen Dual Fuel Engine Equipped with EGR, through Operational Parameter Optimization using RSM-MOGA Technique. *Int. J. Hydrogen Energy*, 78, 1157–1172, 2024.
49. Singh, B., Yadav, A.K., Saha, A., Fabrication and Characterization of AA6082-T6 Produced by Friction Stir Additive Manufacturing. *Surf. Rev. Lett.*, 2024, DOI:10.1142/S0218625X25500222.
50. Equbal, A., Equbal, A., Ahmad, R., Kumar, M., Khan, O., Yadav, A.K., RSM based statistical approach to enhance the compressive performance of ABS P400 parts fabricated *via* FDM Technology. *J. Surf. Rev. Lett.*, 2024, DOI:10.1142/S0218625X25500210.
51. Dewangan, A., Mallick, A., Yadav, A.K., Islam, S., Saleel, C.A., Shaik, S., Ağbulut, U., Production of oxy-hydrogen gas and the impact of its usability on CI engine combustion, performance, and emission behaviors. *Energy*, 278, 127937, 2023.
52. Khan, O., Parvez, M., Kumari, P., Yadav, A.K., Akram, W., Ahmad, S., Idrisi, M.J., Modelling of compression ignition engine by soft computing techniques (ANFIS-NSGA-II and RSM) to enhance the performance characteristics for leachate blends with nano-additives. *Sci. Rep.*, 13, 1, 15429, 2023.
53. Ahmad, A., Yadav, A.K., Singh, A., Singh, D.K., Multi-response optimization of a microalgae-spirulina-fueled VCR diesel engine: a comprehensive RSM-GA approach, in: *Environment, Development and Sustainability*, pp. 1–29, 2023.
54. Ahmed, A., Yadav, A.K., Singh, A., Process optimization of spirulina microalgae biodiesel synthesis using RSM coupled GA technique: a performance study of a biogas powered dual fuel engine. *Int. J. Environ. Sci. Technol.*, 21, 169–188, 2023.
55. Khan, O., Ali, V., Parvez, M., Alhodaib, A., Yahya, Z., Yadav, A.K., Ağbulut, U., Exploring the performance of biodiesel-hydrogen blends with diverse nanoparticles in diesel engine: A hybrid machine learning K-means clustering approach with weighted performance metrics. *Int. J. Hydrogen Energy*, 78, 547–563, 2024.
56. Ahmed, A., Yadav, A.K., Singh, A., Application of machine learning and genetic algorithm for prediction and optimization of Biodiesel Yield from Waste Cooking Oil. *Korean J. Chem. Eng.*, 40, 2941–2956, 2023.
57. Ahmed, A., Yadav, A.K., Singh, A., Enhancement of Biogas Yield from Dual Organic Waste Using Hybrid Statistical Approach and Its Effects on Ternary Fuel Blend (Biodiesel/n-butanol /Diesel) Powered Diesel Engine. *Environ. Prog. Sustain. Energy*, 42, 5, e14163, 2023, doi.org/10.1002/ep.14163.
58. Dewangan, A., Ahmad, A., Yadav, A.K., Innovative research on waste tire recycling for sustainable biofuel production: Assessment of its usability on multi-cylinder diesel engine employing constant injection of oxyhydrogen and biogas through a premixing device. *Int. J. Hydrogen Energy*, 84, 155–163, 2024.

59. Ahmed, A., Yadav, A.K., Singh, A., Optimization of Cavitation-Assisted Biodiesel Production and Fuel Properties from Neochloris Oleoabundans Microalgae oil using Genetic Algorithm and Response Surface Methodology. *J. Process Mech. Eng. Part E*, 238, 4, 1821–1835, 2023, doi.org/10.1177/09544089231159832.
60. Yadav, M., Karimi, M.N., Yadav, A.K., Effect of Graphene Oxide Nanoparticles Dispersed Biodiesel on Combustion, Performance, and Emission Characteristics of a Diesel Engine. *J. Process Mech. Eng.: Part E*, 2023, (in press), https://doi.org/10.1177/09544089231209139.
61. Dewangan, A., Mallick, A., Yadav, A.K., Ahmad, A., Alqahtani, D., Islam., S., Combined Effect of Operating Parameters and Nano Particles on Performance of a Diesel Engine: Response Surface Methodology coupled Genetic algorithm Approach. *ACS Omega*, 8, 27, 24586–24600, 2023.
62. Ahmed, A., Yadav, A.K., Singh, A., Optimization of Biogas Yield from Anaerobic Co-digestion of Dual waste for environmental sustainability: ANN, RSM, and GA, Approach. *Int. J. Oil Gas Coal Technol.*, 33, 1, 75, 2023.
63. Dwivedi, S.P., Yadav, A.K., Saxena, A., Dwivedi, V.K., Tribo-mechanical, physical and thermal behaviour of Al/Si3N4 composite with and without the addition of Cu, Ni and Cr entropy elements. *Proc. Inst. Mech. Eng., Part E: J. Process Mech. Eng.*, 2023, 09544089231189663.
64. Ahmed, A., Yadav, A.K., Singh, A., An environmental impact assessment and optimization study of biodiesel production from microalgae. *Int. J. Global Warming*, 31, 3, 294–313, 2023.
65. Ahmed, A., Yadav, A.K., Singh, A., Biodiesel yield optimization from a third-generation feedstock (microalgae spirulina) using a hybrid statistical approach. *Int. J. Ambient Energy*, 44, 1, 1202–1213, 2023.
66. Dubey, A.M., Kumar, A., Yadav, A.K., Wear behaviour of friction stir weld joint of cast Al (4–10%) Cu alloy welded at different operating parameters. *J. Mater. Process. Technol.*, 240, 87–97, 2017.
67. Ahmed, A., Yadav, A.K., Singh, A., Modelling and optimisation of VCR diesel engine parameters fuelled with microalgae methyl-ester: a RSM-coupled Taguchi approach. *Int. J. Ambient Energy*, 44, 1, 2498–2506, 2023.
68. Yadav, A.K., Vinay, Singh, B., Optimization of Biodiesel Production from Annona Squamosa Seeds Oil Using Response Surface Methodology and Its Characterization. *Energy Sources Part A*, 40, 9, 1051–1059, 2018.
69. Ahmed, A., Yadav, A.K., Singh, A., Biodiesel Production from Mahua Oil: Characterization, Optimization, and Modeling with a Hybrid Statistical Approach. *Int. J. Ambient Energy*, 44, 1, 2600–2609, 2023.
70. Khan, T.A., Khan, T.A., Yadav, A.K., Khan, M.E., Experimental Studies on Utilization of Neochloris Oleoabundans Microalgae Biodiesel as an Alternative Fuel for Diesel Engine. *Int. J. Ambient Energy*, 44, 1, 115–123, 2023.
71. Singh, D., Yadav, A.K., Kumar, A., Samsher, Energy matrices and life cycle conversion analysis of N-identical hybrid double slope solar distiller unit

using Al_2O_3 nanoparticle. *J. Water Environ. Nanotechnol.*, 8, 3, 267–284, 2023.
72. Khan, T.A., Yadav, A.K., Khan, T.A., A hydrodynamic cavitation-assisted system for optimization of biodiesel production from green microalgae oil using a genetic algorithm and response surface methodology approach. *Environ. Sci. Pollut. Res.*, 29, 49465–49477, 2022.
73. Dewangan, A., Yadav, A.K., Mallick, A., Singh, S., Comparative study of Manilkara Zapota and Karanja based Biodiesel Properties and Its Effect on Multi-Cylinder Diesel Engine Characteristics. *Energy Sources Part A*, 44, 2, 5143–5153, 2022.
74. Yadav, A.K., Khan, T.A., Khan, T.A., Kumar, S., Methyl ester of Gmelina arborea oil as a substitute for petroleum diesel: an experimental study on its performance and emissions in a diesel engine. *Energy Sources Part A*, 43, 11, 1307–1314, 2021.
75. Khan, O., Yadav, A.K., Khan, M.E., Parvez, M., Characterization of Bio-Ethanol Obtained from Eichhornia Crassipes Plant; Its Emission and Performance Analysis on CI Engine. *Energy Sources Part A*, 43, 14, 1793–1803, 2021.
76. Khan, I.A., Singh, S.K., Pal, A., Yadav, A.K., Efficient Production of Biodiesel from Cannabis Sativa oil using Intensified Transesterication (Hydrodynamic Cavitation) Method. *Energy Sources Part A*, 42, 20, 2461–2470, 2020.
77. Dewangan, A., Yadav, A.K., Mallick, A., Tyagi, A., Optimisation of biodiesel production and engine performance from Simarouba oil in compression ignition engine. *Int. J. Oil Gas Coal Technol.*, 25, 3, 357–376, 2020.
78. Khan, M.E., Khan, T.A., Yadav, A.K., Pal, A., A comprehensive review on performance, emission and combustion characteristics of diesel engines fuelled with algae oil and methyl ester. *WEENTECH Proc. Energy*, 6, 2, 160–168, 2020.
79. Yadav, A.K., Khan, M.E., Pal, A., Custard Apple Seed (Annona Squamosa) Oil as a Promising Feedstock for Biodiesel Production using Advanced Techniques and Its Experimental Investigations on an Agricultural Diesel Engine. *Int. J. Oil Gas Coal Technol.*, 20, 4, 473–492, 2019.
80. Yadav, A.K., Pal, A., Ghosh, U., Gupta, S.K., Comparative Study of Biodiesel Production Methods from Yellow Oleander Oil and its Performance Analysis on an Agricultural Diesel Engine. *Int. J. Ambient Energy*, 40, 2, 152–157, 2019.
81. Khan, I.A., Singh, S.K., Yadav, A.K., Sharma, D., Enhancement in the Performance of a Diesel Engine Fueled with Pongamia Methyl Ester and n-Butanol as Oxygenated Additive. *Int. J. Ambient Energy*, 40, 8, 842–846, 2019.
82. Yadav, A.K., Khan, O., Khan, M.E., Utilization of High FFA Landfill Waste (Leachates) as a Feedstock for Sustainable Biodiesel Production: Its Characterization and Engine Performance Evaluation. *Environ. Sci. Pollut. Res.*, 25, 32, 32312–32320, 2018.

83. Yadav, A.K., Dewangan, A., Mallick, A., Effect of n-Butanol and Diethyl Ether Additives on Performance and Emission Characteristics of a Diesel Engine Fuelled with Diesel-Pongamia Biodiesel Blends. *J. Energy Eng.*, 144, 6, 04018062, 2018.
84. Vinay, Singh, B., Yadav, A.K., Optimization of Performance and Emission Characteristics of CI Engine Fuelled with Mahua Oil Methyl Ester-Diesel Blend using Response Surface Methodology. *Int. J. Ambient Energy*, 44, 6, 674–85, 2020.
85. Yadav, A.K., Dewangan, A., Mallick, A., Synthesis and Stability Study of Biodiesel from Kachnar Seed Oil. *J. Energy Eng.*, 144, 5, 04018053, 2018.
86. Yadav, A.K., Pal, A., Dubey, A.M., Experimental Studies on Utilization of Prunus Armeniaca L. (Wild Apricot) Biodiesel as an Alternative Fuel for CI Engine. *Waste Biomass Valorization*, 9, 10, 1961–1969, 2018.
87. Yadav, A.K., Khan, M.E., Pal, A., Ultrasonic Assisted Production of Biodiesel from Karabi Oil using Heterogeneous Catalyst. *Biofuels*, 9, 1, 101–112, 2018.
88. Dewangan, A., Yadav, A.K., Mallick, A., Current Scenario of Biodiesel Development in India: Prospects and Challenges. *Energy Sources Part A*, 40, 20, 2494–2501, 2018.
89. Yadav, A.K., Khan, M.E., Pal, A., Performance and Emission Characteristics of a Stationary Diesel Engine fuelled by Schleichera Oleosa oil Methyl Ester (SOME) Produced through Hydrodynamic Cavitation Process. *Egypt. J. Pet.*, 27, 1, 89–93, 2018.
90. Yadav, A.K., Khan, M.E., Pal, A., Performance, Emission and Combustion Characteristics of an Indica Diesel Engine Operated with Yellow Oleander (Thevetia Peruviana) Oil Biodiesel Produced through Hydrodynamic Cavitation Method. *Int. J. Ambient Energy*, 39, 4, 365–371, 2018.
91. Dewangan, A., Mallick, A., Yadav, A.K., Effect of Metal Oxide Nanoparticles and Engine Parameters on the Performance of a Diesel Engine: A Review. *Mater. Today: Proc.*, 21, 1722–27, 2020.
92. Yadav, A.K., Khan, M.E., Pal, A., Dubey, A.M., Experimental Investigations of Performance and Emissions Characteristics of Kusum (Schleichera Oleosa) Biodiesel in a Multi-Cylinder Transportation Diesel Engine. *Waste Biomass Valoriz.*, 8, 4, 1331, 2017.
93. Dubey, A.M., Ram, R., Yadav, A.K., Ion cell performance using single particle representation of battery electrode. *J. Electroanal. Chem.*, 768, 11–17, 2016.
94. Yadav, A.K., Khan, M.E., Pal, A., Biodiesel Production from Oleander (Thevetia Peruviana) Oil and Its Performance Testing on a Diesel Engine. *Korean J. Chem. Eng.*, 34, 2, 340–345, 2017.
95. Khan, O., Khan, M.E., Yadav, A.K., Sharma, D., The Ultrasonic Assisted Optimization of Biodiesel Production from Eucalyptus Oil. *Energy Sources Part A*, 39, 13, 1323–1331, 2017.
96. Dewangan, A. and Yadav, A.K., Wax Deposition during Production of Waxy Crude Oil and Its Remediation. *Pet. Sci. Technol.*, 35, 18, 1831–1838, 2017.

97. Yadav, A.K., Khan, M.E., Pal, A., Kaner biodiesel production through hybrid reactor and its performance testing on a CI engine at different compression ratios. *Egypt. J. Pet.*, 26, 2, 525–532, 2017.
98. Yadav, A.K., Khan, M.E., Pal, A., Biodiesel Production from Nerium Oleander (Thevetia Peruviana) Oil through Conventional and Ultrasonic Irradiation Methods. *Energy Sources Part A*, 38, 23, 3447–3452, 2016.
99. Sharma, G., Dewangan, A.K., Yadav, A.K., Ahmed, A., Current status of research on hydrogen generation, storage and transportation technologies: A state-of-the-art reviewtowards sustainable energy. *Process Saf. Environ. Prot.*, 191, Part A, 1445–1460, 2024.
100. Yadav, A.K., Khan, M.E., Pal A., A., Comparative Study on Ultrasonic Cavitation and Mechanical Stirring Method towards Efficient Production of Biodiesel from Non-Edible Oils and Performance Testing on a C.I. Engine. *Int. J. Environ. Waste Manag.*, 18, 4, 349–367, 2016.
101. Dewangan, A., Mallick, A., Yadav, A.K., Kumar, R., Combustion-generated pollutions and strategy for its control in CI engines: A review. *Mater. Today: Proc.*, 21, 1728–1733, 2020.
102. Yadav, A.K., Khan, M.E., Pal, A., Sharma, D., Optimization of biodiesel production from bitter groundnut oil using Taguchi method and its performance and emissions characteristics on a 4-cylinder Tata Indica engine. *Int. J. Sustain. Agric. Manag. Inform.*, 1, 285–300, 2015.
103. Khan, F., Khan, O., Parvez, M., Ahmad, S., Yahya, Z., Alhodaib, A., Yadav, A.K., Ağbulut, U., K-means clustering optimization of various quantum dots and nanoparticles-added biofuels for engine performance, emission, vibration, and noise characteristics. *Therm. Sci. Eng. Prog.*, 54, 102815, 2024.
104. Yadav, A.K., Khan, M.E., Pal, A., Dubey, A.M., Performance and emission characteristics of a transportation diesel engine operated with non-edible vegetable oils biodiesel. *Case Stud. Therm. Eng.*, 8, 236–244, 2016.
105. Khan, O., Parvez, M., Yadav, A.K., *et al.*, Investigation of the Thermal Properties of Lauric Acid Melting as a Phase Change Material Using Priority based K-Means Clustering Method. *Energy Storage*, 6, 6, e70026, 2024, doi:10.1002/est2.70026.
106. Singh, S., Singh, B., Kumar, S., Yadav, A.K., Temperature-dependent dynamic hysteresis scaling of ferroelectric hysteresis parameters of lead free [(Ba 0.825 + xCa0.175-x) (Ti1- x Snx)O3]ceramics. *Ferroelectrics*, 551, 1, 133–142, 2019.

17

A Comparative Analysis of Carbon Footprints in Hybrid, Diesel, and Petrol Vehicles: An MCDM Approach Using VIKOR and Entropy Methods

Mohammad Seraj[1]*, Mohammad Ashad Ghani Nasim[1], Piyush Bhatnagar[2], Azhar Equbal[2], Osama Khan[2] and Ashok Kumar Yadav[3]

[1]*Department of Mechanical Engineering, Al Falah University, Faridabad, Haryana, India*
[2]*Department of Mechanical Engineering, Jamia Millia Islamia, New Delhi, India*
[3]*Department of Mechanical Engineering, Symbiosis Institute of Technology, Symbiosis International (Deemed University), Pune, India*

Abstract

In this research, a comparative analysis of three distinct vehicles: hybrid, diesel, and petrol are conducted based on their emissions of CO_2, CO and UBHC. The comparison encompasses various operational parameters such as load, efficiency losses, and torque. Employing the VIKOR and entropy methods, this study evaluates the environmental performance of these vehicles under different conditions. The experimental findings revealed that the hybrid vehicle exhibits the most favorable emissions profile, characterized by low CO_2, CO and UBHC levels. Consequently, it emerges as the vehicle with the lowest carbon footprint, followed by diesel and petrol counterparts. These results underscore the environmental advantages of hybrid vehicles over conventional diesel and petrol models. In conclusion, this study underscores the significance of considering emissions criteria

**Corresponding author*: seraj.hvac@gmail.com; ORCID: 0009-0006-3624-4324
Mohammad Ashad Ghani Nasim: ORCID: 0000-0001-9615-4407
Azhar Equbal: ORCID: 0000-0002-7725-1356
Osama Khan: ORCID: 0000-0002-8094-7573
Ashok Kumar Yadav: ORCID: 0000-0002-7350-7641

Ashwani Kumar, Mukesh Kumar Awasthi, Nitesh Dutt, Yogesh Kumar Singla and Sivasakthivel Thangavel (eds.) Thermal Battery Management System for Hybrid and Electric Vehicles, (337–366) © 2025 Scrivener Publishing LLC

in vehicle selection and highlights the potential of hybrid technology in mitigating environmental impacts associated with transportation.

***Keywords*:** Hybrid vehicle, carbon footprint, fuels, energy efficiency, ranking

17.1 Introduction

In recent decades, global carbon emissions from transportation have surged, contributing to approximately 14% of total greenhouse gas emissions worldwide [1, 2]. Petrol and diesel vehicles remain major contributors, with over 1 billion cars on the road globally, emitting an estimated 9.3 billion metric tons of CO_2 annually [3, 4]. Despite advancements in fuel efficiency, the transportation sector accounts for nearly one-fourth of global energy-related CO_2 emissions (International Energy Agency, 2021). Rising concerns about climate change and air pollution have intensified the search for sustainable transportation alternatives, driving interest in hybrid vehicles as a potential solution to mitigate carbon footprints and reduce environmental impact [5–10]. Hybrid vehicles present a compelling case for their superiority over traditional fuel-powered or fully electric motors because of their precise mixture of features. One of the number one benefits lies of their potential to significantly lessen fuel consumption and emissions in comparison to traditional fuel cars. By integrating each inner combustion engines and electric powered vehicles, hybrids obtain greater gas efficiency, making them environmentally friendly alternatives [9, 11–19]. Additionally, hybrid cars provide versatility by means of seamlessly switching between strength resources, thereby extending their using range and lowering dependence on fossil fuels. Their regenerative braking structures similarly enhance efficiency through shooting energy for the duration of braking and converting it into usable strength [20–31]. Moreover, hybrids often boast quieter operation and smoother acceleration compared to standard automobiles, enhancing the general driving enjoy. Furthermore, the lower maintenance necessities of hybrid automobiles make contributions to decreased long-time period possession charges, making them financially tremendous [32–39]. With their evolving technology and growing infrastructure support, hybrid cars constitute a pivotal step toward sustainable transportation answers within the modern-day era. Hybrid engines stand out for his or her capacity to combination a couple of strength assets, harnessing the advantages of both internal combustion engines and electric automobiles to optimize gas efficiency and reduce emissions [40–49]. This unique configuration presents hybrid cars greater flexibility and adaptability as compared to conventional fuel or fully electric powered motors.

Hybrid vehicles face numerous demanding situations regardless of their promise of decreased gas intake and emissions. The complexity of hybrid structures, which combine internal combustion engines with electric powered motors, often ends in higher preservation prices and deters capacity customers because of their preliminary buy price [7, 50–59]. Battery degradation through the years poses another trouble, necessitating highly priced replacements and affecting performance. Limited electric variety calls for reliance on inner combustion engines for longer journeys, impacting fuel performance profits. Establishing a strong charging infrastructure for plug-in hybrids remains a project in certain areas [60–67]. The added weight of hybrid components influences automobile performance and managing, even as the reliability of regenerative braking structures may additionally vary. Uncertainty about lengthy-term reliability and renovation prices can also impact resale fee, influencing marketplace enchantment [68–77]. Addressing these demanding situations *via* technological advancements and infrastructure development is critical for maximizing the potential of hybrid vehicles in sustainable transportation. Automobiles can triumph over their inherent demanding situations and retain to evolve as sustainable transportation solutions for the future [78–83].

Incorporating mechatronics into hybrid vehicles considerably contributes to lowering carbon footprints *via* numerous key mechanisms [84–90]. The optimized manage and coordination enabled *via* mechatronics among the inner combustion engine and electric motor beautify gasoline performance and reduce emissions all through driving. Mechatronics additionally performs a critical position in enhancing regenerative braking systems by using intelligently taking pictures kinetic strength during braking and converting it into electrical power, lowering the want for conventional friction braking and thereby reducing energy wastage and carbon emissions [91–96]. Furthermore, mechatronic algorithms optimize energy control within hybrid automobiles with the aid of balancing electricity necessities and battery utilization in actual time, making sure efficient operation whilst decreasing carbon emissions. Predictive preservation facilitated *via* mechatronics detects capability issues early, scheduling upkeep proactively and ensuring peak performance, which in turn minimizes emissions associated with inefficient components [97]. Also, the implementation of adaptive drive modes, enabled by way of mechatronics, optimizes automobile performance based on riding situations, efficaciously lowering electricity intake and carbon emissions at the same time as keeping overall performance standards.

Furthermore, mechatronics integration notably complements hybrid vehicle performance and sustainability, making them a possible answer in efforts to mitigate environmental impact and reduce carbon footprints [98–101].

Zhao *et al.* [102] examine how PHEVs, integrated with AI, reduce life cycle costs and GHG emissions, addressing challenges like range limitations and pollution from energy sources, hinting at enhanced fuel efficiency through state of charge adjustments. Cao *et al.* [103] analyze HEVs, highlighting energy management and torque distribution for performance optimization, serving as a key resource for hybrid vehicle research. Limon *et al.* [104] emphasize the need to shift from fossil fuels for environmental preservation. They suggest exploring HEV adoption drivers in developing countries as a promising path for sustainable transportation and emission reduction. Liu *et al.* [105] examines the importance of PEM strategies in PHEVs for fuel consumption optimization with real-time data and advanced algorithms. Their study employs DP, PMP, and ECMS to assess the fuel consumption benefits compared to traditional EMS in diverse real-world scenarios. Pielecha *et al.* [106] evaluate car energy use and emissions, prioritizing urban impact for EU climate goals. Their findings rank electric vehicles as most efficient, followed by plug-in hybrids, promoting hybrid and electric tech for sustainability.

It's crucial to determine carbon-based emissions from hybrid electric vehicles to assess their environmental impact accurately. Understanding operating parameters is essential for optimizing efficiency, ensuring effective integration of electric and diesel engines to minimize emissions and enhance performance. The aim of the research is aimed to compare the carbon footprint of various vehicles with this hybrid, while also developing an optimization model to suffice environmental concerns and enhance efficiency in transportation. The novelty of the study lies in its comprehensive comparison of carbon footprints across vehicle types and the development of an optimization model tailored to address environmental concerns and enhance transportation efficiency. Based on the literature the objectives of the research are summarized as:

- Compare carbon footprints of hybrid electric vehicles with diesel and biodiesel vehicles.
- Develop an optimization model to minimize environmental impact and enhance efficiency.
- Assess the operating parameters influencing hybrid vehicle efficiency.
- Provide insights for policymakers and stakeholders to promote sustainable transportation practices.

17.2 Materials and Experimental Setup

17.2.1 Equipment and Material

Understanding the materials and equipment utilized in a hybrid vehicle engine is paramount due to its direct influence on performance, efficiency, and environmental impact. The selection of materials affects factors such as weight, durability, and energy consumption, thereby impacting the overall efficiency and lifespan of the vehicle. Moreover, choosing appropriate equipment ensures seamless integration of electric and combustion engine components, optimizing their synergy and enhancing overall performance. Additionally, a thorough understanding of materials and equipment facilitates the development of sustainable practices, enabling manufacturers to minimize emissions, reduce resource consumption, and meet regulatory standards. Therefore, comprehending the intricacies of these components is essential for advancing hybrid vehicle technology towards greater efficiency, sustainability, and environmental responsibility.

17.2.2 Material Selection

In the assessment of carbon emissions from a diesel engine, exhaust gas analyzers play a crucial role by providing real-time measurements of various pollutants released during combustion. Diesel engines emit carbon emissions primarily in the form of carbon dioxide (CO_2) and carbon monoxide (CO), among other pollutants. The exhaust gas analyzer is strategically positioned to analyze the composition of exhaust gases exiting the engine. By sampling the exhaust stream, the analyzer quantifies the concentration of CO_2 and CO present in the emissions, among other gases. Through this process, the diesel engine's combustion efficiency and emission levels are monitored and assessed. By accurately measuring the concentration of carbon emissions, the exhaust gas analyzer aids in evaluating the environmental impact of diesel engines and facilitates the development of strategies to mitigate carbon emissions for environmental sustainability and regulatory compliance.

In the estimation of carbon emissions from a petrol engine, the utilization of a gas analyzer is instrumental in providing real-time measurements of pollutants released during combustion. A petrol engine, when filled with petrol, undergoes combustion processes that generate emissions comprising carbon dioxide (CO_2), carbon monoxide (CO), and other pollutants. Gas analyzers are strategically positioned to sample and analyze the composition of exhaust gases emitted from the engine [107–113]. Through this analysis, the concentration of CO_2, CO, and other gases in the exhaust stream is quantified, facilitating the assessment of the engine's combustion efficiency and emission levels. By accurately measuring carbon emissions, gas analyzers aid in evaluating the environmental impact of petrol engines and play a crucial role in the development of strategies aimed at reducing emissions for environmental sustainability and regulatory compliance.

A hybrid engine, combining both electric battery systems and traditional petrol engines, operates by seamlessly switching between power sources based on driving conditions and energy demands [114–117]. During low-speed urban driving or when cruising, the electric battery system powers the vehicle, emitting minimal to zero emissions. In contrast, during high-speed or heavy-load scenarios, the petrol engine engages to provide additional power and support, typically recharging the battery system in the process. This dual-engine configuration optimizes fuel efficiency and minimizes carbon emissions by harnessing the strengths of both power sources. Additionally, hybrid engines incorporate regenerative braking systems that capture kinetic energy during deceleration and convert it into electrical energy, further enhancing efficiency and reducing environmental impact. By intelligently integrating petrol and electric propulsion systems, hybrid engines offer a sustainable and adaptable solution to modern transportation needs.

17.2.3 Engine Setup

Hybrid cars come in various types, with each having its unique specifications that influence their operation and environmental impact. One common type is the series-parallel hybrid, which combines both series and parallel hybrid architectures. In this configuration (Figure 17.1), the vehicle can operate using the electric motor, internal combustion engine, or a combination of both depending on driving conditions. The electric motor powers the vehicle at low speeds, while the internal combustion engine engages during higher speeds or when additional power is required. This flexibility allows series-parallel hybrids to achieve excellent fuel efficiency in both city and highway driving scenarios. Additionally, regenerative braking systems capture kinetic energy during deceleration, converting it into electrical energy to recharge the battery and improve overall efficiency. Another type is the plug-in hybrid electric vehicle (PHEV), which offers extended electric-only driving range compared to conventional hybrids. PHEVs feature larger battery packs that can be charged from an external power source, such as a standard electrical outlet or a dedicated charging station. This enables drivers to commute significant distances solely on

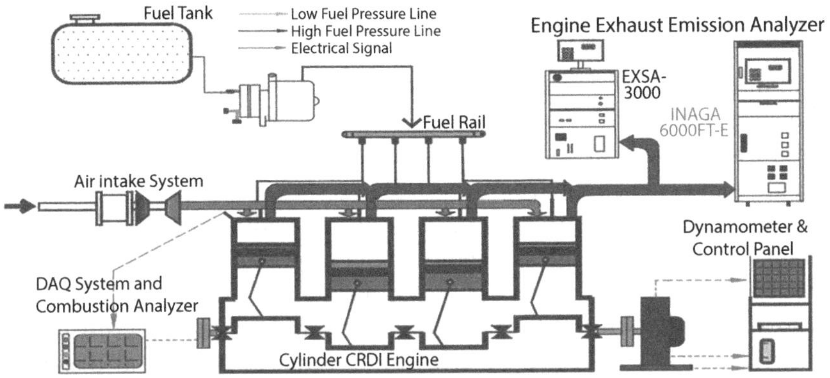

Figure 17.1 Diesel engine setup.

electric power, reducing emissions and reliance on fossil fuels. PHEVs also utilize internal combustion engines for longer trips or when the battery charge is depleted, providing greater versatility, and reducing range anxiety associated with pure electric vehicles. With customizable charging options and sophisticated energy management systems, PHEVs offer consumers a practical and eco-friendly transportation solution tailored to their driving needs and environmental concerns. Understanding the specifications and capabilities of different hybrid car types is essential for consumers to make informed decisions regarding vehicle selection and environmental impact.

17.2.4 Operating Parameters and Their Levels

When evaluating the performance and efficiency of vehicles, particularly hybrids and electric cars, selecting key input parameters like load, efficiency losses, and torque is crucial for comprehensive analysis. Load represents the demand placed on the vehicle's powertrain during operation, encompassing factors like acceleration, uphill driving, and payload. Mathematically, load can be expressed as the force exerted on the vehicle. The significance of load lies in understanding how different driving conditions impact the vehicle's energy consumption, emissions, and overall performance. By examining load across a range of scenarios from low to very high, researchers gain insights into the vehicle's versatility and efficiency across diverse driving environments. Efficiency losses quantify the energy dissipation incurred by the vehicle's powertrain during operation. These losses result from factors such as mechanical friction, heat generation, and energy conversion inefficiencies. Mathematically, efficiency losses can be calculated as the difference between the energy input and the useful work output, expressed as a percentage of the input energy. Understanding efficiency losses allows engineers to optimize vehicle design and operation to minimize wasted energy and improve overall efficiency. By exploring efficiency losses across varying load conditions, researchers can identify areas for improvement and develop strategies to enhance the vehicle's performance and energy efficiency.

Torque represents the rotational force generated by the vehicle's powertrain, crucial for propulsion and acceleration. In electric and hybrid vehicles, torque from the electric motor plays a significant role in driving performance, particularly during acceleration and overcoming resistance. Mathematically, torque is defined as the product of force and distance. Choosing torque as an input parameter allows researchers to assess the vehicle's ability to deliver sufficient power across different driving conditions, ensuring optimal performance and responsiveness. By examining torque within a range of values, engineers can design powertrains capable of meeting diverse driving demands while maintaining energy efficiency and minimizing emissions.

17.3 Methodology

The entropy method, a pivotal tool in multi-criteria decision-making (MCDM), serves to calculate the weightage of criteria by quantifying the inherent uncertainty or variability within the dataset. It operates on the principle that higher entropy values indicate greater dispersion among criteria, thereby necessitating a more balanced allocation of weights. Through entropy, each criterion's contribution to decision-making is objectively assessed, enabling the identification of key factors that significantly influence the outcome. By analyzing the spread of information across criteria, the entropy method facilitates the establishment of robust decision frameworks that account for diverse perspectives and preferences. Its application fosters transparency and objectivity in the weightage determination process, enhancing the reliability and effectiveness of MCDM models across various domains and contexts.

Step 1: Normalize the decision matrix:

$$r_{ij} = \frac{x_{ij}}{\sum_{i=1}^{m} x_{ij}} \tag{17.1}$$

Where:
r_{ij} is the normalized element.
x_{ij} is the original element.

Step 2: Compute Entropy (e_j):

$$e_j = -\frac{1}{\ln(m)}\sum_{i=1}^{n} r_{ij} * \ln(r_{ij})$$

$$j = 1, 2, 3, \ldots n \tag{17.2}$$

m = number of alternatives.

e_j is the entropy

Step 3: Compute the weightage vector (W_j):

$$d_j = 1 - e_j \tag{17.3}$$

$$W_j = \frac{d_j}{\Sigma_{j=1}^{n} d_j} \tag{17.4}$$

Where:
W_j is the entropy vector.

The VIKOR (VIsekriterijumsko KOmpromisno Rangiranje) approach is a multi-criteria selection-making technique that allows the rating and selection of alternatives in complex choice environments. Originating from operations studies and control science, VIKOR gives a scientific method to deal with choice issues characterized by multiple conflicting criteria. Unlike conventional choice-making techniques, VIKOR considers each the overall performance and the degree of compromise amongst alternatives, making it especially suitable for conditions in which decision-makers are trying to find a balance between conflicting goals. The method entails a step-with the aid of-step technique that consists of criteria normalization, dedication of the weights of standards, calculation of the values of a compromise measure, and ranking of options based on their closeness to the proper answer. VIKOR's potential to address more than one criterion concurrently and pick out compromise solutions makes it a valuable tool for numerous programs in fields which include engineering, economics, environmental control, and strategic making plans.

Step 1: Determination of best and worst value:

- For non-beneficial criteria:

$$\text{Best } (X_i^+) = \min (X_{ij})$$

$$\text{Worst } (X_i^-) = \max (X_{ij})$$

- For beneficial criteria:

$$\text{Best } (X_i^+) = \max (X_{ij})$$

$$\text{Worst } (X_i^-) = \min (X_{ij})$$

Step 2: Normalization of S_i and R_i as presented in Eq. (17.5) and Eq. (17.6):

$$S_i = \sum_{j=1}^{m} (W_j * \frac{X^+_i - X_{ij}}{X^+_i - X^-_i}) \tag{17.5}$$

$$R_i = \max (W_j * \frac{X^+_i - X_{ij}}{X^+_i - X^-_i}) \tag{17.6}$$

Where:
$S^* = \min S_i$
$S^- = \max S_i$
$R^* = \min R_i$
$R^- = \max R_i$
W_j = Weightage

Step 3: Computation of Q_j for group utility function as given in Eq. (17.7):

$$Q_i = \left(\vartheta * \frac{S_i - S^*}{S^- - S^*} \right) + ((1-\vartheta) * \frac{R_i - R^*}{R^- - R^*}) \tag{17.7}$$

Where $\vartheta = 0.5$

Step 4: Ranking the alternatives:
Sorting of R_i, S_i and Q_i are made from their minimum value. Hence the ranking list is obtained.

17.3.1 Outcome Parameters Selection

When evaluating hybrid engines, it's essential to consider emissions such as carbon monoxide (CO), carbon dioxide (CO_2), and unburned hydrocarbons (UBHC) as outcome parameters due to their direct impact on environmental and human health, as well as their correlation with engine efficiency and combustion characteristics. Carbon monoxide (CO) is a byproduct of incomplete combustion in internal combustion engines, including those found in hybrid vehicles. It poses health risks, such as reducing the blood's ability to carry oxygen, leading to symptoms like headaches, dizziness, and even death in high concentrations. Monitoring CO emissions from hybrid engines provides insight into the combustion process efficiency and the effectiveness of emission control systems, enabling engineers to optimize engine design and reduce harmful emissions.

The formula for calculating carbon monoxide emissions (CO) typically involves measuring the mass of CO emitted per unit distance travelled by the vehicle. It can be expressed as Eq. (17.8):

$$\text{CO emissions} = \frac{m_{co}}{d} \tag{17.8}$$

Where:

- m_{co} is the mass of carbon monoxide emitted (in grams),
- d is the distance travelled by the vehicle (in kilometres or miles).

Carbon dioxide (CO_2) is a greenhouse gas emitted during fuel combustion, including the operation of hybrid engines. While CO_2 emissions are inherent to the combustion of hydrocarbon fuels, hybrid engines can mitigate them by supplementing the internal combustion engine with electric power, improving fuel efficiency, and reducing overall emissions. Understanding CO_2 emissions from hybrid engines is crucial for assessing their environmental impact and evaluating the effectiveness of hybridization in reducing greenhouse gas emissions. By monitoring CO_2 levels across various driving conditions, researchers can identify opportunities for further emissions reduction and develop strategies for enhancing hybrid vehicle sustainability.

Carbon dioxide emissions (CO_2) from a vehicle are calculated based on the amount of fuel consumed and the carbon content in the fuel. The formula for CO_2 emissions is expressed in Eq. (17.9):

$$\mathrm{CO_2\ emissions} = \frac{m_{fuel}}{FE} * C_{CO_2} \tag{17.9}$$

Where:

- m_{fuel} is the mass of fuel consumed (in kilograms or pounds),
- C_{co2} is the carbon content factor specific to the fuel (in kilograms or pounds of CO_2 emitted per kilogram or pound of fuel burned),
- FE is the fuel efficiency of the vehicle (in kilometers or miles or gallon).

Unburned hydrocarbons (UBHC) represent fuel that escapes combustion and remains in the exhaust gases emitted by hybrid engines. UBHC emissions contribute to air pollution and smog formation, leading to adverse environmental and health effects. Hybrid engines aim to minimize UBHC emissions by optimizing combustion processes, enhancing fuel-air mixing, and implementing advanced emission control technologies. Monitoring UBHC levels helps assess engine performance, combustion efficiency, and the effectiveness of emission control systems in hybrid vehicles. By controlling UBHC emissions, hybrid engines can achieve compliance with stringent emissions regulations, improve air quality, and promote sustainable transportation solution.

Calculating unburned hydrocarbons (UBHC) emissions involves measuring the mass of hydrocarbons that escape combustion and are released into the atmosphere per unit distance travelled by the vehicle. Formula for UBHC emissions can be represented as Eq. (17.10):

$$\mathrm{UBHC\ emissions} = \frac{m_{ubhc}}{d} \tag{17.10}$$

Where:

m_{UBHC} is the mass of unburned hydrocarbons emitted (in grams), d is the distance travelled by the vehicle (in kilometers or miles).

17.4 Results

17.4.1 Weightage Analysis of Vehicles

The analysis revealed that the highest QI (Quality Index) score was attributed to petrol vehicles, followed by diesel vehicles, with hybrid vehicles ranking last, as illustrated in Figure 17.2. This ranking highlights the varying performance of each vehicle type concerning the parameters assessed. Petrol vehicles demonstrated the highest overall quality index, indicating superior performance across the evaluated criteria. Diesel vehicles secured the second position, suggesting favorable attributes but falling short of petrol vehicles in certain aspects. Conversely, hybrid vehicles obtained the lowest QI score, indicating areas where improvements may be needed to enhance their overall performance and efficiency. This ranking provides valuable insights into the comparative strengths and weaknesses of each vehicle type, aiding stakeholders in making informed decisions regarding vehicle selection and future development efforts.

This combined result can serve as a basis for decision-making in selecting the most suitable alternative considering the quality index criterion across different loads.

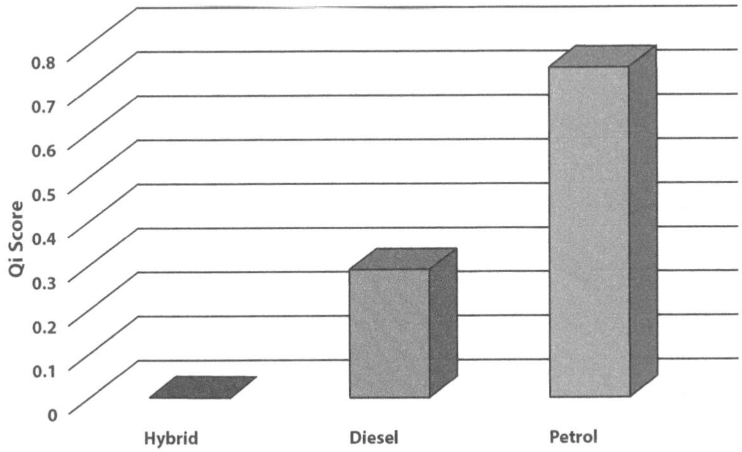

Figure 17.2 Q_i score of hybrid, diesel, and petrol vehicles.

17.4.2 Impact of Input Parameters on Outcomes

Carbon emissions, including unburned hydrocarbons (UBHC), carbon monoxide (CO), and carbon dioxide (CO_2), exhibit varying trends in response to changes in load, torque, and efficiency parameters. The load significantly impacts carbon emissions, with heavier loads generally resulting in increased emissions due to higher fuel consumption and energy demands. As load increases, engines operate under greater stress, leading to higher combustion temperatures and increased emissions of UBHC and CO. Additionally, heavier loads require more energy for propulsion, resulting in increased CO_2 emissions as more fuel is burned to overcome the added resistance. Conversely, lighter loads typically correspond to reduced carbon emissions, as engines operate under less strain and require less fuel for propulsion as shown in Figure 17.3.

Efficiency plays a crucial role in determining carbon emissions, with higher efficiency generally associated with lower emissions across all parameters. Improved engine efficiency leads to more complete combustion of fuel, resulting in reduced emissions of UBHC, CO, and CO_2. Additionally, efficient propulsion systems require less fuel to achieve the same level of performance, further contributing to emissions reductions. By optimizing load, torque, and efficiency parameters, vehicle manufacturers and operators can minimize carbon emissions and promote sustainable transportation practices, ultimately mitigating environmental impact and addressing climate change concerns as shown in Figure 17.4.

Torque variations also influence carbon emissions, albeit to a lesser extent than load. Higher torque requirements generally result in increased

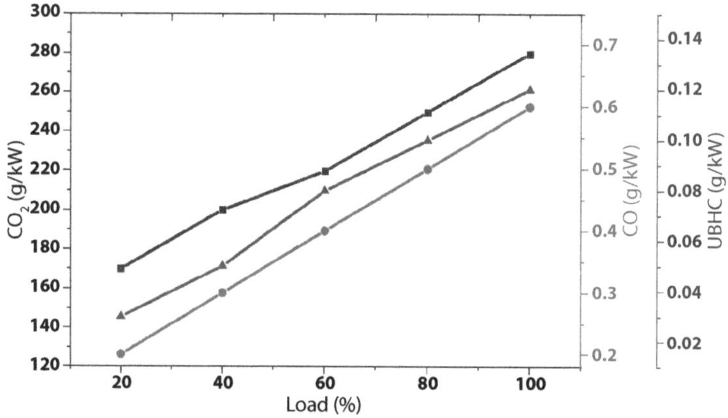

Figure 17.3 Load vs. emissions.

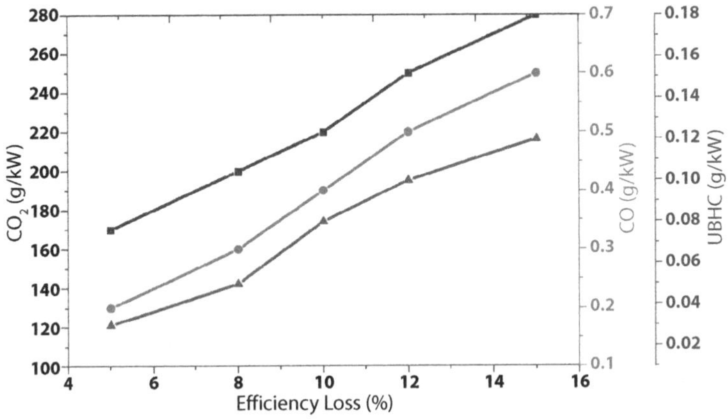

Figure 17.4 Diesel efficiency loss vs. emissions.

emissions of UBHC, as more intense combustion processes are required to generate the necessary power. However, the impact on CO and CO_2 emissions may vary depending on engine efficiency and combustion characteristics. In some cases, higher torque levels may lead to more efficient combustion, resulting in lower CO and CO_2 emissions. Conversely, inefficient combustion processes under high torque conditions may result in increased CO emissions, despite potential reductions in CO_2 emissions due to more complete combustion as shown in Figure 17.5.

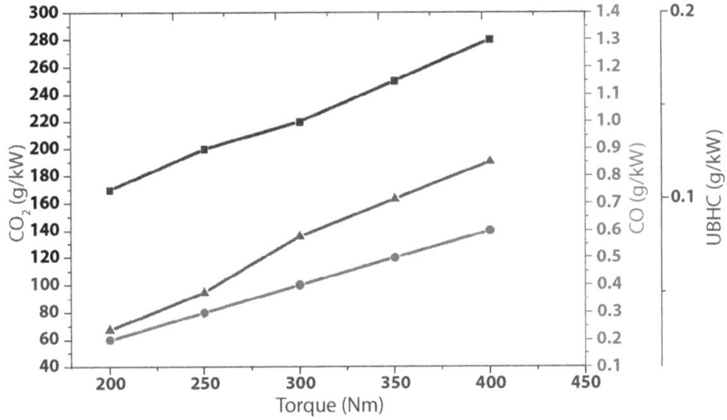

Figure 17.5 Diesel torque vs. emissions.

17.4.3 Heat Map Similarity Analysis for Outcomes

Load exerts a significant influence on carbon emissions for all vehicles, with particular attention to diesel engines as depicted in Figure 17.6. As load increases, diesel engines experience heightened fuel consumption and energy demands to meet the additional propulsion requirements. This elevated workload results in higher combustion temperatures and increased emissions of carbon dioxide (CO_2), unburned hydrocarbons (UBHC), and carbon monoxide (CO). Diesel engines, known for their robust performance under heavy loads, are especially prone to emissions spikes during high-load operations. The combustion process in diesel engines relies on air compression to ignite the fuel, and under heavy loads, this compression becomes more intense, leading to higher combustion temperatures and increased emissions. Additionally, the combustion efficiency of diesel engines may decrease under heavy loads, leading to incomplete combustion and higher emissions of UBHC and CO. Therefore, load management strategies and engine optimization techniques are crucial for mitigating

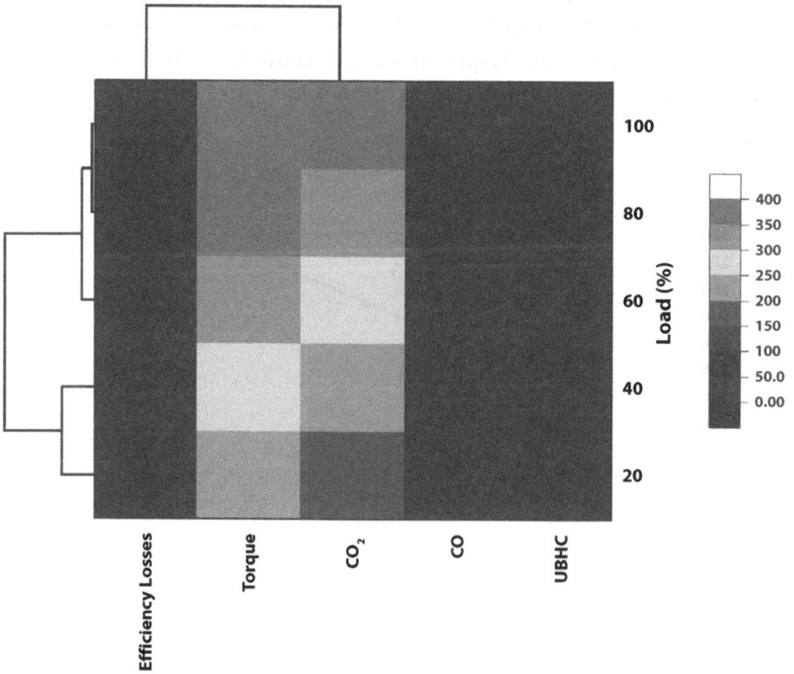

Figure 17.6 Diesel heat map.

carbon emissions from diesel vehicles, ensuring sustainable and environmentally friendly transportation practices.

17.5 Analysis and Discussion

The VIKOR method is utilized to determine the most suitable alternative among the given options for various loads, considering criteria such as CO_2 emissions, CO emissions, UBHC emissions, Si (Silicon) emissions, Ri, and Qi (Quality index). The results for hybrid, diesel and petrol are shown in Figure 17.7.

Table 17.1 – Table 17.5 presents the ranking of the different vehicles at different load percentages of 20%, 40%, 60%, 80% and 100% for different outputs as mentioned in the methodology section. The combined result reflects the overall ranking of the alternatives based on their quality index (Qi) across all loads. The Hybrid option consistently holds the 1st rank

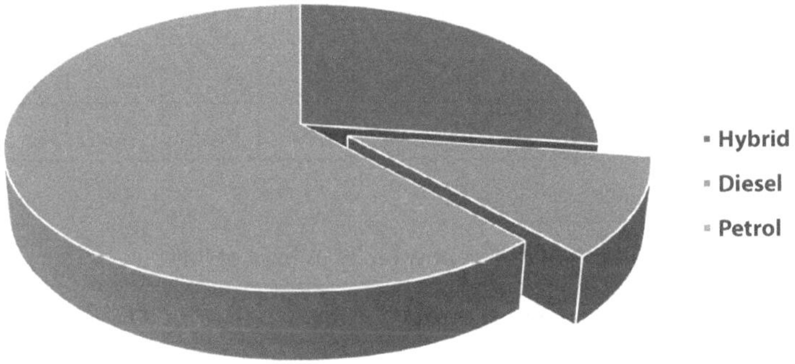

Figure 17.7 Weightages of hybrid, diesel, and petrol vehicles at 60% load.

Table 17.1 Ranking of different vehicles at load of 20%.

Vehicles	CO_2	CO	UBHC	S_i	R_i	Q_i	Rank
Hybrid	0	0	0	0	0	0	1
Diesel	0.24	0.22	0	0.46	0.24	0.51	2
Petrol	0.17	0.44	0.32	0.93	0.44	1	3
S*, R*				0	0		
S-, R-				0.93	0.44		

Table 17.2 Ranking of different vehicles at load of 40%.

Vehicles	CO_2	CO	UBHC	S_i	R_i	Q_i	Rank
Hybrid	0	0	0	0	0	0	1
Diesel	0.25	0.08	0	0.33	0.25	0.39	2
Petrol	0.19	0.16	0.59	0.94	0.58	1	3
S*, R*				0	0		
S-, R-				0.94	0.59		

Table 17.3 Ranking of different vehicles at load of 60%.

Vehicles	CO_2	CO	UBHC	S_i	R_i	Q_i	Rank
Hybrid	0	0	0	0	0	0	1
Diesel	0.27	0.05	0	0.33	0.27	0.30	2
Petrol	0.19	0.11	0.62	0.92	0.62	0.75	3
S*, R*				0	0		
S-, R-				0.92	0.62		

Table 17.4 Ranking of different vehicles at load of 80%.

Vehicles	CO_2	CO	UBHC	S_i	R_i	Q_i	Rank
Hybrid	0	0	0	0	0	0	1
Diesel	0.25	0.03	0	0.28	0.25	0.33	2
Petrol	0.18	0.06	0.69	0.93	0.69	1	3
S*, R*				0	0		
S-, R-				0.93	0.69		

followed by Diesel in the 2nd and Petrol in the 3rd rank at all load percentages. This consolidated ranking provides a clear indication of the preferred alternatives based on their overall performance across the considered criteria. It emphasizes the stability of Hybrid performance and the relatively lower ranking of Petrol compared to Diesel and Hybrid.

Table 17.5 Ranking of different vehicles at load of 100%.

Vehicles	CO_2	CO	UBHC	S_i	R_i	Q_i	Rank
Hybrid	0	0	0	0	0	0	1
Diesel	0.26	0.02	0	0.28	0.26	0.34	2
Petrol	0.16	0.04	0.70	0.90	0.70	1	3
S*, R*				0	0		
S-, R-				0.90	0.70		

17.6 Conclusion

Hybrid vehicles are poised to come to be dominant vehicles in the car panorama of the brand new international because of their pivotal role in addressing urgent environmental worries and assembly stringent emissions rules. As technological improvements keep beautifying battery efficiency and lower manufacturing expenses, hybrid cars offer a compelling pathway towards decreasing carbon footprints and mitigating fossil gas dependence. Moreover, with the developing emphasis on sustainability and the electrification of transportation, hybrid cars constitute a crucial transitional solution that mixes the blessings of internal combustion engines with electric powered propulsion systems. As research and development efforts push the limits of hybrid era, their big adoption appears inevitable, heralding a cleanser and greater sustainable future for transportation.

The VIKOR method offers a structured approach to compare the environmental performance of hybrid, diesel, and petrol vehicles based on their CO_2, CO, and UBHC emissions across different loads.

- The VIKOR method ranks the vehicles in terms of their environmental impact, with the hybrid consistently receiving the top rank (1), followed by diesel (2), and petrol (3).
- By evaluating emissions across various load scenarios, the VIKOR method enables decision-makers to identify the most environmentally friendly option among the three vehicle types.
- This systematic analysis supports informed decision-making in sustainable transportation planning, emphasizing the importance of reducing emissions and promoting eco-friendly vehicle technologies.

Hybrid vehicles still rely partially on internal combustion engines, limiting their potential for zero-emission driving. Additionally, their complexity can lead to higher maintenance costs compared to traditional vehicles. The future scope for hybrid vehicles lies in advancing battery technology to extend electric range and reducing reliance on internal combustion engines. Integration of renewable energy sources and smart charging infrastructure will further enhance their efficiency and environmental benefits.

References

1. Khan, O., Parvez, M., Seraj, M., Yahya, Z., Devarajan, Y., Nagappan, B., Optimising building heat load prediction using advanced control strategies and Artificial Intelligence for HVAC system. *Therm. Sci. Eng. Prog.*, 49, 102484, 2024.
2. Khan, F., Alsaduni, I., Khan, O., Parvez, M., Yadav, A. K., Agbulut, Ü., Enhancing biogas/biohydrogen utilization in dual-fuel engines using advanced machine learning algorithms. *Int. J. Hydrogen Energy*, 112, 81–90, 2025.
3. Seraj, M., Parvez, M., Khan, O., Yahya, Z., Optimizing smart building energy management systems through industry 4.0: A response surface methodology approach. *Green Technol. Sustain.*, 2, 100079, 2024.
4. Ahamad, T., Parvez, M., Lal, S., Khan, O., Idrisi, M.J., 4-E analysis and multiple objective optimizations of a novel solar-powered cogeneration energy system for the simultaneous production of electrical power and heating. *Sci. Rep.*, 13, 1, 22246, 2023.
5. Khan, O., Mufazzal, S., Sherwani, A.F., Khan, Z.A., Parvez, M., Idrisi, M.J., Experimental investigation and multi-performance optimization of the leachate recirculation based sustainable landfills using Taguchi approach and an integrated MCDM method. *Sci. Rep.*, 13, 1, 19102, 2023.
6. Khan, F., Khan, O., Parvez, M., Almujibah, H., Pachauri, P., Yahya, Z., Ağbulut, Ü., Innovative hydrogen production from waste bio-oil *via* steam methane reforming: An advanced ANN-AHP-k-means modelling approach using extreme machine learning weighted clustering. *Int. J. Hydrogen Energy*, 105, 1080–1091, 2025.
7. Khan, O., Parvez, M., Kumari, P., Yadav, A.K., Akram, W., Ahmad, S., Idrisi, M.J., Modelling of compression ignition engine by soft computing techniques (ANFIS-NSGA-II and RSM) to enhance the performance characteristics for leachate blends with nano-additives. *Sci. Rep.*, 13, 1, 15429, 2023
8. Seraj, M., Parvez, M., Ahmad, S., Khan, O., Sustainable energy transition and decision-making for enhancing the performance of building equipment in diverse climatic conditions. *Green Technol. Sustain.*, 1, 3, 100043, 2023.

9. Khan, F., Karimi, M.N., Khan, O., Exploring the scalability and commercial viability of biosynthesized nanoparticles for cooling panels with the help of artificial intelligence and solar energy systems. *Green Technol. Sustain.*, 1, 3, 100036, 2023.
10. Akram, W., Parvez, M., Rafi, K.M., Khan, O., First and Second Law Analyses of an Organic Rankine Cycle to Recover the Waste Heat from the Solar-Operated Combined Cycle Power Plants with Grid, in: *International Conference on Nanotechnology: Opportunities and Challenges*, Springer Nature Singapore, Singapore, pp. 369–376, 2022.
11. Parvez, M., Khan, O., Howari, H., Investigation of (3E) energy, exergy, and environmental analysis of syngas production for combined power and cooling plant by testing six different biomasses. *Biomass Convers. Biorefin.*, 14, 1–15, 2023.
12. Khan, O., Alsaduni, I., Parvez, M., & Yadav, A. K., Enhancing hydrogen production using solar-driven photocatalysis with biosynthesized nanocomposites: A hybrid machine learning approach towards enhanced performance and sustainable environment. *Int. J. Hydrogen Energy*, 102, 609–625, 2025.
13. Khan, O., Khan, M.Z., Bhatt, B.K., Alam, M.T., Tripathi, M., Multi-objective optimization of diesel engine performance, vibration and emission parameters employing blends of biodiesel, hydrogen and cerium oxide nanoparticles with the aid of response surface methodology approach. *Int. J. Hydrogen Energy*, 48, 56, 21513–21529, 2023.
14. Bhatt, B.K., Akram, W., Khan, O., Parvez, M., Ahmad, S., A Methodology for Multi-criteria Assessment of Chiller Plants for Building Setup Utilising Waste Heat and Solar Energy. *Sustain. Anal. Model.*, 3, 100022, 2023.
15. Nasim, M.A.G., Khan, O., Parvez, M., Bhatt, B.K., Optimizing ultrasonic reactor operating variables using intelligent soft computing models for increased biodiesel production. *Green Technol. Sustain.*, 1, 100033, 2023, https://doi.org/10.1016/j.grets.2023.100033.
16. Prakash, V., Saraswat, V., Dewangan, A., Yadav, A. K., Kumar, A., Production and utilization of hydrogen as a future fuel for electric vehicles: A review. *Clean Energy*, 259–281, CRC Press, 2024.
17. Nasim, M.A.G., Khan, O., Parvez, M., Chouhan, P.S., Maximizing biodiesel production: An intelligent software-driven optimization of ultrasonic reactor operating variables. *Bioresour. Technol. Rep.*, 22, 101474, 2023, https://doi.org/10.1016/j.biteb.2023.101474.
18. Khan, O., Parvez, M., Kumari, P., Parvez, S., Ahmad, S., The future of pharmacy: How AI is revolutionizing the industry. *Intell. Pharm.*, 1, 1, 32–40, 2023, https://doi.org/10.1016/j.ipha.2023.04.008.
19. Fatima, M., Sherwani, N.U.K., Singh, V., Mindful medicine: Exploring the benefits of mindfulness-based stress reduction for doctors using statistical techniques. *Intell. Pharm.*, 1, 1, 41–47, 2023, https://doi.org/10.1016/j.ipha.2023.04.009.

20. Howari, H., Parvez, M., Khan, O., Alhodaib, A., Mallah, A., Yahya, Z., Multi-Objective Optimization for Ranking Waste Biomass Materials Based on Performance and Emission Parameters in a Pyrolysis Process—An AHP-TOPSIS Approach. *Sustainability*, 15, 4, 3690, 2023, https://doi.org/10.3390/su15043690.
21. Khan, O., Khan, M.Z., Alam, M.T., *et al.*, Comparative Study of Soft Computing and Metaheuristic Models in Developing Reduced Exhaust Emission Characteristics for Diesel Engine Fueled with Various Blends of Biodiesel and Metallic Nanoadditive Mixtures: An ANFIS–GA–HSA Approach. *ACS Omega*, 8, 7344–7367, 2023, https://doi.org/10.1021/acsomega.2c05246.
22. Patra, B., Nema, P., Khan, M.Z., Khan, O., Optimization of solar energy using MPPT techniques and industry 4.0 modelling. *Sustain. Oper. Comput.*, 4, 22–28, 2023, https://doi.org/10.1016/j.susoc.2022.10.001.
23. Ahmad, S., Parvez, M., Khan, T.A., Khan, O., A Hybrid Approach Using AHP–TOPSIS Methods for Ranking of Soft Computing Techniques Based on their Attributes for Prediction of Solar Radiation. *Environ. Challenges*, 9, 100634, 2022, https://doi.org/10.1016/j.envc.2022.100634.
24. Ahmad, S., Parvez, M., Khan, T.A., Siddiqui, S.A., Khan., O., Performance comparison of solar powered cogeneration and trigeneration systems *via* energy and exergy analyses. *Int. J. Exergy*, 39, 4, 395–409, 2022.
25. Islam, S.M.U., Khan, S., Ahmad, H., Rahman, M.A.U., Tomar, S., Khan, M.Z., Assessment of challenges and problems in supply chain among retailers during COVID-19 epidemic through AHP-TOPSIS hybrid MCDM technique. *Internet Things Cyber-Phys. Syst.*, 2, 180–193, 2022, https://doi.org/10.1016/j.iotcps.2022.10.001.
26. Ahmad, S., Alam, M.T., Bilal, M., Khan, O., Khan, M.Z., Analytical Modelling of HVAC-IoT Systems with the Aid of UVGI and Solar Energy Harvesting, in: *Energy Harvesting*, pp. 65–80, 2022.
27. Khan, O., Khan, M.Z., Khan, M.E., Goyal, A., Bhatt, B.K., Khan, A., Parvez, M., Experimental analysis of solar powered disinfection tunnel mist spray system for coronavirus prevention in public and remote places. *Mater. Today: Proc.*, 46, 15, 6852–6858, 2021.
28. Zeeshan, M., Alam, M.R., Khan, O., Parvez, M., Azad, A.S., Thermal load calculation of building envelope set up with the help of a drone setup. *2021 First International Conference on Advances in Computing and Future Communication Technologies (ICACFCT)*, IEEE, pp. 186–190, 2021.
29. Parvez, M., Khalid, F., Khan, O., Thermodynamic performance assessment of solar-based combined power and absorption refrigeration cycle. *Int. J. Exergy*, 31, 3, 232–248, 2020.
30. Seraj, M.M., Khan, O., Khan, M.Z., Parvez, M., Bhatt, B.K., Ullah, A., Alam, M.T., Analytical Research of Artificial Intelligent Models for Machining Industry Under Varying Environmental Strategies: An Industry 4.0 Approach. *Sustain. Oper. Comput.*, 3, 176–187, 2022.

31. Parvez, M. and Khan, O., Parametric simulation of biomass integrated gasification combined cycle (BIGCC) power plant using three different biomass materials. *Biomass Convers. Biorefin.*, 10, 4, 803–812, 2020.
32. Ahmad, A., Yadav, A.K., Hasan, S., Enhanced production of methane enriched biogas through intensified co-digestion process and its effective utilization in a biodiesel/biohydrogen fueled engine with duel injection strategies: ML-RSM based an efficient optimization approach. *Int. J. Hydrogen Energy*, 65, 671–686, 2024.
33. Meraj, M., Khan, M.E., Tiwari, G.N., Khan, O., Optimization of Electrical Power of Solar Cell of Photovoltaic Module for a Given Peak Power and Photovoltaic Module Area. *Advances in Fluid and Thermal Engineering*, pp. 417–433, 2019.
34. Khan, S., Tomar, S., Fatima, M., Khan, M.Z., Impact of artificial intelligent and industry 4.0 based products on consumer behaviour characteristics: A meta-analysis-based review. *Sustain. Oper. Comput.*, 3, 218–225, 2022.
35. Fatima, M., Sherwani, N.U.K., Khan, S., Khan, M.Z., Assessing and predicting operation variables for doctors employing industry 4.0 in health care industry using an adaptive neuro-fuzzy inference system (ANFIS) approach. *Sustain. Oper. Comput.*, 3, 286–295, 2022.
36. Khan, O., Khan, M.E., Parvez, M., Ahmed, K.A.A.R., Ahmad., I., Extraction and Experimentation of Biodiesel Produced from Leachate Oils of Landfills Coupled with Nano-Additives Aluminium Oxide and Copper Oxide on Diesel Engine, in: *Nanomaterials for Innovative Energy Systems and Devices*, pp. 319–332, 2022.
37. Ahmed, A., Yadav, A.K., Singh, A., Singh, D.K., A machine learning-response surface optimization approach to enhance the performance of diesel engine using novel blends of aloe vera biodiesel with MWCNT nanoparticles and hydrogen. *Process Saf. Environ. Prot.*, 186, 738–755, 2024, doi: 10.1016/j.psep.2024.04.013.
38. Parvez, M., Khan, M.E., Khalid, F., Khan, O., Akram, W., A Novel Energy and Exergy Assessments of Solar Operated Combined Power and Absorption Refrigeration Cogeneration Cycle, in: *Electric Vehicles*, pp. 213–229, 2021.
39. Ahmed, A. and Yadav, A.K., Parametric Analysis of Wastewater Electrolysis for Green Hydrogen Production: A Combined RSM, Genetic Algorithm, and Particle Swarm Optimization Approach. *Int. J. Hydrogen Energy*, 59, 51–62, 2024.
40. Ahmed, A., Yadav, A.K., Singh, A., Singh, D.K., A machine learning–genetic algorithm based models for optimization of intensification process through microwave reactor: A new approach for rapid and sustainable synthesis of biodiesel from novel Hiptage benghalensis seed oil. *Fuel*, 374, 132470, 2024.
41. Khan, O., Khan, M.P., Yadav, A.K., A Comparative Analysis of Various Nanocomposites for the Remediation of Heavy Metals from Biomass Liquid Digestate using Multi-Criteria Decision Methods. *Biomass Bioenergy*, 186, 107281, 2024.

42. Khan, O., Alsaduni, I., Equbal, A., Parvez, M., Yadav, A.K., Performance and Emission Analysis of Biodiesel Blends Enriched with Biohydrogen and Biogas in Internal Combustion Engines. *Process Saf. Environ. Prot.*, 183, 1013–1037, 2024.
43. Ahmed, A., Yadav, A.K., Singh, D.K., Singh, A., A Comprehensive Machine Learning-Coupled Response Surface Methodology Approach for Predictive Modeling and Optimization of Biogas Potential in Anaerobic Co-Digestion of Organic Waste. *Biomass Bioenergy*, 180, 106995, 2024.
44. Dewangan, A., Mallick, A., Yadav, A.K., Ahmad, A., Alqahtani, D., Islam., S., Combined Effect of Operating Parameters and Nano Particles on Performance of a Diesel Engine: Response Surface Methodology coupled Genetic algorithm Approach. *ACS Omega*, 8, 27, 24586–24600, 2023.
45. Ahmad, A., Yadav, A.K., Singh, A., Singh, D.K., Ağbulut, U., A hybrid RSM-GA-PSO approach on optimization of process intensification of linseed biodiesel synthesis using an ultrasonic reactor: Enhancing biodiesel properties and engine characteristics with ternary fuel blends. *Energy*, 288, 129077, 2024.
46. Ashraf, F., Shajar, M., Khan, M. E., Equbal, A., Khan, O., Yadav, A. K., Parvez, M., Comparative evaluation of carbon emissions from various vehicles: Integrating AHP techniques for ranking, in: *Clean Energy*, pp. 282–303, CRC Press, 2024.
47. Ahmed, A., Yadav, A.K., Hasan, S., Enhancing Efficiency and Environmental Impacts of a Nano-Enhanced Oleander Biodiesel-Biohydrogen Dual Fuel Engine Equipped with EGR, through Operational Parameter Optimization using RSM-MOGA Technique. *Int. J. Hydrogen Energy*, 78, 1157–1172, 2024.
48. Khan, O., Parvez, M., Yahya, Z., Alhodaib, A., Yadav, A.K., Hoang, A.T., Ağbulut, U., Waste-to-energy power plants: Multi-objective analysis and optimization of landfill heat and methane gas by recirculation of leachate. *Process Saf. Environ. Prot.*, 186, 957–968, 2024.
49. Ahmed, A., Yadav, A.K., Singh, A., Enhancing waste cooking oil biodiesel yield and characteristics through machine learning , response surface methodology , and genetic algorithm for optimal utilization in CI Engines. *Int. J. Green Energy*, 21, 1345–1365, 2024.
50. Singh, A.A., Singh, R.K., Yadav, A.K., Saha, A., Dewangan, A., Singh, B., Experimental investigations on the electrochemical machining of D3 die steel material and multi-objective parameter optimization. *Surf. Rev. Lett.*, 31, 2450074, 2024.
51. Ahmed, A., Yadav, A.K., Hasan, S., Biogas as a Sustainable and Viable Alternative Fuel for Diesel Engines: A Comprehensive Review of Production, Purification, Economic Analysis, and Performance Evaluation. *J. Process Mech. Eng. Part E*, 2024, https://doi.org/10.1177/09544089241255930.
52. Khan, F., Karimi, M.N., Khan, O., Yadav, A.K., Alhodaib, A., Gürel, A.E., Ağbulut, U., A hybrid MCDM optimization for utilization of novel set of

biosynthesized nanofluids on thermal performance for solar thermal collectors. *Int. J. Thermofluids*, 22, 100686, 2024.
53. Alhodaib, A., Yahya, Z., Khan, O., Equbal, A., Equbal, M.S., Parvez, M., Idrisi, M.J., Sustainable coatings for green solar photovoltaic cells: performance and environmental impact of recyclable biomass digestate polymers. *Sci. Rep.*, 14, 1, 11221, 2024.
54. Singh, B., Yadav, A.K., Saha, A., Fabrication and Characterization of AA6082-T6 Produced by Friction Stir Additive Manufacturing. *Surf. Rev. Lett.*, 2024, DOI:10.1142/S0218625X25500222.
55. Equbal, A., Equbal, A., Ahmad, R., Kumar, M., Khan, O., Yadav, A.K., RSM based statistical approach to enhance the compressive performance of ABS P400 parts fabricated *via* FDM Technology. *J. Surf. Rev. Lett.*, 2024, DOI:10.1142/S0218625X25500210.
56. Dewangan, A., Ahmad, A., Yadav, A.K., Innovative research on waste tire recycling for sustainable biofuel production: Assessment of its usability on multi-cylinder diesel engine employing constant injection of oxyhydrogen and biogas through a premixing device. *Int. J. Hydrogen Energy*, 84, 155–163, 2024.
57. Dewangan, A., Mallick, A., Yadav, A.K., Islam, S., Saleel, C.A., Shaik, S., Ağbulut, U., Production of oxy-hydrogen gas and the impact of its usability on CI engine combustion, performance, and emission behaviors. *Energy*, 278, 127937, 2023.
58. Ahmed, A., Yadav, A.K., Singh, A., Enhancement of Biogas Yield from Dual Organic Waste Using Hybrid Statistical Approach and Its Effects on Ternary Fuel Blend (Biodiesel/n-butanol /Diesel) Powered Diesel Engine. *Environ. Prog. Sustain. Energy*, 42, 5, e14163, 2023, doi.org/10.1002/ep.14163.
59. Ahmed, A., Yadav, A.K., Singh, A., Optimization of Cavitation-Assisted Biodiesel Production and Fuel Properties from Neochloris Oleoabundans Microalgae oil using Genetic Algorithm and Response Surface Methodology. *J. Process Mech. Eng. Part E*, 238, 4, 1821–1835, 2023, doi.org/10.1177/09544089231159832.
60. Ahmad, A., Yadav, A.K., Singh, A., Singh, D.K., Multi-response optimization of a microalgae-spirulina-fueled VCR diesel engine: a comprehensive RSM-GA approach, in: *Environment, Development and Sustainability*, pp. 1–29, 2023.
61. Ahmed, A., Yadav, A.K., Singh, A., Process optimization of spirulina microalgae biodiesel synthesis using RSM coupled GA technique: a performance study of a biogas powered dual fuel engine. *Int. J. Environ. Sci. Technol.*, 21, 169–188, 2023.
62. Khan, O., Ali, V., Parvez, M., Alhodaib, A., Yahya, Z., Yadav, A.K., Ağbulut, U., Exploring the performance of biodiesel-hydrogen blends with diverse nanoparticles in diesel engine: A hybrid machine learning K-means clustering approach with weighted performance metrics. *Int. J. Hydrogen Energy*, 78, 547–563, 2024.

63. Ahmed, A., Yadav, A.K., Singh, A., Application of machine learning and genetic algorithm for prediction and optimization of Biodiesel Yield from Waste Cooking Oil. *Korean J. Chem. Eng.*, 40, 2941–2956, 2023.
64. Yadav, M., Karimi, M.N., Yadav, A.K., Effect of Graphene Oxide Nanoparticles Dispersed Biodiesel on Combustion, Performance, and Emission Characteristics of a Diesel Engine. *J. Process Mech. Eng.: Part E*, 2023, https://doi.org/10.1177/09544089231209139.
65. Dwivedi, S.P., Yadav, A.K., Saxena, A., Dwivedi, V.K., Tribo-mechanical, physical and thermal behaviour of Al/Si3N4 composite with and without the addition of Cu, Ni and Cr entropy elements. *Proc. Inst. Mech. Eng., Part E: J. Process Mech. Eng.*, 09544089231189663.
66. Ahmed, A., Yadav, A.K., Singh, A., Biodiesel yield optimization from a third-generation feedstock (microalgae spirulina) using a hybrid statistical approach. *Int. J. Ambient Energy*, 44, 1, 1202–1213, 2023.
67. Ahmed, A., Yadav, A.K., Singh, A., Modelling and optimisation of VCR diesel engine parameters fuelled with microalgae methyl-ester: a RSM-coupled Taguchi approach. *Int. J. Ambient Energy*, 44, 1, 2498–2506, 2023.
68. Das, D.K., Ahmad, A., Singh, A., Singh, D.K., Yadav, A.K., Numerical simulation of the combustion chamber and performance analysis of microalgae-based biodiesel/diesel-powered CI engine: An axisymmetric flow approach. *Proc. Inst. Mech. Eng., Part E: J. Process Mech. Eng.*, 2023, 09544089231218121.
69. Singh, D., Singh, S., Yadav, A.K., Khan, O., Dewangan, A., Alkahtani, M.Q., Islam, S., From Theory to Practice: A Sustainable Solution to Water Scarcity by Using a Hybrid Solar Distiller with a Heat Exchanger and Aluminum Oxide Nanoparticles. *ACS Omega*, 37, 8, 33543–33553, 2023.
70. Ahmed, A., Yadav, A.K., Singh, A., Optimization of Biogas Yield from Anaerobic Co-digestion of Dual waste for environmental sustainability: ANN, RSM, and GA, Approach. *Int. J. Oil Gas Coal Technol.*, 33, 1, 75, 2023.
71. Ahmed, A., Yadav, A.K., Singh, A., An environmental impact assessment and optimization study of biodiesel production from microalgae. *Int. J. Global Warming*, 31, 3, 294–313, 2023.
72. Ahmed, A., Yadav, A.K., Singh, A., Biodiesel Production from Mahua Oil: Characterization, Optimization, and Modeling with a Hybrid Statistical Approach. *Int. J. Ambient Energy*, 44, 1, 2600–2609, 2023.
73. Khan, T.A., Khan, T.A., Yadav, A.K., Khan, M.E., Experimental Studies on Utilization of Neochloris Oleoabundans Microalgae Biodiesel as an Alternative Fuel for Diesel Engine. *Int. J. Ambient Energy*, 44, 1, 115–123, 2023.
74. Singh, D., Yadav, A.K., Kumar, A., Samsher, Energy matrices and life cycle conversion analysis of N-identical hybrid double slope solar distiller unit using Al2O3 nanoparticle. *J. Water Environ. Nanotechnol.*, 8, 3, 267–284, 2023.

75. Khan, T.A., Yadav, A.K., Khan, T.A., A hydrodynamic cavitation-assisted system for optimization of biodiesel production from green microalgae oil using a genetic algorithm and response surface methodology approach. *Environ. Sci. Pollut. Res.*, 29, 49465–49477, 2022.
76. Yadav, A.K., Khan, M.E., Pal, A., Performance, Emission and Combustion Characteristics of an Indica Diesel Engine Operated with Yellow Oleander (Thevetia Peruviana) Oil Biodiesel Produced through Hydrodynamic Cavitation Method. *Int. J. Ambient Energy*, 39, 4, 365–371, 2018.
77. Dewangan, A., Yadav, A.K., Mallick, A., Singh, S., Comparative study of Manilkara Zapota and Karanja based Biodiesel Properties and Its Effect on Multi-Cylinder Diesel Engine Characteristics. *Energy Sources Part A*, 44, 2, 5143–5153, 2022.
78. Yadav, A.K., Khan, T.A., Khan, T.A., Kumar, S., Methyl ester of Gmelina arborea oil as a substitute for petroleum diesel: an experimental study on its performance and emissions in a diesel engine. *Energy Sources Part A*, 43, 11, 1307–1314, 2021.
79. Khan, O., Yadav, A.K., Khan, M.E., Parvez, M., Characterization of Bio-Ethanol Obtained from Eichhornia Crassipes Plant; Its Emission and Performance Analysis on CI Engine. *Energy Sources Part A*, 43, 14, 1793–1803, 2021.
80. Khan, I.A., Singh, S.K., Pal, A., Yadav, A.K., Efficient Production of Biodiesel from Cannabis Sativa oil using Intensified Transesterication (Hydrodynamic Cavitation) Method. *Energy Sources Part A*, 42, 20, 2461–2470, 2020.
81. Yadav, A.K., Vinay, Singh, B., Optimization of Biodiesel Production from Annona Squamosa Seeds Oil Using Response Surface Methodology and Its Characterization. *Energy Sources Part A*, 40, 9, 1051–1059, 2018.
82. Dewangan, A., Yadav, A.K., Mallick, A., Tyagi, A., Optimisation of biodiesel production and engine performance from Simarouba oil in compression ignition engine. *Int. J. Oil Gas Coal Technol.*, 25, 3, 357–376, 2020.
83. Khan, M.E., Khan, T.A., Yadav, A.K., Pal, A., A comprehensive review on performance, emission and combustion characteristics of diesel engines fuelled with algae oil and methyl ester. *WEENTECH Proc. Energy*, 6, 2, 160–168, 2020.
84. Yadav, A.K., Khan, M.E., Pal, A., Custard Apple Seed (Annona Squamosa) Oil as a Promising Feedstock for Biodiesel Production using Advanced Techniques and Its Experimental Investigations on an Agricultural Diesel Engine. *Int. J. Oil Gas Coal Technol.*, 20, 4, 473–492, 2019.
85. Yadav, A.K., Pal, A., Ghosh, U., Gupta, S.K., Comparative Study of Biodiesel Production Methods from Yellow Oleander Oil and its Performance Analysis on an Agricultural Diesel Engine. *Int. J. Ambient Energy*, 40, 2, 152–157, 2019.
86. Khan, I.A., Singh, S.K., Yadav, A.K., Sharma, D., Enhancement in the Performance of a Diesel Engine Fueled with Pongamia Methyl Ester and n-Butanol as Oxygenated Additive. *Int. J. Ambient Energy*, 40, 8, 842–846, 2019.

87 Yadav, A.K., Khan, O., Khan, M.E., Utilization of High FFA Landfill Waste (Leachates) as a Feedstock for Sustainable Biodiesel Production: Its Characterization and Engine Performance Evaluation. *Environ. Sci. Pollut. Res.*, 25, 32, 32312–32320, 2018.
88. Yadav, A.K., Dewangan, A., Mallick, A., Effect of n-Butanol and Diethyl Ether Additives on Performance and Emission Characteristics of a Diesel Engine Fuelled with Diesel-Pongamia Biodiesel Blends. *J. Energy Eng.*, 144, 6, 04018062, 2018.
89. Vinay, Singh, B., Yadav, A.K., Optimization of Performance and Emission Characteristics of CI Engine Fuelled with Mahua Oil Methyl Ester-Diesel Blend using Response Surface Methodology. *Int. J. Ambient Energy*, 44, 6, 674–85, 2020.
90. Yadav, A.K., Dewangan, A., Mallick, A., Synthesis and Stability Study of Biodiesel from Kachnar Seed Oil. *J. Energy Eng.*, 144, 5, 04018053, 2018.
97. Yadav, A.K., Pal, A., Dubey, A.M., Prunus Armeniaca, L., (Wild Apricot) Biodiesel as an Alternative Fuel for CI Engine. *Waste Biomass Valorization*, 9, 10, 1961–1969, 2018.
92. Yadav, A.K., Khan, M.E., Pal, A., Ultrasonic Assisted Production of Biodiesel from Karabi Oil using Heterogeneous Catalyst. *Biofuels*, 9, 1, 101–112, 2018.
93. Dewangan, A., Yadav, A.K., Mallick, A., Current Scenario of Biodiesel Development in India: Prospects and Challenges. *Energy Sources Part A*, 40, 20, 2494–2501, 2018.
94. Yadav, A.K., Khan, M.E., Pal, A., Performance and Emission Characteristics of a Stationary Diesel Engine fuelled by Schleichera Oleosa oil Methyl Ester (SOME) Produced through Hydrodynamic Cavitation Process. *Egypt. J. Pet.*, 27, 1, 89–93, 2018.
95. Dewangan, A., Mallick, A., Yadav, A.K., Effect of Metal Oxide Nanoparticles and Engine Parameters on the Performance of a Diesel Engine: A Review. *Mater. Today Proc.*, 21, 1722–27, 2020.
96. Yadav, A.K., Khan, M.E., Pal, A., Dubey, A.M., Experimental Investigations of Performance and Emissions Characteristics of Kusum (Schleichera Oleosa) Biodiesel in a Multi-Cylinder Transportation Diesel Engine. *Waste Biomass Valorization*, 8, 4, 1331, 2017.
97. Yadav, A.K., Khan, M.E., Pal, A., Biodiesel Production from Oleander (Thevetia Peruviana) Oil and Its Performance Testing on a Diesel Engine. *Korean J. Chem. Eng.*, 34, 2, 340–345, 2017.
98. Dubey, A.M., Kumar, A., Yadav, A.K., Wear behaviour of friction stir weld joint of cast Al (4–10%) Cu alloy welded at different operating parameters. *J. Mater. Process. Technol.*, 240, 87–97, 2017.
99. Khan, O., Khan, M.E., Yadav, A.K., Sharma, D., The Ultrasonic Assisted Optimization of Biodiesel Production from Eucalyptus Oil. *Energy Sources Part A*, 39, 13, 1323–1331, 2017.
100. Dewangan, A. and Yadav, A.K., Wax Deposition during Production of Waxy Crude Oil and Its Remediation. *Pet. Sci. Technol.*, 35, 18, 1831–1838, 2017.

101. Yadav, A.K., Khan, M.E., Pal, A., Kaner biodiesel production through hybrid reactor and its performance testing on a CI engine at different compression ratios. *Egypt. J. Pet.*, 26, 2, 525–532, 2017.
102. Zhao, J., Xi, X.I., Na, Q.I., Wang, S., Kadry, S.N., Kumar, P.M., The technological innovation of hybrid and plug-in electric vehicles for environment carbon pollution control. *Environ. Impact Assess. Rev.*, 86, 106506, 2021.
103. Cao, Y., Yao, M., Sun, X., An overview of modelling and energy management strategies for hybrid electric vehicles. *Appl. Sci.*, 13, 10, 5947, 2023.
104. Limon, M.H., Debnath, B., Bari, A.M., Exploration of the drivers influencing the growth of hybrid electric vehicle adoption in the emerging economies: Implications towards sustainability and low-carbon economy. *Sustain. Oper. Comput.*, 4, 76–87, 2023.
105. Liu, X., Deng, Y., Xiong, J., Wu, X., Srivastava, V., Pan, W., Sun, J., Real-time Multi-layer Predictive Energy Management for a Plug-in Hybrid Vehicle based on Horizon and Navigation Data (No. 2024-01-2773). SAE Technical Paper, 2024.
106. Pielecha, J., Skobiej, K., Kurtyka, K., Exhaust emissions and energy consumption analysis of conventional, hybrid, and electric vehicles in real driving cycles. *Energies*, 13, 23, 6423, 2020.
107. Khan, O., Khan, M.Z., Khan, E., Bhatt, B.K., Afzal, A., Ağbulut, U., Shaik, S., An enhancement in diesel engine performance, combustion, and emission attributes fueled with Eichhornia crassipes oil and copper oxide nanoparticles at different injection pressures. *Energy Sources Part A: Recovery Util. Environ. Eff.*, 44, 3, 6501–6522, 2022.
108. Sharma, G., Dewangan, A.K., Yadav, A.K., Ahmed, A., Current status of research on hydrogen generation, storage and transportation technologies: A state-of-the-art review towards sustainable energy. *Process Saf. Environ. Prot.*, 191, Part A, 1445–1460, 2024.
109. Yadav, A.K., Khan, M.E., Pal, A., Dubey, A.M., Performance and emission characteristics of a transportation diesel engine operated with non-edible vegetable oils biodiesel. *Case Stud. Therm. Eng.*, 8, 236–244, 2016.
110. Yadav, A.K., Khan, M.E., Pal, A., Biodiesel Production from Nerium Oleander (Thevetia Peruviana) Oil through Conventional and Ultrasonic Irradiation Methods. *Energy Sources Part A*, 38, 23, 3447–3452, 2016.
111. Dubey, A.M., Ram, R., Yadav, A.K., Ion cell performance using single particle representation of battery electrode. *J. Electroanal. Chem.*, 768, 11–17, 2016.
112. Yadav, A.K., Khan, M.E., Pal A., A., Comparative Study on Ultrasonic Cavitation and Mechanical Stirring Method towards Efficient Production of Biodiesel from Non-Edible Oils and Performance Testing on a C.I. Engine. *Int. J. Environ. Waste Manag.*, 18, 4, 349–367, 2016.
113. Dewangan, A., Mallick, A., Yadav, A.K., Kumar, R., Combustion-generated pollutions and strategy for its control in CI engines: A review. *Mater. Today: Proc.*, 21, 1728–1733, 2020.

114. Yadav, A.K., Khan, M.E., Pal, A., Sharma, D., Optimization of biodiesel production from bitter groundnut oil using Taguchi method and its performance and emissions characteristics on a 4-cylinder Tata Indica engine. *Int. J. Sustain. Agric. Manag. Inform.*, 1, 285–300, 2015.
115. Khan, O., Parvez, M., Yadav, A.K., *et al.*, Investigation of the Thermal Properties of Lauric Acid Melting as a Phase Change Material Using Priority based K-Means Clustering Method. *Energy Storage*, 6, 6, e70026, 2024, doi: 10.1002/est2.70026.
116. Singh, S., Singh, B., Kumar, S., Yadav, A.K., Temperature-dependent dynamic hysteresis scaling of ferroelectric hysteresis parameters of lead free [(Ba 0.825 + xCa0.175-x) (Ti1- x Snx)O3]ceramics. *Ferroelectrics*, 551, 1, 133–142, 2019.
117. Khan, F., Khan, O., Parvez, M., Ahmad, S., Yahya, Z., Alhodaib, A., Yadav, A.K., Ağbulut, Ü., K-means clustering optimization of various quantum dots and nanoparticles-added biofuels for engine performance, emission, vibration, and noise characteristics. *Therm. Sci. Eng. Prog.*, 54, 102815, 2024.

18

Safety Assessments of Battery Thermal Management Systems (BTMS) in Electric Vehicles (EVs)

Sainu Baliyan

Department of Chemistry, School of Basic Sciences & Technology, IIMT University Meerut, Uttar Pradesh, India

Abstract

As electric vehicles (EVs) become increasingly popular, ensuring their safety and efficiency is critical, especially regarding Battery Thermal Management Systems (BTMS). This paper examines the safety evaluations of BTMS, which are essential for regulating battery temperature and preventing dangerous thermal runaway events. BTMSs are responsible for maintaining optimal battery temperatures, which enhances performance and extends battery life. The evaluation process focuses on the system's ability to manage extreme thermal conditions, its response to potential failures, and its overall dependability in keeping temperatures within safe limits. Key aspects of the safety assessment include thermal modeling, simulating thermal incidents, and testing under diverse environmental conditions. The study reviews different BTMS technologies, such as active and passive cooling methods, and assesses their effectiveness in reducing risks. It also considers the role of advanced sensors and control systems in real-time temperature monitoring and management. By pinpointing potential weaknesses and developing effective strategies to address them, this paper aims to support the development of safer and more reliable BTMS designs, thereby improving the overall safety of electric vehicles.

***Keywords*:** Battery thermal management systems (BTMS), electric vehicles (EVs)

Email: baliyanji97@gmaiil.com

Ashwani Kumar, Mukesh Kumar Awasthi, Nitesh Dutt, Yogesh Kumar Singla and Sivasakthivel Thangavel (eds.) Thermal Battery Management System for Hybrid and Electric Vehicles, (367–386) © 2025 Scrivener Publishing LLC

18.1 Introduction

The shift towards electric vehicles (EVs) has gained momentum due to growing environmental concerns and advances in battery technology. The safety and efficiency of these vehicles are heavily dependent on the performance of their BTMSs. These systems are important to regulate the thermal process of the cell pack, keeping it within safe operating limits and preventing thermal runaway—a dangerous condition that can lead to severe battery failures, including fires and explosions. These days, the world's most pressing problems are the worldwide energy issue and pollution in the atmosphere, with traditional cars that run on petrol and diesel fuel having a direct correlation with these two problems. In order to address the climate change issue and difficulties with pollution of the environment, electric automobiles and battery-electric vehicles with clean, high energy utilization, and environmental protection are being created. Li-ion batteries, which have a significant amount of energy and power density, are becoming more and more popular as EV and HEV power sources. Nevertheless, a number of issues, including thermal safety, declining discharge efficiency, and cycle life degradation of Li-ion batteries, can pose significant barriers to the continued commercialization of EVs and HEVs [1–3].

18.1.1 The Role of BTMS in EVs

Many HEVs and pure electric vehicles with different mixing levels are available on the current auto market. Depending on the degree of blending, EVs have a variety of battery cell types, sizes, and quantities implanted. Compared to conventional fuel, battery cells have stricter criteria for the operating environment. They are incredibly sensitive to temperature. BTMS and battery cells will need to be paired to provide an optimally thermally efficient environment. Thus, it's essential to comprehend the prerequisites for a battery's proper operation as well as the kinds of management techniques that can satisfy them. With this basis, the efficiency and lifespan of the electrical system in an EV may be optimized [4, 5].

Investment in electric motor vehicles (EVs), such as electric hybrids (HEVs), plug-in hybrid electric automobiles (PHEVs), and battery-powered vehicles (BEVs), has grown as environmental regulations on emissions of greenhouse gases (GHG) have become more stringent. The biggest challenge for EVs is locating the appropriate energy source. High-performance storage system, rapid charging, and long range [6]. Because of their better density of energy and specific power, increased recyclable nature, increased lightness, lower self-discharge rates in comparison with other rechargeable power sources like lead-acid, and longer cycle life, rechargeable Li-ion batteries are considered the greatest alternative for storing energy. They gain from not having any memory impact as well [7].

But when it comes of performance, longevity, and safety, these lithium-ion batteries are very temperature-sensitive. Because commercial batteries use a variety of materials for electrodes and electrolyte mixtures, it is difficult to design a clear and complete system that lowers the safety and efficacy of Li-ion batteries. All cell materials have an anomalous temperature range, however it is obvious that nickel-cadmium and Ni-MH batteries, which are composed of nickel and cadmium, exhibit lower performance and stability than Li-ion batteries. Because they are affected by external influences and release heat as a result of various chemical reactions associated to charge and discharge, batteries often undergo temperature variations. Numerous articles have made significant contributions to the design of the battery system and the creation of a cooling mechanism to the battery system. Continuous monitoring of the battery system's thermal model, charge, and health is the responsibility of the battery thermal management system [8]. To keep the battery within its functioning range and lower its temperature, the cooling system plays a crucial role. The air the cooling process, inactive, and active cooling systems are the main focus of this battery thermal management system, and the observations gained from this project enable us to create an effective battery cooling system.

Lithium-ion batteries, which are widely used in EVs, are highly sensitive to temperature changes. Operating the battery outside its optimal temperature range can result in accelerated degradation, reduced capacity, or, in severe cases, thermal runaway. The BTMS is designed to maintain the battery's temperature within a safe range during charging, discharging, and storage by employing various cooling methods, such as air cooling, liquid cooling, phase change materials, and heat pipes, each with its own benefits and drawbacks [9–11].

18.2 Evaluating BTMS Safety

The safety of BTMS is rigorously evaluated through extensive testing and simulations under a range of operational conditions. These evaluations focus on several key aspects:

1. Thermal Runaway Prevention

Thermal runaway is a critical safety concern, where an uncontrollable increase in temperature can lead to fires or explosions. The BTMS must be equipped to detect early signs of thermal runaway and implement mitigation strategies to prevent catastrophic failures. Electric cars and trucks frequently employ batteries made of lithium ion (LiBs) due to their elevated density of energy and extended lifespan. It is essential to design a battery thermal management system (BTMS) that stops the spread of runaway thermal energy (TR) in the event of abusive accidents. To stop TR propagation in the power source module, this study aims to construct a unique hybrid BTMS featuring liquid as well as passive cooling features. To study the transverse repulsion propagation induced by a single cell, a computational framework for a battery system (16 spherical 18 650 cells) was created using the multiphysics program COMSOL. Passive cooling was achieved using copper-based foam and expandable graphite-paraffin (EG-PCM) composite material [12–14].

2. Overheating Protection

Overheating can significantly reduce battery life and performance. The BTMS must include robust overheating protection mechanisms, utilizing sensors and algorithms to monitor and regulate temperatures within safe limits. For both electric and hybrid vehicles, the power module's thermal characterization is a crucial component that affects the inverter's dependability and efficiency. A three-phase PWM switching inverter produces the power absorption of conductivity and switching losses in IGBTs and diodes. Junction temperature (Tj) rises as a result of this power dissipation, which is dependent upon the DC-Link voltage, modulation index, phase current frequency, and magnitude. The Tj must not exceed the specified limit, no matter what driving circumstances, including fast traveling, regenerative braking, and hill-holding. In order to determine the highest junction temperature under various driving situations, the PWM inverter's power loss computation and thermal model are created in this study. A HEV or EV's lifetime is ensured and overheating is safely prevented using the expected junction temperature. NTC, where a temperature sensor inside the battery module, is additionally required to operate the electrical appliances as long as the maximum permissible junction temperature has not exceeded the limit. This is because the ones suggested thermal energy models are no longer valid in abnormal conditions, such as a lack of water or a failure of the current sensor [15–17].

3. Cooling Efficiency

Effective cooling is vital during periods of high power demand or environmental stress. The BTMS should be designed to dissipate heat efficiently, using either passive or active cooling systems, depending on the vehicle's needs. With the right cooling technology, battery packs can withstand more negative temperature changes, have higher power battery efficiency, be safer to use, age more slowly, and have longer service lives. This context examines battery thermal management systems (BTMS), including the change in phase material-based the BTMS process, heat pipe-based BTMS, and the thermoelectric system element-based BTMS in the new battery thermal management system; and the air cooling, liquid cooling, and refrigeration fluid direct cooling BTMS in the conventional battery thermal management system. The six energy battery cooling technologies mentioned above are discussed in order to lessen the detrimental effects of high temperatures on battery packs and to look for practical approaches for BTMS systems in future development. Provide an overview of the current research priorities and advancements for each BTMS. Analyze each BTMS's benefits and drawbacks with objectivity. After taking into account each BTMS's financial benefits, installation viability, and existing working conditions, appropriate solutions should be discussed, and future development patterns should be forecast with reasonable accuracy. Lastly, examine and talk about the distinctions and shortcomings between the new and traditional BTMS. supplying a model for creating the ideal BYMS system. Maintaining the battery in the ideal operating temperature range, enhancing battery conversion efficiency, and preserving the BTMS stable operation are all important ways to contribute to the future BTMS research [18, 19].

4. Insulation Integrity

Proper insulation within the BTMS prevents electrical short circuits and limits thermal propagation between cells. The choice of insulation materials and their integrity are essential for maintaining overall system safety. An essential part of electric car technology, the cooling system for the lithium-iron phosphate batteries plays a major role in the safe and effective functioning of these vehicles. Cell cooling is one of the most important issues that electric vehicles must overcome to guarantee peak performance, longer battery life, and increased safety. Excessive heat produced when a battery is operating can cause performance issues as well as possible safety risks. The cooling capacity, dimensions, and complications of traditional cooling systems, like liquid or air cooling, are all constrained. Thus, it's necessary to investigate cutting-edge cooling strategies that can successfully lessen the thermal problems related to lithium-ion battery cells for electric vehicles. The use of phase transition substances at the crucial thickness and a thermal insulating layer appears to be a novel strategy for reducing battery cell surface temperature. Throughout the charging phase, the highest temperatures of the naked cell, Teflon-insulated cell, and paraffin wax-coated battery were observed to be 69 °C, 57 °C, and 53.3 °C, respectively. It was shown that the Teflon-coated cell reduced peak temperature by 17.39% when compared to the naked cell, but the layer of paraffin wax-coated cell reduced peak temperature by a more significant 23.18%. Through the battery cell's discharging phase, a similar trend of temperature drop is seen. Reduced surface temperatures improve battery cell safety and lengthen its lifespan when insulating or phase-changing materials with a required thickness are used [20–23].

5. Temperature Uniformity

Maintaining a consistent temperature across all battery cells is crucial to prevent uneven aging or degradation. Some papers examined a thermal management technique based on a tiny heat sink integrated with PCM in order to further improve the thermal uniformity of the power battery. The main elements influencing the performance of thermal management are the thermophysical characteristics and dispersion of PCM. Carefully considered was a comparative analysis of the cell's thermal management system's cooling and homogenization capabilities under various thermal conductivities, PCM distributions, and melting temperatures. The primary findings indicate that the battery's maximum operating temperature and thermal standard deviation (STD) can be lowered thanks to the heat sink's PCM-embedded functionality. The battery's maximum temperature and STD can be lowered in comparison to the air cooling approach. Thermal conductivity's impact might be disregarded. It is possible to lower the highest temperature and STD by reducing the temperature at which things melt and raising latent heat. It might be a practical strategy to enhance the effectiveness of thermal management [24–26].

6. Fault Detection and Diagnostics

Rapid detection and diagnosis of faults, such as sensor failures or wiring issues, are critical for BTMS reliability. The system must incorporate advanced diagnostics to identify and address issues swiftly. To guarantee the dependability and safety of electric vehicles (EVs), fault detection and diagnosis, or FDD, is crucial. Critical parts of an electric vehicle (EV) that are prone to various failures include the battery system and electric motor drive, which together make up the energy storage and power train. EV problems and even disastrous accidents may result from a failure to identify and fix these issues on time. Fixed Magnetic Synchronise Motors and battery packs containing lithium-ion have attracted a lot of interest in the field of electric vehicle applications. As a result, defect detection techniques for batteries made of lithium-ion and their engines have gained prominence in the field of study. Precision, acceleration, sensitivities, and value for money are essential characteristics of a successful FDD strategy. Conventional fault-finding and diagnosis (FDD) techniques comprise model-based and signal-based approaches. On the other hand, data-driven approaches—such as machine learning-based approaches—have gained popularity recently because of their promising fault detection capabilities. This paper covers the most recent state-of-the-art research on EV fault detection and attempts to offer a thorough overview of possible defects in electric vehicle motors and battery systems. The data presented here can be a useful resource for future research in this area [27–30].

7. Fire Suppression Systems

Incorporating fire suppression systems into the BTMS enhances safety by providing immediate response to potential fires. These systems should be well-integrated and effective in extinguishing or containing fires within the battery pack. Electric vehicles (EVs) have revolutionized the global automotive industry, mostly due to the rapid advancement of lithium-ion battery technology. However, a significant safety worry for EVs is the flames risk and hazard related to this high-energy battery. The most recent EV fire safety concerns, such as thermal runaway and fire in Li-ion batteries, are reviewed in detail. Extreme misuse circumstances, which might as malfunctioning equipment or traffic accidents, can lead to thermal runaway or fire. An explosion, combustion, rocket flames, and the emission of poisonous fumes could then follow the battery's failure. Suppression agent must be used excessively to decrease the temperature the battery, put out the fire, and stop it from starting again. By addressing these issues, the study hopes to support academics and businesses involved in battery, electric vehicle (EV), and fire safety engineering. It also hopes to foster active research partnerships and draw attention to further research and development aimed at enhancing the general safety of EVs in the future. It won't be until then that EVs and conventional cars are as comfortable for society as they are [31–33].

8. Redundancy and Fail-Safes

Redundant systems and fail-safes are essential to ensure that the BTMS remains operational even if the primary system fails. This redundancy is key to maintaining vehicle safety and performance. A fail-safe approach for the sensor arrays of autonomous cars is presented in this study. The three components of the proposed fault diagnosis process are (1) in-vehicle sensor fail-safes; (2) fault isolating by potential area prediction; and (3) fault detection *via* a duplication-comparison method. The unique fault isolation approach using target predictions and the use of duplicated external sensor pairs that estimate the same target and its outcomes have been assessed in finding the problem by an enhanced duplication-comparison method are the major ideas. It is possible to identify the location of defects and distinguish between those caused by software errors and sensor-related issues through the comparison of the estimations of nearby vehicles with the raw measurement data. The algorithm's performance has been evaluated by offline MATLAB vehicle data analysis. The efficacy of the proposed approach was assessed using the duration of the gap between injection and fault identification in many fault injection trials [34–36].

9. Battery Cooling System Reliability

The reliability of the BTMS cooling system, including components like pumps, fans, and coolant, must be thoroughly tested to ensure consistent performance and prevent failures under various operating conditions. The most recent developments in various BTMS, including air cooling, liquid cooling, and heat pipe-based cooling, were covered in the review work that is currently underway. Air-cooled BTMS systems are straightforward to design, reliable, and safe, but they have lower heat capacities and thermal efficiencies in addition to having restricted functionality when it comes to thermal management systems. Because of the ventilation patterns within the battery pack of an electric car, forced air-cooled battery thermal management systems (BTMS) that use the airflow pathways inside battery packs effectively cool the battery [37–40].

10. Vibration and Shock Resistance

Electric vehicles are subject to mechanical vibrations and shocks during operation. The BTMS must be designed to withstand these forces without compromising its functionality or safety. In many engineering systems, the retention of the energy of vibration by mechanical damping poses a serious challenge. To lessen vibration in machinery and airplanes, high-damping substances are employed. Advanced material durable strength, as well as decreases in mass and expenses, are advantages of damping treatment. Only a small number of research publications discuss dampened composite structures and shock-resistant nanomaterials, despite the possibility of controlling shock waves that propagate in the material and greatly increasing impact energy dissipation by utilizing the features of nanostructures [41–43].

11. Overcharge and Overdischarge Protection

To prevent damage, the BTMS must include mechanisms that protect against overcharging and overdischarging, both of which can lead to reduced battery capacity or even safety hazards. The reactions of commercially available lithium-ion phosphate batteries during overcharging and over-discharge scenarios were examined through testing. Batteries got charged at 1C progressively during overcharge experiments until a certain visual sign, like gas explosion or vent burst, occurred. Cells were discharged and maintained at zero volts or even negative polarity during overdischarge testing. To analyze the abuse process, variables including internal temperature, cell voltage, and ac impedance were tracked and recorded. Tested cells displayed varying degrees of failure. This work establishes an early precaution strategy for battery failures in order to ensure the safety of applications that use batteries. It does this by connecting the symptoms, impacts, and effects of overcharged and overdischarge phenomena, in addition to the electrochemical analysis and post-disassembly evaluation [44, 45].

12. Electrical Insulation and Isolation

Effective electrical insulation and isolation within the BTMS are critical to preventing short circuits and leakage currents. This ensures that the system remains safe and functional under various conditions. For hybrid electric vehicles, a new kind of online monitor for insulation resistance has been developed (HEV). It is primarily made up of a processing module for data sampling and a proportional circuit. By measuring the voltage between the positive line and the chassis and the DC bus voltage, the gadget measures insulation resistance. An electrically controlled unit (ECU) is used to control the voltage sample. The opto-coupler isolation amplifier isolates these voltages before sending them to the ECU. The ECU computes the insulating resistance value. The formulas for calculating insulating resistance, software realization, and hardware circuit are provided. The gadget has undergone laboratory testing, and the findings indicate that it is capable of quickly and correctly detecting any type of insulation state and sounding an alert whenever the insulating material is compromised. The gadget uses an electrically powered switch to regulate sampling and does not require a voltage or current sensor. As a result, it lowers costs and facilitates engineering realization [46, 47].

13. Environmental Resistance

The BTMS must be resilient to environmental factors such as moisture, dust, and extreme temperatures. Ensuring that the system is protected against these elements is vital for long-term reliability. Li-ion batteries are extremely sensitive to temperature, so temperature management is a serious problem that needs to be solved. Electric car batteries generate a great deal of heat, which lowers the vehicle's performance and efficiency. Electric cars (EVs) are currently necessary because traditional internal combustion engines (IC engines) generate a significant amount of pollution that negatively impacts the environment, whereas EVs produce a very tiny amount of pollution. Many individuals are now recommending and using it. However, the electric vehicle's battery module overheating is causing some serious issues. These days, a device known as a BTMS controls the temperature of batteries. EVs of today have both passive and active cooling systems. Battery architecture is improved through thermal management in an effort to increase autonomy or speed up charging. Future thermal management challenges will require the use of liquid or air cooling. The vehicle's power, performance, energy preservation, recharging, and discharges are all adversely affected by the battery overheating; therefore, a dependable battery thermal management system is necessary to address these issues. The issues with temperature and safety are covered first in some investigations. The advantages and disadvantages of battery pack temperature sensors (BTMSs), which have been utilized to maintain battery packs at appropriate temperatures, are listed below. Finally, a review of the developments made in the area of temperature control systems for next-generation batteries is provided. It has been suggested to perform a thorough evaluation and categorization of both current and prospective battery management systems as a preliminary look at prospective BTMSs for locomotive operations [48].

14. Cooling System Leak Detection

Detecting leaks in liquid cooling systems early is crucial to prevent potential failures. The BTMS should incorporate reliable leak detection mechanisms to maintain the integrity of the cooling system. The growing concern of protecting the environment from pollutants is focusing special attention on the electric vehicle. Research on the effects of valve opening and refrigerant charge quantity on the operation of the sustainable CO2 automobile heat pump system and its individual components performance is still lacking, particularly in cases where the system incorporates an accumulator heat exchanger. Furthermore, research on oil circulating rate in CO2 vehicle systems is still lacking. This study carefully investigated the effects of oil circulation performance and refrigerant charging quantity and opening of valves on system/component performance and cooling mode. When the low-pressure conditions prevail accumulator inlet port is in the superheated state due to low refrigerant charge levels, the oil distribution rate tends to decrease as valve opening increases; however, when the inlet is in the two-phase state due to high refrigerant charge levels, the oil circulation evaluate curve exhibits a growing trend. The engineering community will profit from this study by having a thorough understanding of the thermal management of electric vehicle systems, which will preserve vehicle dependability and enhance comfort and safety [49, 50].

15. Material Compatibility

The materials used in the BTMS must be compatible with the battery chemistry to avoid adverse reactions. Careful selection and testing of materials are essential for ensuring the system's safety and longevity. Fossil fuel consumption and the need for energy are growing in the modern world, particularly in nations experiencing rapid economic expansion. Moreover, the transportation sector's emissions of atmospheric greenhouse gases and other pollutants have serious negative effects on the environment and public health, which makes the creation of new, renewable energy sources necessary. In many nations, the auto industry is progressively shifting to battery-powered cars in an effort to reduce pollution and reach the objective of zero emissions. The Li-ion battery is one of the most common kinds of battery packs in cars because of its high specified energy and power, extended life span per cycle, and relatively low self-discharge rate. Therefore, improving Li-ion battery capacity and cycle life is essential to raising the effectiveness of electric vehicles. Li-ion batteries are sensitive to temperature, though. Li-ion batteries operate best in a temperature range of 15 to 40 °C, with a temperature differential of no more than 5 °C between various battery components. Long-term occurrence of a wrong temperature zone and variation in temperature range in the lithium-ion battery will result in a number of issues. The Li-ion battery's capacity and charging and discharging voltages vary dramatically in cold environments. Low-temperature lithium coating on a battery made of lithium-ion will lower the battery's electrical capacity and cause significant deterioration. When the battery functions at a high temperature, there is a greater chance of heat runaway or even explosion [51–53].

16. Emergency Shutdown Procedures

In the event of a critical fault, the BTMS must have clear protocols for an emergency shutdown to protect the battery and vehicle. These procedures should be designed for rapid and safe execution. This study suggests a functional modeling technique for identifying safety goals and immediate shutdown (ESD) system functions during the design stage that allows for safe emergency shutdown: multilevel flow modeling (MFM). First, an analysis is done on the data needed to develop safety goals and tasks for an emergency shutdown. Process topology, the process's starting point, the shutdown goals, and additional limitations including unit operations' operational and design bounds as well as environmental and hazardous constraints are all included in the data. Second, utilizing MFM, a procedure is suggested based on the necessary data. The process includes the following steps: 1) The modeling methods are followed in order to build the MFM visualization of the procedure in normal operations. 2) When a process unexpectedly shuts down, an objective-function graph within MFM modeling is used to describe initial states and identify changing conditions. 3) The shutdown operation's objectives are specified, 4) Applying MFM causal reasoning, the first security duty during shutdown process is determined after taking into account the shutdown operational goals and changing situations. 5) The remaining safety functions are found and the ESD processes are carried out by applying MFM risk logic upon the very first verified protection value described in the MFM model. Designing safety features for an emergency shutdown of a saltwater deaeration technique in a saltwater injection system serves as an example of this approach. The outcomes demonstrate the viability of the emergency shutdown protocol developed using the suggested approach and its ability to be verified against the official operating protocol [24].

17. Battery Cell Balancing

The BTMS should ensure that all battery cells are evenly charged, preventing imbalances that can shorten battery life or reduce performance. Effective cell balancing enhances the efficiency or safety of that cell. An automobile battery management unit must do more than just monitor and regulate the voltages, currents, and temperatures of each individual battery cell in the battery unit of an electric car. Mechanisms to rectify intracellular imbalances and the kinds and causes of these imbalances will be discussed in various research studies. Moreover, safety features of BMS concerning operational safety according to ISO 26262, standard safety devices and measures of automotive battery systems, and HV and IT safety/security are discussed. Battery state estimate with cell models is also covered. Finally, sources of more information and prospective trends are provided [25].

18. Thermal Propagation Prevention

Preventing thermal propagation, where heat spreads from one cell to adjacent cells, is critical for safety. The BTMS should be designed to contain and manage heat within individual cells to prevent wider system failures. One of the biggest problems with battery packs is the propagation of thermal breakdown, which often exacerbates thermal dangers and leads to more catastrophic accidents. Several studies on thermal failure were carried out to investigate how the level of charge number of heaters, location of failure, and pack size affect the battery pack's ability to spread thermal failure. The findings indicate that clear domino effected battery pack's temperature failure. Usually, there are multiple stages to a battery pack's thermal failure. The battery pack's propagation is hampered by the increased SOC. In addition, the pack with more heaters experiences the thermal failure early. The problem of failure propagate will be exacerbated if the battery near the pack's center has a thermal failure. Furthermore, it is demonstrated that as pack size increases, the propagation of thermal failure would be prolonged longer when it reaches the outermost portion of the batteries [26].

19. Load and Stress Testing

Testing the BTMS under maximum load and stress conditions is essential to assess its performance and reliability. These tests help identify potential weaknesses and ensure the system can withstand real-world demands. Consensus documents known as standards allow an innovation or practice to be homologated. Some research provides a review of the industry standards used by the EV (electric vehicle) battery business. It also lists the tests that are carried out to evaluate the battery's age and lifespan, normal working conditions, and instances of abuse or malfunction. The majority of standards are created and suggested by testing laboratories, battery manufacturers, device integrators, automakers, and government agencies. The standards are continuously evaluated to ensure that they remain applicable in light of new breakthroughs in technology. Commonly carried out testing include dynamic current simulations of EV consumption, charging and discharging capacity checks, thermal tests within the battery's stated operating range, and pulse current tests. The most popular tests for battery safety screening include controlled crush, indentation or nail penetration, overvoltage, short-circuit, and simulated fire exposure [27].

20. System Integration and Compatibility

The BTMS must be seamlessly integrated with other vehicle systems to ensure smooth operation. Compatibility and effective communication between systems are crucial for maintaining overall vehicle safety and performance. Should the government step in and require compatibility when businesses invest in complementary commodities or technical standards that are incompatible? Vigorous policy and antitrust discussions are sparked by this dilemma in a variety of industries, including manufacturing and digital markets. A move in the direction of compatibility allows customers to access the pooled investments of all companies, perhaps improving convenience, variety, or other quality indicators for them. Benefits for customers, however, can be outweighed by a drop in business investment. The concept of competition amongst firms is significantly altered by compatibility, which transforms investments made by firms from being competitors' takeovers to complements that benefit other enterprises. Because private profits from business theft contribute little to social surplus, businesses may invest excessively under incompatibility, and they may invest insufficiently under compatibility, as private incentives fail to internalize the positive spillovers. According to theoretical literature, private incentives for compatibility can be excessively large or low in comparison to social incentives. Empirical analysis can be used to determine how a compatibility policy affects welfare [28].

18.3 Challenges and Future Prospects

Despite considerable advancements, challenges persist in developing BTMS. Balancing efficiency, weight, and cost remains a significant challenge. Additionally, integrating BTMS with other vehicle systems, such as power electronics and cabin cooling, requires further innovation. Future research is focused on developing more advanced materials and cooling technologies, as well as enhancing simulation models to better predict and prevent failures. The advancement of Battery Thermal Management Systems (BTMS) in electric vehicles (EVs) presents a range of challenges and opportunities for the future. Key among these challenges is enhancing the early detection and prevention of thermal runaway, which could be improved through the integration of AI and machine learning technologies. Another concern is cooling efficiency; future developments are expected to focus on creating more compact and energy-efficient cooling solutions. As battery technologies advance, BTMS must evolve to accommodate new materials, necessitating innovative design approaches.

Efforts to reduce weight and size are crucial for improving vehicle efficiency, requiring the use of advanced materials and design techniques. Enhancing fault detection and integrating BTMS with autonomous vehicle systems will be essential for future safety and performance. Additionally, addressing environmental durability and scaling BTMS for mass production are ongoing challenges.

Future BTMS must also improve thermal propagation prevention, energy efficiency, and adhere to evolving regulatory standards while minimizing costs. Lifecycle management, including recycling and end-of-life processing, will become increasingly important. Real-time data analytics, leveraging cloud computing and IoT, will be vital for ongoing monitoring and predictive maintenance. Lastly, adapting BTMS to meet diverse global market needs will require versatile and customizable solutions.

18.4 Conclusion

This chapter emphasizes the critical role of BTMS in ensuring the safety and efficiency of electric vehicle batteries. By addressing these 20 key considerations, manufacturers can design BTMS that protect battery systems and meet stringent safety standards. Safety evaluations of BTMS in EVs are crucial for ensuring the reliable and safe operation of electric vehicles. These systems are vital in preventing battery-related incidents, boosting consumer confidence, and promoting the wider adoption of EVs. Ongoing innovation and thorough testing are essential to overcome existing challenges and ensure that BTMS can meet the evolving demands of the EV industry. In summary, conducting a thorough safety assessment of Battery Thermal Management Systems (BTMS) is essential to ensure the reliability, longevity, and safety of batteries, particularly in electric vehicles. The 20 key safety assessments outlined offer a solid framework for evaluating the effectiveness and durability of BTMS across various operational conditions.

Critical areas of focus include the system's capability to prevent and manage thermal runaway, maintain consistent temperature distribution, and protect against overheating—each vital for averting catastrophic incidents such as fires or explosions. The significance of fault detection, diagnostic tools, and emergency shutdown procedures cannot be understated, as these features enable quick responses to potential problems, thereby minimizing risks and potential damage.

Additionally, the system's structural resilience under mechanical stress, its resistance to environmental factors, and its ability to withstand electrical faults ensure reliable, safe operation under even the most challenging conditions. The integration of fire suppression systems, redundant components, and thermal propagation prevention mechanisms further strengthens the safety profile of BTMS, providing multiple layers of protection.

The BTMS's ability to balance battery cells, prevent overcharge and overdischarge, and ensure material compatibility with battery chemistry is crucial for optimizing battery performance and extending its operational life. Effective cooling systems, equipped with reliable leak detection and cooling efficiency, are also critical in maintaining optimal temperatures and protecting the battery from thermal stress.

Ultimately, a well-designed and thoroughly evaluated BTMS is vital for the safe and efficient operation of battery systems. By addressing these critical safety factors, manufacturers and operators can greatly mitigate the risks associated with battery use, thereby improving the safety and reliability of modern energy storage systems.

References

1. Fleckenstein, M., *et al.*, Thermal Management of Lithium-Ion Batteries: Modeling and Simulation, Springer, 2019.
2. Pesaran, A., Battery Thermal Models for Hybrid Vehicle Simulations. *J. Power Sources*, *110*, 2, 377–382, 2001.
3. Andre, D., *et al.*, Characterization of High Power Lithium-Ion Batteries by Electrochemical Impedance Spectroscopy II: Modelling. *J. Power Sources*, *196*, 12, 5349–5356, 2014.
4. Raj, A. and Dhoble, A.S., Battery Thermal Management System in Electric Vehicles: A Review. *J. Therm. Anal. Calorim.*, *135*, 1, 451–471, 2019.
5. Tu, H., Feng, H., Srdic, S., Lukic, S., Extreme rapid charging of electric vehicles: A technical review. *IEEE Trans. Transp. Electrif.*, 5, 4, 861–878, December 2019, doi: 10.1109/TTE.2019.2958709.
6. Park, C. and Jaura, A., Dynamic Thermal Model of Li-Ion Battery for Predictive Behavior in Hybrid and Fuel Cell Vehicles. *SAE Tech. Pap. Ser.*, 2003-01-2286, 2003, https://doi.org/10.4271/2003-01-2286.
7. Kumar, A., Prasad, A., Kumar, G., Sustainable Mobility: Policies, Challenges and Advancements. John Wiley & Sons, Inc., USA, 2024. https://doi.org/10.1002/9781394166831. ISBN: 9781394166831.
8. Jeon, S.U., Digital Object Identifier Study of Electric Vehicle Battery Charging Strategy Considering Battery Capacity. *IEEE Access*, June 21, 2021, 10.1109/ACCESS.2021.3090763.
9. Sun, Z., Guo, Y., Zhang, C., Xu, H., Zhou, Q., Wang, C., A novel hybrid battery thermal management system for prevention of thermal runaway propagation. *IEEE Trans. Transp. Electrif.*, 9, 4, 5028–5038, 2022.
10. Lee, J., Jang, H., Shin, S., Jang, K., Jung, J., Over temperature protection in power module for hybrid and electric vehicle, in: *2016 IEEE Transportation Electrification Conference and Expo, Asia-Pacific (ITEC Asia-Pacific)*, IEEE, pp. 432–435, 2016, June.

11. Lu, M., Zhang, X., Ji, J., Xu, X., Zhang, Y., Research progress on power battery cooling technology for electric vehicles. *J. Energy Storage*, 27, 101155, 2020.
12. Kadam, S., Gunjavate, P., Bhise, S., Enhancing battery cell safety and lifespan through strategic application of insulating and phase change materials utilizing critical thickness for enhanced heat transfer. *Eng. Res. Express*, 6, 2, 025503, 2024.
13. Zhao, J., Wu, C., Rao, Z., Investigation on the cooling and temperature uniformity of power battery pack based on gradient phase change materials embedded thin heat sinks. *Appl. Therm. Eng.*, 174, 115304, 2020.
14. Khaneghah, M.Z., Alzayed, M., Chaoui, H., Fault detection and diagnosis of the electric motor drive and battery system of electric vehicles. *Machines*, 11, 7, 713, 2023.
15. Sun, P., Bisschop, R., Niu, H., Huang, X., A review of battery fires in electric vehicles. *Fire Technol.*, 56, 4, 1361–1410, 2020.
16. Shin, D., Park, K.M., Park, M., Development of fail-safe algorithm for exteroceptive sensors of autonomous vehicles. *Electronics*, 9, 11, 1774, 2020.
17. Thakur, A.K., Prabakaran, R., Elkadeem, M.R., Sharshir, S.W., Arıcı, M., Wang, C., Saidur, R., A state of art review and future viewpoint on advance cooling techniques for Lithium–ion battery system of electric vehicles. *J. Energy Storage*, 32, 101771, 2020.
18. Kireitseu, M., Hui, D., Tomlinson, G., Advanced shock-resistant and vibration damping of nanoparticle-reinforced composite material. *Compos. Part B Eng.*, 39, 1, 128–138, 2008.
19. Wu, C., Sun, J., Zhu, C., Ge, Y., Zhao, Y., Research on overcharge and over-discharge effect on lithium-ion batteries, in: *2015 IEEE Vehicle Power and Propulsion Conference (VPPC)*, IEEE, pp. 1–6, 2015, October.
20. Zhen-jun, W. and Li-Fang, W., A novel insulation resistance monitoring device for hybrid electric vehicle, in: *2008 IEEE Vehicle Power and Propulsion Conference*, IEEE, pp. 1–4, 2008, September.
21. Chidambaranathan, B., Rajendran, A.K., Vijaya Kumar, A., Velayutham, B.K., Rajendran, A.K., Mohan, A., Sundaram, M., Thermal management system in electric vehicle batteries for environmental sustainability. *Environ. Qual. Manage.*, 33, 2, 159–167, 2023.
22. Li, G., Tang, Z., Zou, H., Zhang, R., Experimental investigation of cooling performance of a CO2 heat pump system with an integrated accumulator heat exchanger for electric vehicles: Impact of refrigerant charge and valve opening. *Appl. Therm. Eng.*, 224, 120077, 2023.
23. Luo, J., Zou, D., Wang, Y., Wang, S., Huang, L., Battery thermal management systems (BTMs) based on phase change material (PCM): A comprehensive review. *Chem. Eng. J.*, 430, 132741, 2022.
24. Wu, J., Song, M., Zhang, X., Lind, M., Identifying Safety Objectives and Functions for Emergency Shutdown in the Design Phase by Using Functional Modelling. *Chem. Eng. Trans.*, 86, 967–972, 2021.

25. Hauser, A. and Kuhn, R., Cell balancing, battery state estimation, and safety aspects of battery management systems for electric vehicles, in: *Advances in Battery Technologies for Electric Vehicles*, pp. 283–326, Woodhead Publishing, New Delhi, India, 2015.
26. Ouyang, D., Liu, J., Chen, M., Weng, J., Wang, J., An experimental study on the thermal failure propagation in lithium-ion battery pack. *J. Electrochem. Soc.*, 165, 10, A2184, 2018.
27. Castillo, E.C., Standards for electric vehicle batteries and associated testing procedures, in: *Advances in Battery Technologies for Electric Vehicles*, pp. 469–494, Woodhead Publishing, New Delhi, India, 2015.
28. Li, J., Compatibility and investment in the us electric vehicle market. Unpublished manuscript, MIT, 2019.
29. Kumar, B.A., Jyothi, B., Singh, A.R., Bajaj, M., Rathore, R.S., Berhanu, M., A novel strategy towards efficient and reliable electric vehicle charging for the realisation of a true sustainable transportation landscape. *Sci. Rep.*, 14, 1, 3261, 2024, 10.1038/s41598-024-53214-w.
30. Singh, A.R., Vishnuram, P., Alagarsamy, S., Bajaj, M., Blazek, V., Damaj, I., Othman, K.M., Electric vehicle charging technologies, infrastructure expansion, grid integration strategies, and their role in promoting sustainable e-mobility. *Alexandria Eng. J.*, 105, 300–330, 2024, 10.1016/j.aej.2024.06.093.
31. Gupta, A., Singh, R.K., Paul, J., Sharma, M., Joshi, S., Battery-Operated Electric Vehicles (BOEVs) Adoption in India: Analysis of Barriers and Strategies. *IEEE Trans. Eng. Manage.*, 71, 7076–7087, 2023, http://dx.doi.org/10.1109/TEM.2023.3253621.
32. Aggarwal, S., Singh, A.K., Rathore, R.S., Bajaj, M., Gupta, D., Revolutionizing load management: A novel technique to diminish the impact of electric vehicle charging stations on the electricity grid. *Sustain. Energy Technol. Assess.*, 65, 103784, 2024, 10.1016/j.seta.2024.103784.
33. Rekioua, D., Mokrani, Z., Kakouche, K., Rekioua, T., Oubelaid, A., Logerais, P.O., Ghoneim, S.S., Optimization and intelligent power management control for an autonomous hybrid wind turbine photovoltaic diesel generator with batteries. *Sci. Rep.*, 13, 1, 21830, 2023, 10.1038/s41598-023-49067-4.
34. Ramakrishnan, V., Vishnuram, P., Yang, T., Bajaj, M., Rathore, R.S., Zaitsev, I., Design and implementation of a high misalignment-tolerance wireless charger for an electric vehicle with control of the constant current/voltage charging. *Sci. Rep.*, 14, 1, 13165, 2024, 10.1038/s41598-024-63952-6.
35. Dimri, S.C., Indu, R., Bajaj, M., Rathore, R.S., Blazek, V., Dutta, A.K., Alsubai, S., Modeling of traffic at a road crossing and optimization of waiting time of the vehicles. *Alexandria Eng. J.*, 98, 114–129, 2024, 10.1016/j.aej.2024.04.050.
36. Kanouni, B., Badoud, A.E., Mekhilef, S., Bajaj, M., Zaitsev, I., Advanced efficient energy management strategy based on state machine control for multi-sources PV-PEMFC-batteries system. *Sci. Rep.*, 14, 1, 7996, 2024, 10.1038/s41598-024-58785-2.

37. Badri, Y.I., Sudabattula, S.K., Hussain, I., Malik, H., Márquez, F.P.G., Ultracapacitor as selectable energy buffer in electric vehicle application. *J. Energy Storage*, 73, 109200, 2023, http://dx.doi.org/10.1016/j.est.2023.109200.
38. Kasri, A., Ouari, K., Belkhier, Y., Bajaj, M., Zaitsev, I., Optimizing electric vehicle powertrains peak performance with robust predictive direct torque control of induction motors: A practical approach and experimental validation. *Sci. Rep.*, 14, 1, 14977, 2024, 10.1038/s41598-024-65988-0.
39. Afzal, M.Z., Aurangzeb, M., Iqbal, S., Pushkarna, M., Rehman, A.U., Kotb, H., Bereznychenko, V., A Novel Electric Vehicle Battery Management System Using an Artificial Neural Network-Based Adaptive Droop Control Theory. *Int. J. Energy Res.*, 2023, 1, 2581729, 2023, 10.1155/2023/2581729.
40. Almutairi, A., Albagami, N., Almesned, S., Alrumayh, O., Malik, H., Electric vehicle load estimation at home and workplace in Saudi Arabia for grid planners and policy makers. *Sustainability*, 15, 22, 15878, 2023, http://dx.doi.org/10.3390/su152215878.
41. Sahri, Y., Tamalouzt, S., Belaid, S.L., Bajaj, M., Ghoneim, S.S., Zawbaa, H.M., Kamel, S., Performance improvement of Hybrid System based DFIG-Wind/PV/Batteries connected to DC and AC grid by applying Intelligent Control. *Energy Rep.*, 9, 2027–2043, 2023, 10.1016/j.egyr.2023.01.021.
42. Bagga, P., Das, A.K., Wazid, M., Rodrigues, J.J., Choo, K.K.R., Park, Y., On the design of mutual authentication and key agreement protocol in internet of vehicles-enabled intelligent transportation system. *IEEE Trans. Veh. Technol.*, 70, 2, 1736–1751, 2021, 10.1109/TVT.2021.3050614.
43. Bhatt, A., Ongsakul, W., Pawar, J., Optimal energy management system for carbon–neutral microgrid integrating second-life batteries and crypto mining devices. *Sustain. Energy Technol. Assess.*, 64, 103686, 2024, 10.1016/j.seta.2024.103686.
44. Amoussou, I., Paddy, E.Y., Agajie, T.F., Ibrahim, F.S., Agajie, E.F., Nsanyuy, W.B., Mohammadi, S.A.D., Enhancing residential energy access with optimized stand-alone hybrid solar-diesel-battery systems in Buea, Cameroon. *Sci. Rep.*, 14, 1, 15543, 2024, 10.1038/s41598-024-66582-0.
45. Razaviyn, Z., Heidari, M., Thangavel, S., Verma, V., Kumar, A., Yadav, A.K., Marine Energy Harvesting from Fluid Flow *via* Vortex Induced Vibrations. *Int. J. Thermofluids*, 25, 101015, 2025, https://doi.org/10.1016/j.ijft.2024.101015.
46. Thapliyal, S., Wazid, M., Singh, D.P., Das, A.K., Islam, S.H., Robust authenticated key agreement protocol for internet of vehicles-envisioned intelligent transportation system. *J. Syst. Archit.*, 142, 102937, 2023, 10.1016/j.sysarc.2023.102937.
47. Das, S., Thakur, P., Singh, A.K., Singh, S.N., Optimal management of vehicle-to-grid and grid-to-vehicle strategies for load profile improvement in distribution system. *J. Energy Storage*, 49, 104068, 2022, http://dx.doi.org/10.1016/j.est.2022.104068.

48. Singh, P.P., Das, S., Wen, F., Palu, I., Singh, A.K., Thakur, P., Multi-objective planning of electric vehicles charging in distribution system considering priority-based vehicle-to-grid scheduling. *Swarm Evol. Comput.*, 77, 101234, 2023, http://dx.doi.org/10.1016/j.swevo.2023.101234.
49. Oubelaid, A., Kakouche, K., Belkhier, Y., Khosravi, N., Taib, N., Rekioua, T., Tuka, M.B., New coordinated drive mode switching strategy for distributed drive electric vehicles with energy storage system. *Sci. Rep.*, 14, 1, 6448, 2024, 10.1038/s41598-024-56209-9.
50. Ravindran, M.A., Nallathambi, K., Vishnuram, P., Rathore, R.S., Bajaj, M., Rida, I., Alkhayyat, A., A novel technological review on fast charging infrastructure for electrical vehicles: Challenges, solutions, and future research directions. *Alexandria Eng. J.*, 82, 260–290, 2023, 10.1016/j.aej.2023.10.009.
51. Jena, N.K., Sahoo, S., Sahu, B.K., Naik, A.K., Bajaj, M., Misak, S., Prokop, L., Impact of a redox flow battery on the frequency stability of a five-area system integrated with renewable sources. *Energies*, 16, 14, 5540, 2023, 10.3390/en16145540.
52. Hamed, S.B., Hamed, M.B., Sbita, L., Bajaj, M., Blazek, V., Prokop, L., Ghoneim, S.S., Robust optimization and power management of a triple junction photovoltaic electric vehicle with battery storage. *Sensors*, 22, 16, 6123, 2022, 10.3390/s22166123.
53. Rawat, T., Niazi, K.R., Gupta, N., Sharma, S., Multi-objective techno-economic operation of smart distribution network integrated with reactive power support of battery storage systems. *Sustain. Cities Soc.*, 75, 103359, 2021, http://dx.doi.org/10.1016/j.scs.2021.103359.

About the Editors

Ashwani Kumar, PhD is a senior lecturer at the Technical Education Department of Uttar Pradesh. He has over 260 publications, including more than 20 books and numerous book chapters and articles. His research interests include vibration analysis, finite element analysis, finite element modelling, and renewable energy technologies.

Mukesh Kumar Awasthi, PhD is an assistant professor in the Department of Mathematics at Babasaheb Bhimarao Ambedkar University. He has over 170 publications, including several books, patents, and articles in international journals. His research interests include heat and mass transfer, fluid mechanics, and computational fluid dynamics.

Nitesh Dutt, PhD is an associate professor and the Head of the Department of Mechanical Engineering at COER University with over ten years of teaching experience. He has authored three books, four patents, and several articles in international journals. His research focuses on computational fluid dynamics, solar energy, heat transfer, and nuclear energy.

Yogesh Kumar Singla, PhD is a research associate in the Department of Engineering at the University of Idaho. He has over 40 publications to his credit, including five books and numerous book chapters and journal articles. His research interests include welding, manufacturing, surface engineering, and tribology.

Sivasakthivel Thangavel, PhD is a senior lecturer in the Department of Mechanical Engineering at the Global College of Engineering and Technology. He has published over 30 articles in international journals of repute. His research interests include pyrolosis, gasification and combustion, hydrogen energy, and industrial safety analysis.

About the Editors

Index

Also of Interest

Check out these other related titles from Scrivener Publishing

Solar Electric, Water and Air Tribrid Auto Engineering, Edited by Chandan Deep Singh, Kanwaljit Singh, Davinder Singh, Talwinder Singh, Er. Jasvinder Singh, and Er. Rajdeep Singh, ISBN: 9781394166169. *Solar Electric, Water and Air Tribrid Auto Engineering* is a must-have for anyone in the automotive industry, as it offers a comprehensive analysis of cutting-edge technologies that could revolutionize vehicle design and fuel efficiency, paving the way for a more sustainable future.

Digital Convergence in Intelligent Mobility Systems, Edited by Rathishchandra R. Gatti & Chandra Singh, ISBN: 9781394275243. *Digital Convergence in Intelligent Mobility Systems* gives a comprehensive understanding of how digital technologies are revolutionizing transportation, equipping you with the insights needed to navigate the future of intelligent mobility systems.

AI for Autonomous Vehicles: The Future of Driverless Technology, Edited by Sathiyaraj Rajendran, Munish Sabharwal, Yu-Chen Hu, Rajesh Kumar Dhanaraj, and Balamurugan Balusamy, ISBN: 9781119847465. This title presents the research prospects and security challenges present in the exciting new field of autonomous vehicles by comprehensively introducing advanced concepts of artificial intelligence (AI) in driverless technology.

Introduction to Automotive Engineering, By R. Sakthivel, Faisal O. Mahroogi, S. Narayan, S. Abubakar, M. U. Kaisan, and Youssef Alammari, ISBN: 9781119479802. A brief introduction to automotive engineering, this volume is a useful reference as well, covering not just the basics, but also some of the latest innovations, such as electric and self-driving vehicles.

AUTONOMOUS VEHICLES VOLUME 1: Using Machine Intelligence, Edited by Romil Rawat, A. Mary Sowjanya, Syed Imran Patel, Varshali Jaiswal, Imran Khan, and Allam Balaram. ISBN: 9781119871958. Addressing the current challenges, approaches and applications relating to autonomous vehicles, this groundbreaking new volume presents the research and techniques in this growing area, using Internet of Things, Machine Learning, Deep Learning, and Artificial Intelligence.

AUTONOMOUS VEHICLES VOLUME 2: Smart Vehicles for Communication, Edited by Romil Rawat, Purvee Bhardwaj, Upinder Kaur, Shrikant Telang, Mukesh Chouhan, and K. Sakthidasan Sankaran, ISBN: 9781394152254. The companion to *Autonomous Vehicles Volume 1: Using Machine Intelligence,* this second volume in the two-volume set covers intelligent techniques utilized for designing, controlling and managing vehicular systems based on advanced algorithms of computing like machine learning, artificial Intelligence, data analytics, and Internet of Things with prediction approaches to avoid accidental damages, security threats, and theft.

Hydrogen Electrical Vehicles, Edited by Mehmet Sankir and Nurdan Demirci Sankir, ISBN: 9781394166381. Hydrogen electrical vehicles are an essential component of the "Green New Deal" and this book covers cutting-edge technologies designed for fuel-cell-powered cars.

Smart Charging Solutions for Hybrid and Electric Vehicles, Edited by Sulabh Sachan, Sanjeevikumar Padmanaban, and Sanchari Deb, ISBN: 9781119768951. Written and edited by a team of experts in the field, this is the most comprehensive and up to date study of smart charging solutions for hybrid and electric vehicles for engineers, scientists, students, and other professionals.

Printed in the USA/Agawam, MA
September 16, 2026

910661.129